Irrigated Agriculture and the Environment

The Management of Water Resources

Series Editor: Charles W. Howe

Professor Emeritus of Economics and Professional Staff, Environment and Behavior Program, Institute of Behavioral Science, University of Colorado at Boulder, USA

1. Irrigated Agriculture and the Environment
   James S. Shortle and Ronald C. Griffin

Future titles will include:

Water Resources and Climate Change
Kenneth D. Frederick

The Economics of Industrial Water Use
Steven Renzetti

Water Resources and Economic Development
R. Maria Saleth

Conflict Prevention and Resolution in Water Systems
Aaron T. Wolf

Wherever possible, the articles in these volumes have been reproduced as originally published using facsimile reproduction, inclusive of footnotes and pagination to facilitate ease of reference.

For a list of all Edward Elgar published titles visit our site on the World Wide Web at
http://www.e-elgar.co.uk

Irrigated Agriculture and the Environment

Edited by

James S. Shortle

Professor of Agricultural and Environmental Economics
Pennsylvania State University, USA

and

Ronald C. Griffin

Professor of Natural Resource Economics
Texas A&M University, USA

THE MANAGEMENT OF WATER RESOURCES

An Elgar Reference Collection
Cheltenham, UK • Northampton, MA, USA

Published by
Edward Elgar Publishing Limited
Glensanda House
Montpellier Parade
Cheltenham
Glos GL50 1UA
UK

Edward Elgar Publishing, Inc.
136 West Street
Suite 202
Northampton
Massachusetts 01060
USA

A catalogue record for this book is available from the British Library.

Library of Congress Cataloguing in Publication Data

Irrigated agriculture and the environment / edited by James S. Shortle and Ronald C. Griffin.
p.cm. – (The Management of water resources ; 1) (An Elgar reference collection)
Includes bibliographical references.
1. Irrigation farming. 2. Irrigation farming–Environmental aspects. 3. Water quality management. 4. Irrigation farming–United States. 5. Irrigation farming–Environmental aspects–United States. 6. Water quality management–United States. I. Shortle, J. S. (James S.) II. Griffin, Ronald C. III. Series. IV. Series: An Elgar referemce collection

S613 .I66 2001
333.91'3'0973–dc21

2001040961

ISBN 1 84064 503 2

Printed and bound in Great Britain by MPG Books Ltd, Bodmin, Cornwall

Contents

Acknowledgements

The editors and publishers wish to thank the authors and the following publishers who have kindly given permission for the use of copyright material.

Academic Press for articles: Kathleen Segerson (1988), 'Uncertainty and Incentives for Nonpoint Pollution Control', *Journal of Environmental Economics and Management*, **15** (1), March, 87–98; Erik Lichtenberg, David Zilberman and Kenneth T. Bogen (1989), 'Regulating Environmental Health Risks Under Uncertainty: Groundwater Contamination in California', *Journal of Environmental Economics and Management*, **17**, 22–34; Richard Cabe and Joseph A. Herriges (1992), 'The Regulation of Non-Point-Source Pollution Under Imperfect and Asymmetric Information', *Journal of Environmental Economics and Management*, **22**, 134–46; J.F. Booker and R.A. Young (1994), 'Modeling Intrastate and Interstate Markets for Colorado River Water Resources', *Journal of Environmental Economics and Management*, **26**, 66–87; Richard D. Horan, James S. Shortle and David G. Abler (1998), 'Ambient Taxes When Polluters Have Multiple Choices', *Journal of Environmental Economics and Management*, **36**, 186–99.

American Geophysical Union and the Copyright Clearance Center, Inc. for articles: Neal S. Johnson and Richard M. Adams (1988), 'Benefits of Increased Streamflow: The Case of the John Day River Steelhead Fishery', *Water Resources Research*, **24** (11), November, 1839–46; Bonnie G. Colby (1990), 'Enhancing Instream Flow Benefits in an Era of Water Marketing', *Water Resources Research*, **26** (6), June, 1113–20; Bruce A. McCarl, Carl R. Dillon, Keith O. Keplinger and R. Lynn Williams (1999), 'Limiting Pumping from the Edwards Aquifer: An Economic Investigation of Proposals, Water Markets, and Spring Flow Guarantees', *Water Resources Research*, **35** (4), April, 1257–68.

Blackwell Publishers Ltd for articles: Ronald C. Griffin and Daniel W. Bromley (1982), 'Agricultural Runoff as a Nonpoint Externality: A Theoretical Development', *American Journal of Agricultural Economics*, **64** (3), August, 547–52; James S. Shortle and James W. Dunn (1986), 'The Relative Efficiency of Agricultural Source Water Pollution Control Policies', *American Journal of Agricultural Economics*, **68** (3), August, 668–77; Scott L. Johnson, Richard M. Adams and Gregory M. Perry (1991), 'The On-Farm Costs of Reducing Groundwater Pollution', *American Journal of Agricultural Economics*, **73** (4), November, 1063–73; Ronald C. Griffin and Shih-Hsun Hsu (1993), 'The Potential for Water Market Efficiency When Instream Flows Have Value', *American Journal of Agricultural Economics*, **75** (1), February, 292–303; Gloria E. Helfand and Brett W. House (1995), 'Regulating Nonpoint Source Pollution Under Heterogeneous Conditions', *American Journal of Agricultural Economics*, **77** (4), November, 1024–32; Donna J. Lee and Richard E. Howitt (1996), 'Modeling Regional Agricultural Production and Salinity Control Alternatives for Water Quality Policy Analysis', *American Journal of Agricultural Economics*, **78** (1), February, 41–53; Marca Weinberg and Catherine

L. Kling (1996), 'Uncoordinated Agricultural and Environmental Policy Making: An Application to Irrigated Agriculture in the West', *American Journal of Agricultural Economics*, **78** (1), February, 65–78.

Kluwer Academic Publishers BV for article: Ariel Dinar, Mark B. Campbell and David Zilberman (1992), 'Adoption of Improved Irrigation and Drainage Reduction Technologies Under Limiting Environmental Conditions', *Environmental and Resource Economics*, **2** (4), 373–98.

Natural Resources Journal for article: Michael R. Moore, Aimee Mulville and Marca Weinberg (1996), 'Water Allocation in the American West: Endangered Fish Versus Irrigated Agriculture', *Natural Resources Journal*, **36** (2), Spring, 319–57.

Western Agricultural Economics Association for article: David B. Willis, Jose Caldas, Marshall Frasier, Norman K. Whittlesey and Joel R. Hamilton (1998), 'The Effects of Water Rights and Irrigation Technology on Streamflow Augmentation Cost in the Snake River Basin', *Journal of Agricultural and Resource Economics*, **23** (1), July, 225–43.

Every effort has been made to trace all the copyright holders but if any have been inadvertently overlooked the publishers will be pleased to make the necessary arrangement at the first opportunity.

In addition the publishers wish to thank the Library of the London School of Economics and Political Science and the Library of Indiana University at Bloomington, USA, for their assistance in obtaining these articles.

Economic Perspectives on Irrigated Agriculture and the Environment

James S. Shortle and Ronald C. Griffin

Irrigated agriculture is the leading human use of water. About 70 percent of the water drawn from rivers, lakes and aquifers is used in agricultural production (Food and Agriculture Organization (FAO) 2000a), and irrigation's lower return flow ratio implies that it accounts for more than 70 percent of total consumptive use. While irrigation is an ancient practice thought to have originated in Mesopotamia 6000 years ago, the amount of agricultural land under irrigation has increased tremendously during the past 100 years. There were approximately 40 million irrigated hectares in 1900 (Field 1990). By 1998, more than 271 million hectares were irrigated, with much of the increase occurring after 1950 (FAO 2000b). Much of the growth in agricultural productivity and world food production over the same period can be attributed to irrigation. Although constituting only about 17 percent of all crop land, irrigated land now produces approximately 40 percent of the world's food.

The contribution of irrigated agriculture to world food production has been accompanied, however, by significant environmental costs. In addition to being a big water user, irrigated agriculture is also a major cause of water-related environmental problems. Reservoirs constructed to supply water to agriculture and other sectors have destroyed significant natural assets by inundation. Dams and diversions present physical barriers to fish migration, alter streamflow regimes and water temperatures, and trap sediments. Consequences include severe degradation of aquatic and riparian habitats, with significant threats to aquatic species. Surface return flows and drainage from irrigated agricultural lands carry salts, fertilizers, pesticides and other pollutants into surface waters, causing harm to fish and wildlife, and impairing water for human uses. Aquifers used for drinking water are also contaminated by agricultural pesticides and nitrogen when irrigation water percolates through the ground into the saturated zone.

Accessible and relatively comprehensive environmental assessment data make it easy to use the United States to illustrate the significance of these problems. The US has the third largest land area under irrigation, with 21.4 million hectares in 1998, representing 12 percent of the nation's crop land (FAO 2000b). The irrigated land is found mostly in the arid West. Presently, scores of species of fish are threatened or endangered in western rivers, where habitat has been degraded by large-scale irrigation projects (Moore, Mulville and Weinberg, Chapter 14 in this volume). A recent assessment by the United States Environmental Protection Agency (USEPA) concluded that irrigated crop land accounted for 89 percent of the water quality-impaired river mileage in eleven states in the West (USEPA 1992a). Irrigated agriculture accounted for 40 percent of the impaired lakes in the US, and was the most common cause of pollution in wildlife refuges. Bird deaths, deformities and reproductive failures at the Kesterson Reservoir in California in 1983 highlighted the risks of agricultural drain water to wildlife. Pesticides and other chemicals applied to farmland have been found in ground water supplies used for drinking water in several irrigated regions of the US (USPA 1992b; Barbash and Resek 1996).

These problems are by no means unique to the United States. For example, water diversion to agriculture and polluted irrigation return flows have contributed to the environmental disaster in the Aral Sea in Central Asia (Tanton and Heaven 1999). On a lesser scale, fish are endangered by irrigation-related habitat degradation in the Murray–Darling River system in Australia. Nor are agriculturally imposed injuries limited to irrigated agriculture. Runoff from agricultural land in humid regions is also an important cause of water pollution.

The environmental consequences of irrigation are drawing increasing attention from scientists, environmental advocates and policy makers:

1. Concerns for the environmental impacts of dam building have become a significant barrier to water development for agriculture and other uses in many countries.
2. Increasingly, scientists and environmental groups are calling for transferring irrigation water reserved for agriculture to the support of freshwater ecosystem services. In the United States, these calls have led to changes in federally operated irrigation projects. For example, in 1992 the Central Valley Project Improvement Act allocated 800,000 acre feet of water to ecosystem maintenance (US Department of the Interior 2000). The Murray–Darling Basin Commission in Australia has also taken steps to limit water withdrawals to protect fish habitats. Several countries in the Aral Sea Basin have agreed in principle that the sea itself ought to be regarded as an independent claimant, to support the sea's ecosystems (Postel 1999).
3. Protecting ground and surface water quality from agriculture has become a major water quality policy issue in many countries. Agriculture in the United States and several countries in Scandinavia and the European Union has become a major target of regulatory initiatives to protect water quality (Organization for Economic Cooperation and Development 1991).

In addition to these impacts on water resources and related ecosystems, irrigated agriculture has an ancient foe: salinization of irrigated soils. Dissolved salts introduced by the application of irrigation waters are progressively concentrated by evaporation and plant transpiration. If these salts are allowed to accumulate in the root zone, agricultural productivity falls. When these salts are leached beyond the root zone, ground water quality and downstream surface water quality can be impaired. Thus salinity constitutes both an on-farm management problem for irrigators and an off-farm externality for other water and environment users.

Policy initiatives to address agriculture's impacts on the environment have stimulated a significant body of economic research. Issues broadly include the effects of existing government interventions in agriculture on the environmental impacts of agriculture, the economic costs and benefits of environmental regulations for agriculture, and the economic and environmental merits of alternative mechanisms for water allocation and water quality protection. In this volume we present a sampling of economic research concerning irrigated agriculture's interfaces with the environment. These eighteen previously published works span the topics of nonpoint pollution control, salinity management and the allocation of water. Articles have been selected to provide both theoretical and empirical guidance for these issues. All have relevance for policy design. We ultimately assume responsibility for which articles are included in this volume (and which are not), but we wish to thank the several water resource economists who responded to our call for suggestions or who took time to comment on our initial efforts.[1]

In the discussion that follows we provide our perspectives on the insights and contributions of each of the selected papers. Due to its abridged nature, these comments underscore only the most significant insights provided by each of these works.

Pollution Control Instruments for Irrigated Agriculture

Agriculture, both irrigated and rain fed, is a major cause of water quality problems in many regions. Yet even in developed countries with extensive regulations for water quality protection, policy measures to control pollution loads from agricultural activities have generally been quite limited by comparison with measures that have been taken to control industrial and municipal point sources of pollution.[2] Although there are exceptions, the principal policy approach to reducing water quality degradation by agricultural drainage and runoff has been 'moral suasion', supplemented to varying degrees by technical and financial assistance to encourage the adoption of agricultural Best Management Practices. The limited results of this voluntary approach, along with increased concern for the water quality impacts of agricultural production, has created a substantial interest in the adoption of more effective policy measures. As with other types and sources of pollution, significant reductions in water pollution from agriculture require the application of either enforceable regulatory approaches or significant changes in the economic environment in which farmers make production and pollution control decisions. The latter might be accomplished through some combination of changes in agricultural price and income policies or the application of economic instruments for pollution control. The appropriate choices among the range of options that fall within these boundaries are the subject of a substantial body of economic literature.

Growing recognition of the magnitude of agriculture's contribution to the water pollution problems in the US and Europe in the 1970s stimulated economic interest in the design of environmental policy instruments for reducing polluting runoff and ground water contamination from agricultural production. Two major questions have emerged in this literature. One is the appropriate basis for defining and measuring compliance with environmental regulations. Conceivable options include farmers' use of inputs and technologies that are closely related to pollution flows from farms (e.g. water, pesticides, fertilizers), measures of pollutant flows from farms (e.g. the volume of salts, pesticides or nutrients leaving fields) and ambient concentrations of pollutants in receiving waters. The second is how best to induce changes in farm production and pollution control practices to achieve water quality objectives. The menu includes direct regulations and various types of economic incentives (e.g. taxes, subsidies, tradeable permits, bonds, liability rules). Initially, researchers looked to the theoretical and empirical literature of environmental economics for guidance. Because this literature highlighted the control of conventional point sources of pollution where emissions are often readily observed, the emphasis was on the economic merits of alternative types of discharge-based economic incentives (e.g. discharge charges/standards, discharge reduction subsidies, transferable discharge permits). Polluting emissions are the economically preferred base for the application of regulatory standards or economic incentives, provided that they can be metered with a reasonable degree of accuracy at a reasonable cost. The literature on discharge-based environmental instruments is, however, of limited relevance to the design of pollution control instruments for agriculture (and other nonpoint sources) because the movement of pollutants from farm fields in runoff or through soil into drains or aquifers generally cannot be so measured. With unobservable pollutant flows, other constructs must be used as performance standards and as a basis for the application of policy instruments. The economics of designing policy instruments for agricultural and other nonpoint pollution externalities are therefore complicated by the fact that choices must be made between suboptimal compliance measures as well as between types of regulations and incentives.

The article by Griffin and Bromley included as Chapter 1 in this volume was the first in a series of three particularly influential articles on the economic theory of nonpoint pollution control applicable to irrigated and rain-fed agriculture. While pollution flows from farm fields are not easily or cheaply measured, hydrological process and statistical models have been developed to assess these flows given measurements of appropriate land characteristics, weather and farm production practices. Griffin and Bromley initiated the development of a theory of nonpoint pollution instruments in which information on the relationship between production choices and emissions provided by such a physical model, which they refer to as a 'nonpoint production function', would replace direct measurement of emissions. They described how the information could be used to construct economically efficient input tax/subsidy schemes or input standards. They also described the construction of an economically efficient tax and standard for an emissions estimate obtained using the nonpoint production function to map from farm production inputs to emissions. Their work enlarged the domain of economic thinking on nonpoint instruments with theoretical support to include input-based instruments and provided a significant foundation for subsequent theoretical and empirical work.

Building on Griffin and Bromley's work, Shortle and Dunn (Chapter 2) presented a model in which nonpoint emissions were unobservable and stochastic. Under this specification, plugging observations of farm inputs into a nonpoint production function is no longer a perfect substitute for measuring emissions without error. Their analysis also incorporated the problem of asymmetric information about the private cost of pollution control. Specifically, in their model environmental policy makers have imperfect information about the costs of compliance with pollution controls, while farmers have perfect information. These departures from Griffin and Bromley's model were shown to have important implications for the choice of base (inputs versus an emissions estimate that is an unbiased prediction of actual emissions) and the type of instrument (economic incentives versus standards) for agricultural or other nonpoint externalities. In particular, they demonstrated that the economically efficient outcome for nonpoint sources could be obtained with properly designed input-based economic incentives, but not by instruments applied to estimated emissions. The key contributions were to introduce uncertainty about the relationship between production choices and emissions into the analysis of nonpoint instruments and demonstrate that this uncertainty had implications for the choice of compliance measures.

Segerson (Chapter 3) takes a very different approach to the nonpoint problem from Griffin and Bromley and Shortle and Dunn. Rather than monitoring the choices of input choices of farmers that are suspected of contributing to environmental degradation, she proposed the use of an ambient-based tax that shifts monitoring from the sources of emissions (i.e. farms) to the receptor (i.e. streams, lakes, estuaries). She devised an ambient tax/subsidy scheme that could achieve the efficient outcome for nonpoint sources as a Nash equilibrium. The scheme pays firm-specific subsidies when the ambient pollution concentration falls below a target and charges firm-specific taxes when the ambient concentration exceeds the target. Horan, Shortle and Abler (Chapter 4) present two ambient taxes for nonpoint polluters that apply under less restrictive conditions. The theoretical framework presented in their article also provides a bridge between Segerson's model of nonpoint pollution and the models used by Griffin and Bromley and Shortle and Dunn.

A crucial feature of ambient tax/subsidy schemes is that farmers are not rewarded or penalized according to their own performance. Rather, rewards and penalties depend on group performance. This makes farmers' expectations (or conjectural variations) about other farmers' behavior, and the regulator's knowledge of these expectations, critical to the design and performance of ambient-based instruments. A second important feature is that ambient taxes shift the burden of information from regulators to farmers. An individual's response to an ambient tax will depend on his or her expectations about the environmental impact of his or her choices, the choices of others and natural events on ambient conditions. In other words, it will depend on the farmer's theory of the ultimate biochemical conversions and movements impinging on runoff contaminants as they progress through the hydrological environment. The importance of expectations for the performance of ambient instruments is demonstrated by Cabe and Herriges (Chapter 5), who explore the consequences of asymmetric prior information about transport and fate. If farmers' prior information about transport and fate differs from the regulator's, then incentives designed on the assumption that they are the same will not have the desired properties. For instance, a farmer may perceive no impact of his or her choices on pollution. In this case, an ambient tax may have either no impact or it will result in a decision to escape the tax entirely by ending the suspect activity. The regulator has a choice between adjusting the incentives for the mismatch, attempting to change the farmer's beliefs, or a combination of both.

This theoretical research on input- and ambient-based instruments demonstrates that, given certain assumptions, including zero transactions costs, either approach can be designed to yield ex-ante efficient solutions. However, like Pigouvian emissions taxes, the transactions costs associated with identifying and implementing optimally designed nonpoint instruments would be prohibitive. Plausible instruments will necessarily be of a second-best type. For instance, input tax/subsidy structures required to achieve a first-best allocation must be farm specific and applied to all inputs that directly affect emissions. However, farm-specific taxes are exceptionally information intensive, and would pose significant enforcement problems because they provide incentives for arbitrage between farms that could defeat the efficiency of the tax system. In practice, uniform rates, at least within regions where such arbitrage could easily occur, would be required. Similarly, while purchases of potentially harmful inputs or investments in pollution control structures can be easily tracked in some instances, many farm and water management decisions having a large impact on the environment are too costly to monitor or verify. These information costs require the tax/subsidy base to be reduced to a subset of choices that are both relatively easy to observe and correlated with ambient impacts.

While theory can help guide the choice and design of second-best instruments, questions about compliance measures, types of instrument and other details are inherently empirical. Experience provides little for economists to work with in evaluating alternatives. There has been only limited use of input-based economic incentives and regulations for agricultural nonpoint pollution control. Accordingly, empirical research on the design and performance of instruments to reduce water pollution from irrigated and rain-fed agriculture is largely conducted using models that simulate economic and ecological impacts. Much of this research is for rain-fed agriculture in humid regions of North America and Western Europe, but there are several noteworthy contributions on environmental instruments for irrigated agriculture in the United States. The empirical literature largely addresses economic incentives or regulations based directly on farm inputs or practices or other farm-specific environmental performance measures (e.g. estimates of pollution flows from particular farms). Ambient-based instruments, while of significant theoretical interest, have received little attention in empirical research or from policy makers.

The article by Johnson, Adams and Perry (Chapter 6) is an outstanding example of the integration of biophysical and economic models to study the costs of reducing pollution flows from irrigated agricultural land. Specifically, they link a plant growth simulation model, a hydrological simulation model and an economic model of farm decision making to examine the costs of reducing nitrate leaching into ground water from irrigated production of wheat, corn and potatoes in the Columbia Basin of Oregon (US). The hydrological simulation model is the empirical analog of Griffin and Bromley's nonpoint production function. Johnson et al. examine the costs, in terms of forgone farm profits, of taxes and restrictions on nitrogen inputs, and taxes and restrictions on estimated leachate rates. Their results show the latter to be much more cost effective in reducing average nitrate leachate rates. It should be noted, however, that this result does not imply, for reasons described in Shortle and Dunn, that a tax or restrictions on estimated leachate are superior to a tax or restrictions on nitrogen applications. They also find that significant reductions in leachate rates could be achieved at modest cost to farmers with appropriate changes in the timing and levels of water and nitrogen applications.

Helfand and House (Chapter 7) compare several instruments for reducing nitrate leaching from two soils used for lettuce production in California's Salinas Valley (US). Like Johnson et al., their research combines an economic model of farm decision making with a physical process model of leaching. Also like Johnson et al., they consider taxes and restrictions on nitrogen inputs, but they also consider taxes and restrictions on irrigation water inputs. Their research is particularly interesting for their comparison of first- and second-best input-based specifications of these instruments. The integrated model is used to solve for taxes or standards that minimize the cost of achieving a 20 percent reduction in nitrogen leaching under several scenarios involving combinations of taxes (or restrictions) on one or both inputs, with the rates (restrictions) being either uniform or differentiated across soils. The lowest cost solution is obtained for the case where a tax is imposed on each input, with the rates optimally differentiated across the soils. This is the first-best policy. The other combinations entail suboptimal but more plausible second-best policies. Like Johnson et al., Helfand and House show that choices about compliance measures are important. In this case, taxes or standards applied to nitrogen alone perform poorly by comparison with the first-best solution and also by comparison with taxes or standards applied to water alone. Their results generally show little difference between the performance of taxes and standards when applied to the same compliance base. Similarly, they find the costs of uniform input taxes (restrictions) as opposed to differentiated taxes (restrictions) to be relatively small. These latter two results most likely reflect a limited degree of heterogeneity in control costs across soils in their model and should not be generalized.[3]

Most economic research on irrigated agriculture and water quality is focused on the costs of reducing the use of polluting inputs or expected emissions from agricultural production. The choice of environmental objectives and the implications for pollution control strategies have received comparatively little attention. The paper by Lichtenberg, Zilberman and Bogen (Chapter 8) is a significant exception. As discussed at length in the articles by Shortle and Dunn, Segerson and Horan et al., uncertainty about environmental outcomes is a key characteristic of agricultural nonpoint pollution problems. These authors demonstrate that the treatment of risk is important to policy design, but do not delve very deeply into the question of how to treat risk. Lichtenberg et al. present a 'safety first' approach to the regulation of environmental risks when decision makers are uncertain about environmental outcomes. They would have decision makers first choose a maximum allowable amount of risk and then select pollution control policies to achieve that risk at least cost to society. The safety-first rule is expressed as a restriction on the probability of an environmental risk measure exceeding a maximum allowable level. Lichtenberg et al. develop and demonstrate the approach, and examine implications for pollution control strategies in a case study of regulations to reduce human health risks caused by pesticide contamination of drinking water. The fundamental concepts and lessons can, however, be applied to other types of environmental risks caused by irrigated agriculture.

Salinity and Water Allocation

In arid and semiarid regions where irrigation is prevalent, the salinity problem is aggravated by naturally high evaporation and transpiration. The lack of precipitation in these areas lessens the environment's assimilative capacity naturally to leach and dilute accumulated salts (National Research Council 1989). Mounting soil salinity eventually alters the production regimes selected by irrigators as they modify crop choices, irrigation quantities, irrigation technologies and drainage activities to combat declining crop yields. Some of these choices work to externalize salinity by shifting salts to off-farm receptors. Inevitably, surface and subsurface runoff is more saline than original irrigation water, and subsequent reuse of these waters by downgradient irrigators multiplies the problem. This is an intrinsically dynamic matter, because salts can be stored in soils and ground water prior to their eventual transport off-farm. The consequent, progressive salinization of waterways has been responsible for habitat and wildlife losses as well as reduced utility for other human uses of water.

Salinization is a significant negative force for agriculture in some areas of the world, so it has produced a noteworthy literature in agricultural economics. Much of this literature emphasizes farm management matters, especially profit-maximizing responses for producers, rather than efficient control of an environmental externality in light of all social costs. Focusing on the externality issue, there is little conceptually to distinguish salinity from the usual nonpoint pollutant. As with other nonpoint pollutants, policy prospects are crucially altered by the limited opportunity to link salinity 'emissions' to the responsible farm. Accountability for rising salinity must be inferred instead of measured, and policy instruments must be based on these inferences. Thus the more general literature of nonpoint pollution theory also guides the selection of salinity policy.

Four articles pertaining to salinity modeling and policy are included in this volume. The first of these (Chapter 9) addresses the choice of irrigation technology in production settings facing salinity. Dinar, Campbell and Zilberman first assemble the theoretical details that underlie producer behavior in selecting from the array of available technologies. These choices are crucial factors in determining eventual water use and runoff salinity. Using farm survey data collected in a heavily irrigated region of California, they estimate farm-level and field-level explanatory models, seeking to relate technology choices to the economic and physical situations faced by farm managers. Arguing that policy makers should understand these choice processes, as well as the inherent diversity of farms and technology selections, they make a unique contribution to the issue of salinity management. It is noteworthy that this paper, like most in the salinity field, emphasizes on-farm decisions advancing private objectives. Still, this represents an important building block for constructing policy. Readers wishing to expand their knowledge in this arena are advised also to examine the many published works of Keith Knapp and Dan Yaron.

An interesting facet of salinity control is the importance of irrigation quantity in determining runoff salinity. By decreasing applied water, the applied mass of salts is also reduced. From producers' perspectives, there are limits to this approach because economic levels of crop water demand must be met and additional water quantities are useful for leaching salts beyond root zones. Socially, the dependence of salinity emissions on water use links the salinity issue to another major interface between irrigated agriculture and the environment: competition for scarce water. Three additional papers are incorporated in this volume because they offer different approaches to examining this two-pronged implication of water use for the environment.

Booker and Young (Chapter 10) empirically investigate the salinity and water availability implications of alternative water allocations along the Colorado River basin. Partitioning this large basin into segments and developing a nonlinear optimization model maximizing net benefits, they inspect the economic implications of policies attending to different goals or allowing different water-marketing scenarios. The spatial model incorporates offstream water demands for irrigation, municipalities and fossil fuel power plants, as well as instream demands for hydropower. Other instream water demands, such as recreation and habitat, are not included. Salinity is modeled and salinity damages are economically incorporated. Among other things, the results demonstrate both the gains available to more flexible reallocation policies (allowing interstate transfers of water) and the economic implications of salinity for water allocation. With respect to the latter item, attention to salinity is shown to recommend substantial downstream transfers of water. That is, optimizing water diversions occur farther downstream (as compared with actual practice) as a consequence of modeling salinity economics. This result is theoretically obtainable, but Booker and Young's empiricism succeeds in demonstrating the magnitude of the gains as well as the extent of the required water reallocation.

Lee and Howitt (Chapter 11) provide an alternative analysis of the same Colorado River problemshed while applying different techniques and emphases. Using salinity-adjusted Cobb–Douglas production functions for irrigated outputs as well as incorporating salinity damages for nonagricultural sectors, they investigate optimizing combinations of water use and federally proposed salinity control projects within the basin. Results clearly underscore the social inefficiency of much of the contemporary irrigation taking place in the upper parts of the basin. Retiring or modifying upper basin irrigation is shown to produce substantial cost savings, especially in relation to the prospective costs of federal control projects.

Weinberg and Kling (Chapter 12) study both the water allocation and salinity issues with an eye toward the theoretical second-best conundrum: policy 'correction' of any single market failure while leaving other market failures in place can possibly lower economic efficiency. Using a California study region where irrigation prices are highly subsidized and salinity problems are advanced, they investigate whether separate policy directed at each of these issues will enhance social net benefits. As with the prior two salinity studies, the central model employs nonlinear optimization. Not surprisingly, coordinated policies simultaneously addressing both salinity and water overuse problems are found to offer the greatest rewards, but single-issue policies are discovered to promise strong efficiency gains as well. As a consequence, confidence is increased regarding the prospective efficiency of piecemeal (i.e. single targeted) policy.

Water Reallocation and the Environment

As the humanly dominant use of water, irrigation not only introduces chemicals, minerals, nutrients and salts into our waters, but it consumes water as well. Naturally flowing waterways were the global rule rather than the exception one century ago. That changed as we used concrete and steel to recast the hydrological system better to suit our own visions of where water should go and how it should be employed. Irrigation and hydropower demands for water largely motivated these conversions, but municipal and industrial demands were sometimes contributing factors. Natural instream water uses came to be regarded as residual claimants, and for a long time some laws did not even acknowledge instream water uses to be a beneficial use of water. During this period, aquatic habitats suffered, species were lost and fishery benefits fell.

Although the domestication of rivers is still under way in some areas of the world, evidence from developed countries suggests that water is overcommitted to irrigation *vis-à-vis* all other water demands. In many places, irrigation development occurred when land was inexpensive, municipal and industrial water demands were slight, water works were subsidized, environmental goods were plentiful, and incomplete scientific information underappreciated the environmental implications of large-scale irrigation. These circumstances no longer exist in much of the world.

Fortunately, two economic tools have emerged to ameliorate tendencies to overinvest in dams and overuse water in irrigation. Cost–benefit analysis has illuminated the inefficiency of many water projects prior to construction, thereby slowing the further appropriation of water from natural systems. Many texts and articles concerning this technique are available. In recent decades water marketing has provided a vehicle for reallocating water from irrigation to growing urban and industrial demand, and water marketing has even fostered transfers of water rights to instream uses. Long championed by resource economists, water marketing has been slowly developing, but there are still obstacles to overcome. Some of the most important hurdles relate to unique features of environmental and instream water demands.

The final six articles of this book deal with various aspects of these problems. Most of these selected papers address economic tradeoffs between irrigation and instream water uses.

Colby's paper (Chapter 13) establishes the setting faced when instream water demands enter contemporary water markets. She observes the types of uses composing instream water demands and the unique character of these demands. The noteworthy water market instruments (e.g. sales, leases, options) are discussed, along with key water market semantics, such as seniority, thinness and transactions costs. Some actual transactions involving western US transfers of irrigation water to instream uses are summarized.

Moore, Mulville and Weinberg (Chapter 14) describe the relationship between endangered fish species and irrigated agriculture in the western US. Two statistical models are estimated to explore the linkages between the number of fish listed as endangered species and irrigation explanatory variables. Their paper dispels any possible arguments that irrigated agriculture may have minimal implications for aquatic environments. Some recovery efforts involving government agencies are also discussed.

Although water resource economists have been zealously promoting water marketing, the requirements for achieving economic efficiency are quite stringent when instream uses enter the picture. Griffin and Hsu (Chapter 15) develop a theory of water markets inclusive of instream flow demand. The theory demonstrates that the economist's usual recommendation of transferable permits in consumptive water rights falls short when the goal is to achieve economic efficiency in water allocation. The unique character of instream water demand requires separate water rights for water diversions and water consumption. This point adds to the difficulties inherent to the nonrivalness of most instream uses. A system of publicly sponsored incentives is observed to be a potential substitute for market participation by instream beneficiaries.

The paper by Johnson and Adams (Chapter 16) demonstrates a procedure for valuing marginal streamflow changes for their capacity to enhance sport fishing benefits. First, streamflow changes are related to changes in the productivity and catchability of a fish species (steelhead). Then, a contingent valuation study is employed to assign recreational value to these changes, thereby quantifying the connection between streamflow and sport fishing benefits. This work permits the economic efficiency of water reallocations between irrigated agriculture and recreation to be analyzed. Although streamflow enhancements of this type certainly produce benefits other than those related to steelhead fishing, the methods employed in this analysis are illustrative.

Willis et al. (Chapter 17) appraise the long-term losses in irrigation profits that would result from irrigators' participation in contingent (option) contracts providing for dry-year water forfeitures. Such contracts are intended to enhance streamflow during periods of low flow so that young salmon can experience higher survival rates during their passage downstream. The hydrological record provides probabilities that the contract terms will be triggered. A profit optimization model estimates consequent profit changes while allowing irrigators to adjust their cropping decisions. It is noteworthy that one of the co-authors, Norm Whittlesey, has long been a student of the salmon/hydropower/irrigation tradeoff in the Columbia River Basin, and has published numerous articles on this topic during his career. The Willis et al. article cites portions of this work, and therefore provides a starting place for researchers undertaking similar studies.

McCarl et al. (Chapter 18) provide an economic study of a situation in which endangered species protection has motivated the adoption of transferable ground water rights. As the study region's largest user of ground water, irrigated agriculture is being granted most of the water rights, and subsequent water marketing will soon play a strong role in the reallocation of water out of irrigation. The authors create a mathematical programming model incorporating fundamental aquifer hydrology and maximizing multisectoral net benefits in the region. Different aggregate amounts of pumping rights are investigated for their impacts on springflows, upon which endangered species depend, and for their impact on efficient water allocation. The results show the impacts and costs of employing ground water marketing to protect biodiversity.

Notes

1. We thank David Abler, Bonnie Colby, Ariel Dinar, Ujjayant Chakravorty, Ziv Bar-Shira, David Zilberman, Frank Ward, Richard McCann, Bill Easter, Ray Huffaker, Micha Gisser, Rich Adams, Gloria Helfand, Rick Horan, Noel Gollehon, Nir Becker and Mahdu Khanna.
2. Pollution sources are generally distinguished as point or nonpoint according to the pathways the pollutants or their precursors follow from the place of origin to the receiving environment. Pollutants from point sources enter at discrete, identifiable locations. Industrial facilities that discharge residuals directly into air or water from the end of a smoke stack or pipe exemplify this class. Pollutants from nonpoint sources follow indirect and diffuse pathways to environmental receptors. Open areas such as farm fields, parking lots and construction sites, from which pollutants move overland in runoff into surface waters or leach through the permeable layer into ground waters, are examples.
3. Most empirical studies show that, transactions costs aside, highly targeted, information-intensive strategies for nonpoint pollution control policies outperform undifferentiated strategies, often by a substantial margin (e.g. Babcock et al. 1997).

References

Babcock, B.A., P.G. Lakshminarayan, J.J. Wu and D. Zilberman (1997), 'Targeting Tools For Purchasing of Environmental Amenities', *Land Economics*, **73**, 325–39.

Barbash, J.E. and E.A. Resek (1996), 'Pesticides in Groundwater: Distribution, Trends and Governing Factors', *Pesticides in Hydrologic Systems Series*, Vol. 2, Ann Arbor Press: Chelsea and Michigan.

Field, W. (1990), 'World Irrigation', *Irrigation and Drainage Systems*, **4**, 91–107.

Food and Agriculture Organization (2000a), *Crops and Drops: Making the Best Use of Land and Water*, FAO: Rome.

Food and Agriculture Organization (2000b), http://apps.fao.org/lim500/nph-wrap.p1? Irrigation& Domain=LUI, accessed 31 May 2000.

National Research Council (1989), *Irrigation-Induced Water Quality Problems: What Can Be Learned from the San Joaquin Valley Experience*, National Academy Press: Washington, DC.

Organization for Economic Cooperation and Development (1991), *State of the Environment*, OECD: Paris.

Postel, S. (1999), *Pillar of Sand: Can the Irrigation Miracle Last?*, W.W. Norton and Company: New York and London.

Tanton, T.W. and S. Heaven (1999), 'Worsening of the Aral Basin Crisis: Can There Be a Solution?', *Journal of Water Resources Planning and Management*, **125**, 363–8.

US Department of the Interior, Bureau of Reclamation Mid-Pacific Region (2000), 'Central Valley Project Improvement Act – Public Law 102-575, Title 34', http://www.mp.usbr.gov/regional/cvpiamain/index.html, accessed 14 June 2000.

US Environmental Protection Agency (1992a), *Managing Nonpoint Source Pollution: Final Report to Congress on Section 319 of the Clean Water Act*, USEPA Office of Water: Washington, DC.

US Environmental Protection Agency (1992b), *Another Look: National Survey of Pesticides in Drinking Water Wells*, Phase II Report, EPA 579/09-91-020, January, USEPA Office of Water: Washington, DC.

Part I
Pollution Control Instruments for Irrigated Agriculture

[1]

Agricultural Runoff as a Nonpoint Externality: A Theoretical Development

Ronald C. Griffin and Daniel W. Bromley

The purpose of this paper is to develop and explore a theory dealing with an important facet of agricultural runoff problems. Agricultural runoff is a nonpoint externality with notable implications for both research and policy. A nonpoint externality exists whenever the externality contributions of individual economic agents cannot be practically measured by direct monitoring. Without monitoring, regulations on emissions cannot be enforced, and charges or subsidies cannot be assessed. Thus, policies which are usually suggested for pollution abatement are not available. There is great need for a theoretical base that can adequately capture these facts. Such a theory yields some important implications even before specific empirical applications.

The economic problem of agricultural runoff can be separated into three distinct categories. First, the sediment, nutrients, and chemicals removed by runoff represent a loss of resources to the individual farmer. These costs are borne privately by farmers.

Second, if the discount rate of the individual farmer is greater than the social discount rate, or if the farmer has a planning horizon which is shorter than society's, the farmer will mine the soil resource at a rate which is more depletive than is socially optimal. This is a temporal externality.

A third category involves the conservation of mass. Physical resources lost by the individual farm must appear elsewhere in the environment. In sufficient quantities, these resources are pollutants. Since water is the primary transport media for these resources, it is water that is polluted by soil, nutrients, and agricultural chemicals. This spatial externality has been the focus of much research. Here we intend to develop a theory for these issues, to employ it to suggest a revised research methodology, and to provide some important conclusions about alternative runoff policies.

Only two of the three categories of agricultural runoff problems involve externalities, and only the spatial externality is considered here.

Ronald C. Griffin is an assistant professor in the Department of Agricultural Economics, Texas A&M University, and Daniel W. Bromley is a professor in the Department of Agricultural Economics, University of Wisconsin.

Texas Agricultural Experiment Station Technical Article No. 17485.

The authors wish to thank John Braden, George Casler, Bart Eleveld, Wesley Seitz, John Stoll, C. Robert Taylor, and Norman Whittlesey for their comments on earlier drafts without implicating them for any errors which may remain.

A Theory of Nonpoint Externalities

The purpose of this section is to develop a nonpoint externality theory. Economically efficient regulatory and incentive policies will be devised and their optimal parameters mathematically specified. The development begins with a simple static model of the traditional point-source externality.

A Point Externality

In this single-period model we assume that it is either impossible or too costly to determine consumers' valuations of marginal benefits from reducing emissions of a point-source pollutant. Instead, we assume that a regional limit on emissions has been politically or bureaucratically resolved and that the objective is to achieve this goal at least cost to the region.

Starting from the Baumol and Oates framework for modeling depletable and detrimental externalities in a least-cost setting, let $\mathbf{y}^j$ be the production bundle of firm j with $y_n{}^j$ being the nth element (positive or negative) of that vector. Positive activities represent outputs; negative ones are inputs. There are J firms. Excluding the pollutant, there are N goods or activities. Pollutant emissions by firm j are nonnegative and are denoted z^j. Total pollutant emissions summed across all firms are limited to Z^*.

Production relationships are allowed to differ among firms. Firm j's production set is given implicitly by $f^j(\mathbf{y}^j, z^j) \leq 0$, although it is assumed that each firm fully exploits its productive abilities. Consequently, these relationships become equalities.

Given this specification, society's problem can be formalized as a desire to maximize total profits given productive abilities and the constraint on emissions. Society's Lagrangian is

$$(1)\quad L = \sum_j \mathbf{p}\mathbf{y}^j - \sum_j \alpha^j f^j(\mathbf{y}^j, z^j) - \mu\left(\sum_j z^j - Z^*\right).$$

As usual, $\mathbf{p}$ represents the price vector ($1 \times N$), and the α^j's and μ are appropriate Lagrange multipliers.

Assuming that the implicit production functions are concave and that all constraints are binding, the following first-order conditions describe the optimal choice of production activities.

$$(2) \quad p_n - \alpha^j f_n^j = 0 \quad \text{for all } j, n, \text{ and}$$

$$(3) \quad -\mu - \alpha^j f_z^j = 0 \quad \text{for all } j.$$

Subscripts on functionals denote partial derivatives and those on vectors denote particular vector elements.

Because only z is an externality quantity, profit maximization by firms will satisfy equation (2) but not (3). To also attain (3), firms theoretically would be confronted with an economic incentive for reducing emissions. Let s represent the per unit incentive on pollutant emissions. This incentive will be a charge at the margin (the externality is detrimental) but can be either a net charge or subsidy to each firm. For this generalization, let Z represent the incentive base level, a predetermined quantity from which greater emissions are charged at rate s and lesser emissions are subsidized at the same rate. Z bears no necessary relationship to Z^* and must be independent of actual emissions. If firm j maximizes profits in the presence of such an incentive, then its Lagrangian is

$$(4) \quad L^j = \mathbf{p}\mathbf{y}^j + s(Z - z^j) - \delta^j f^j(\mathbf{y}^j, z^j).$$

Optimality conditions are then given by equations (5) and (6):

$$(5) \quad p_n - \delta^j f_n^j = 0 \quad \text{for all } n.$$

$$(6) \quad -s - \delta^j f_z^j = 0.$$

Therefore, referring to (3) and (6), if the private value of productive abilities is equivalent to the social value ($\delta^j = \alpha^j$) and the incentive is chosen to equal μ, then the social and private solutions are the same. The first of these conditions is satisfied, but the second can be a major problem. The prescribed method for attaining $s = \mu$ is a trial-and-error procedure in which s is established at some initial level and iteratively adjusted until the standard, Z^*, is just reached. While it is easy to design a mechanism which converges s to μ, the speed of this convergence is very much in doubt. This issue, although important, is not discussed here.

The optimal incentive, s^*, is the same for all firms and is dependent upon the incentive base level. Use of this concept generalizes the model by incorporating the entire charge-subsidy spectrum. A pure charge exists if Z is equal to zero. Choosing each firm's previous externality generalization for Z corresponds to a pure subsidy. The specification employed here recognizes choices lying between these two extremes and permits control over equity and efficiency in externality resolution. Z can be interpreted as an initial endowment which the firm can then sell (collect a subsidy on) or buy more of (pay a charge for).

This analysis is neutral with respect to whether chosen economic incentives are subsidies or charges. The literature on this matter posits asymmetry between subsidies and charges. Subsidies will preserve marginal enterprises and may even induce entry by firms which were previously unprofitable (Baumol and Oates). Charges will do the opposite. Thus, there will be more polluters under subsidies than charges. Moreover, the incentive required with a subsidy will be slightly higher than with a charge achieving the same abatement level. Nonetheless, either policy will be least-cost in a strict sense. The difference is distributional and related to a type of income effect involving entry and exit by firms.

An equally efficient set of regulations can be mathematically stated using profit functions. Firm j's (optimal) profit function, $\pi^j(\mathbf{p}, s)$, specifies optimal profits as a function of the price vector and the incentive level. That is,

$$\pi^j(\mathbf{p}, s) = \text{Max } \mathbf{p}\mathbf{y}^j + s(Z - z^j)$$
$$\text{subject to } f^j(\mathbf{y}^j, z^j) = 0.$$

Given previous assumptions, we need only disallow constant returns to scale in order to guarantee the existence of $\pi^j(\mathbf{p}, s)$. Applying Hotelling's lemma (Varian), the firm's optimal output of pollution is the derivative of the profit function with respect to the cost of emitting the pollutant. Evaluating this partial derivative at the appropriate prices and the optimal incentive, we have an equally optimal regulation for firm j.

$$(7) \quad z^{j*} = z^j(\mathbf{p}, s^*) = \left.\frac{\partial \pi^j(\mathbf{p}, s)}{\partial s}\right|_{\mathbf{p}, s^*} \quad \text{for each } j.$$

Equation (7) describes a set of optimal regulations which, when enforced, will achieve the targeted emission restriction at least cost. Allocatively, the regulations defined by equation (7) are as efficient as the least-cost incentive, $s^* = \mu$. Each of the policies is the dual of the other. This dual relationship guarantees that the allocative efficiency of both programs is equal. The typical argument favoring incentives over regulations is based on informational efficiency. Equation (7) clarifies this issue by indicating that profit functions must be known in order to calculate optimal regulations. Profit functions require knowledge of each firm's implicit production function. Hence, the information needed to determine least-cost regulations is greater than that for least-cost incentives.

Therefore, the point externality can be approached with either of two policies; the first is price guided (an incentive), and the second is quantity guided (regulations). Monitoring and assessment of the incentive or monitoring and enforcement of the regulations will lead to the desired result—least-cost achievement of the aggregate restriction on point-source emissions. Unfortunately, monitoring of nonpoint pollutants is either infeasible or impractical. Hence, the two policies specified above are unworkable because each requires that the pollutant be monitorable.The next section illustrates how some theoretical adjustments can reestablish the usefulness of these approaches.

A Nonpoint Externality

Externality levels have often been linked directly to an output quantity. In such cases, least-cost incentive or regulatory policies can be attached to the output quantity rather than the actual amount of externality. Meade was the first to extend these output-oriented policies to inputs by postulating that externality levels may be dependent on the amount of productive factors employed. Under these conditions, incentive and/or regulatory policies may be applied individually to every factor on which externality generation depends. Of course, the choice of incentives and regulations must all be correct in order to induce least-cost responses by firms.

Whereas it would be technically difficult and prohibitively expensive to measure nonpoint pollutant emissions by individual firms, factors influencing those emissions can be measured at a more reasonable cost. The refinements undertaken below to expand the point externality model so that it accounts for these possibilities are, therefore, valuable. This is true despite the fact that there are many unresolved problems regarding the precise linkages between management choices (and physical land characteristics) and the generation of agricultural runoff. The amended model is completely general because it accommodates a functional relationship between externality levels and outputs, inputs, or some combination of the two.

Assume that every firm is fully utilizing its productive abilities, i.e., $f^j(\mathbf{y}^j, z^j) = 0$ for all j. Applying the implicit function theorem, we have that there exists a neighborhood about $\mathbf{y}^j$ and smooth functions g^j such that $f^j[\mathbf{y}^j, g^j(\mathbf{y}^j)] = 0$ for all j throughout the neighborhood. The only restrictive assumption necessary to apply this theorem requires that $f_z^j \neq 0$. In the presence of the unabated externality, each farm chooses $f_z^j = 0$, so the theorem does not apply. However, externality policy is intended to direct the farm away from this point, so we know that this assumption is indeed valid.

The implication of this theorem is that externality production is expressible as a continuously differentiable function of all inputs and outputs. Hence, this formulation is completely general, accommodating input and/or output determinants of the nonpoint externality. For convenience, assume that the nonpoint production function, $g^j(\mathbf{y}^j)$, does not differ among farms. Hence, the superscript on this function is dropped but may be reinserted with little change in the analysis. This assumption does not imply, for example, that all farms have the same soil types and slopes; these variables are arguments of g.

Use of this functional relationship suggests four distinct policies.[1] The first, a nonpoint incentive, is equivalent to the incentive formulation of the previous section and shall continue to be denoted as s^*. While this incentive is unchanged, policy operation must be revised. Instead of monitoring pollutant emissions, the individual determinants of these emissions are monitored, and the nonpoint production function is used to calculate z^j for each farm. Not every farm input and output needs to be monitored; most will be unrelated to pollution generation and will not enter into g. Under these conditions, farm profits are $\mathbf{p}\mathbf{y}^j + s[Z - g(\mathbf{y}^j)]$.

The second policy, a least-cost system of nonpoint standards, is still expressed by equation (7). Enforcement requires that farm production activities be monitored in order to estimate actual emissions using $g(\mathbf{y}^j)$.

Third, the nonpoint production function can be used to determine individual management incentives for each production activity affecting emissions. Properly chosen, these incentives can induce least-cost efficiency. Let $\boldsymbol{\sigma}$ denote the vector of incentives attached to the elements of $\mathbf{y}_j$. Thus, σ_n is the incentive (either a marginal charge or subsidy) on activity y_n. Letting $\mathbf{Y}$ represent the vector of management incentive base levels (one for each activity), farm profits are $\mathbf{p}\mathbf{y}_j + \boldsymbol{\sigma}(\mathbf{Y} - \mathbf{y}_j)$. Analogous to Z above, each Y_n is a predetermined and arbitrary quantity from which greater activity levels are charged and lesser activity levels are subsidized at the rate σ_n (vice versa if σ_n is negative). Maximizing profits subject to the newly stated technological constraint for each farm yields the following Lagrangian and optimality conditions.

$$(8)\quad L^j = \mathbf{p}\mathbf{y}^j + \boldsymbol{\sigma}(\mathbf{Y} - \mathbf{y}_j) - \delta^j f^j[\mathbf{y}^j, g(\mathbf{y}^j)]$$

$$p_n - \sigma_n - \delta^j(f_n^j + f_z^j g_n) = 0 \qquad \text{for all } n, j.$$

Society's problem is slightly altered because of the removal of z^j as an independent variable. The social Lagrangian and least-cost efficiency are specified by the following relations.

$$(9)\quad L = \sum_j \mathbf{p}\mathbf{y}^j - \sum_j \alpha^j f^j[\mathbf{y}^j, g(\mathbf{y}^j)] - \mu\left[\sum_j g(\mathbf{y}^j) - Z^*\right]$$

$$p_n - \mu g_n - \alpha^j(f_n^j + f_z^j g_n) = 0 \qquad \text{for all } n, j.$$

Comparison of (8) and (9) identifies the least-cost management incentives as

$$(10)\quad \sigma^*_n = \mu g_n = s^* g_n \qquad \text{for all } n.$$

This system of incentives is not necessarily the same for all farms because derivatives of the nonpoint production function are evaluated at different

[1] Much information would be needed to specify precisely a nonpoint production function. To identify accurately the relevant physical relationships, a simulation model may be needed. Hence, the theoretical function may be very complex in actual application. Although the validity of nonpoint production functions must be examined in each setting, their theoretical consequences are significant since they imply interesting and useful policy and research conclusions.

activity levels. Note that each management incentive σ^*_n is a marginal charge or subsidy depending on whether g_n is positive or negative, respectively. It is expected that g_n is zero for many activities.

For a fourth and final runoff policy, there is a system of standards which is the dual to management incentives. Letting $\pi^j(\mathbf{p}, \boldsymbol{\sigma})$ denote the farm's profit function in the presence of management incentives, Hotelling's lemma can be used to specify regulatory constraints on individual farming activities. These constraints are management practices, and are defined by equation (11).

(11)
$$y_n^{j*} = y_n^j(\mathbf{p}, \boldsymbol{\sigma}^*) = \left|\frac{\partial \pi^j(\mathbf{p}, \boldsymbol{\sigma})}{\partial \sigma_n}\right|_{\mathbf{p},\boldsymbol{\sigma}^*} \quad \text{for all } n, j.$$

The absolute-value operator has been added because the derivative of the profit function with respect to an input price results in the optimal supply of the input (which is negative). This slate of management practices will induce the least-cost achievement of the environmental goal. There is, of course, no need to monitor activities for which $g_n = 0$.

Nonpoint Policy

The four policy alternatives for nonpoint pollution are importantly related to one another. With the usual caveats concerning the omission of income effects, transaction costs, and time, every one of these policies induces the allocatively efficient achievement of the target objective. Least-cost parameters for each of these policies are summarized in table 1.

Even though all four policies are allocatively efficient, they differ in many respects. First is the number of parameters to be specified by the controlling authority. There are, at most, one nonpoint incentive, J nonpoint standards, $N \times J$ management incentives, and $N \times J$ management practices needed to accomplish the objective at least cost. In order for the least-cost goal to be attained in two of these programs, the nonpoint incentive and nonpoint standards, individual farmers must have information on the nonpoint production function and its use. Otherwise, they will not be able to choose production plans to maximize profits. If dissemination of this information is a problem, there is justification for pursuing one of the two management policies.

Another major difference among these programs relates to the amount of policy transaction costs that each policy will incur. Policy transaction costs include the costs of initial information for a specific instance of market failure and of deciding whether or not to invoke a nonmarket allocation mechanism, the costs of policy design, the structural costs of the administering agency, variable enforcement costs (for monitoring, assessment, and litigation), and the costs of periodic policy reevaluation. Just

Table 1. Efficient Nonpoint Policy Parameters

Incentives	Regulations	Type
$s^* = \mu$	$z^{j*} = \dfrac{\partial \pi^j(\mathbf{p}, s)}{\partial s}\Big\vert_{\mathbf{p},s^*}$ for each j	Nonpoint Incentives or Standards
$\sigma^*_n = s^* g_n$ for each j and n	$y_n^{j*} = \left\vert\dfrac{\partial \pi^j(\mathbf{p}, \boldsymbol{\sigma})}{\partial \sigma_n}\right\vert_{\mathbf{p},\boldsymbol{\sigma}^*}$ for each j and n	Management Incentives or Practices

as transaction costs are incurred for the operation of markets, the implementation and management of public policies also involve transaction costs. While many of these costs are the same for alternative policies, some of the variable enforcement costs will be different. These differences can become very important for policy selection when the administering agency has a limited budget.

Other differences include the distribution of costs and the ability of each policy to respond to price changes, technological innovation, and entry/exit. Policy adoption is heavily influenced by the distribution of costs between farmers and government as well as among farmers themselves (Sharp and Bromley). In the generalized format of the preceding model the equity of each incentive program is highly variable. The two incentives programs can result in either net charges or net subsidies for farmers depending on the initial choice of incentive base levels (Z or $\mathbf{Y}$). The regulatory policies also are flexible because varying amounts of lump-sum transfers and cost sharing can be engaged.

Implications for Economic Research

A basic tenet of natural resource economics is that both costs and benefits of public investment or intervention should be jointly assessed before the selection of a particular action. Within least-cost (cost-effective) frameworks for pollution control, benefits are improved environmental quality. Therefore, policy selection requires that environmental effectiveness as well as economic impact be known to decision makers. Published research has concentrated on likely economic impacts (i.e., costs) of various actual or proposed nonpoint policies. This kind of research needs to be complemented with analysis involving nonpoint production functions.

It is also important that each nonpoint policy under consideration be specified so that it is a least-cost means of obtaining a given level of environmental improvement. Often this has not been accomplished (Griffin and Bromley). Therefore, policy costs have been overstated. The policy parameters in table 1 can serve as a guide for improved research. Equations (10) and (11) demonstrate that neither management practices nor management incentives can be economically efficient if they are not specified using the appropriate nonpoint production function(s).[2] The correct application of this information will result in the identification of an optimal set of management practices or incentives. Single management practices or incentives will be economically inefficient.

Other implications for economic research are provided by the dualistic relation between least-cost incentives and least-cost regulations. Recognition of this relationship leads to some important conclusions. First, least-cost efficiency can be achieved with regulations. However, such regulations (either nonpoint standards or management practices) will be different for each farm. These differences are apparent in table 1. Therefore, when regulations are constrained to be equal for all farms or for all farms in each land class, incentive programs will be more efficient. Equity concerns and political realities make equivalent regulations more likely. Under these circumstances incentives will be less costly than regulations achieving the same amount of nonpoint emissions (neglecting policy transaction costs). However, this advantage will be concealed by economic models of agricultural pollution unless they properly reflect the differing productive abilities and resource constraints of individual farms (Jacobs and Casler). Since national and regional models are usually based on land classes and soil types rather than individual farms, this problem does exist.

The dualism between regulations and incentives provides insight in designing least-cost parameters for any of the four fundamental policy types. Assuming that we have an appropriate profit-maximization model incorporating several or many farms and a constraint on total emissions, the optimal nonpoint incentive is the shadow price of the environmental constraint; optimal nonpoint standards are the emissions of each farm; and optimal management practices are the activities undertaken by each farm in the solution to this problem. Only the management incentives require additional calculations. The partial derivatives of the nonpoint production function must be evaluated, and the management incentives must be established according to equation (10). Nonpoint externality theory identifies four policy types. Within the context of this theory, all four policies are equally efficient. However, these policies are not equivalent in equity, policy transaction costs, and political and public acceptability. Therefore, the existence of four distinct policies for nonpoint problems expands the number of opportunities for actual programs.

[2] The universal soil loss equation (USLE) is a nonpoint production function. It is a simple relationship expressing soil loss as a multiplicative function of physical and management parameters. While the USLE actually measures erosion not runoff, some studies using it have achieved limited success in identifying economic and environmental effects of various policies.

The flexibility provided by incentive base levels also improves program options. In the theory developed here, the equity of incentive programs becomes a policy parameter. Therefore, if it is found that one of the two incentive policies offers an important opportunity for the efficient achievement of an environmental goal, then that policy need not be dismissed because it is judged to be inequitable.

Conclusions

The nonpoint character of agricultural runoff renders traditional pollution policies inoperative because these policies must identify the externality contributions of each economic agent. This fact has increased the difficulty of economically analyzing alternative runoff policies. It also may have led to the institution of policies without sufficient attention to their economic consequences.

Economic guidance for nonpoint policy can be obtained by relying upon nonpoint production functions. These functional relationships permit the amount of pollution in the runoff of individual farms to be estimated when their respective production activities are known. The essential feature of a nonpoint production function is that it allows economically efficient policies to be based upon those factors which determine pollution rather than the pollutant itself.

The following statements summarize the major consequences of omitting the nonpoint production function from any economic analysis of agricultural pollution. First, only the costs of prospective policies can be identified. Because the linkage between production activities and nonpoint pollution is not explicitly included in the analysis, it will be impossible to identify policy benefits. Second, because any two policies will differ environmentally as well as economically, no conclusions are possible regarding the relative merits of alternative programs. Third, it will be impossible to calculate a set of management practices or management incentives which are truly "best" in terms of the costs of achieving any given level of pollution abatement. Fourth, nonpoint incentives and nonpoint standards are not realistic policy alternatives because emissions cannot be monitored or estimated.

[*Received April 1981; revision accepted February 1982.*]

References

Baumol, William J., and Wallace E. Oates. *The Theory of Environmental Policy*. Englewood Cliffs NJ: Prentice-Hall, 1975.

Griffin, Ronald C., and Daniel W. Bromley. "Agricultural Runoff as a Nonpoint Externality: Theory, Practice, and Policy." Center for Resource Policy Studies

Work. Pap. No. 15, University of Wisconsin, Aug. 1981.

Jacobs, James J., and George L. Casler. "Internalizing Externalities of Phosphorus Discharges from Crop Production to Surface Water: Effluent Taxes versus Uniform Reductions." *Amer. J. Agr. Econ.* 61(1979):309–12.

Meade, J. E. "External Economies and Diseconomies in a Competitive Situation." *Econ. J.* 62(1952):54–67.

Sharp, Basil M. H., and Daniel W. Bromley. "Agricultural Pollution: The Economics of Coordination." *Amer. J. Agr. Econ.* 61(1979):591–600.

Varian, Hal R. *Microeconomic Analysis*. New York: W. W. Norton & Co., 1978.

[2]

The Relative Efficiency of Agricultural Source Water Pollution Control Policies

James S. Shortle and James W. Dunn

This paper examines the relative expected efficiency of four general strategies which have been proposed for achieving agricultural nonpoint pollution abatement. Emphasis is placed on the implications of differential information about the costs of changes in farm management practices, the impracticality of accurate direct monitoring, and the stochastic nature of nonpoint pollution. The possibility of using hydrological models to reduce, but not eliminate, the uncertainty about the magnitude of nonpoint loadings is incorporated into the analysis. The principal result is that appropriately specified management practice incentives should generally outperform estimated runoff standards, estimated runoff incentives, and management practice standards for reducing agricultural nonpoint pollution.

Key words: differential information, efficiency, nonpoint pollution control.

Although much has been written about the relative efficiency of alternative pollution control policy approaches when emissions are nonstochastic and can be monitored accurately by source, the control of nonpoint sources and, therefore, agricultural sources, presents a different and more complex problem. First, flows of pollutants from nonpoint sources cannot be monitored on a continuous and widespread basis with reasonable accuracy or at reasonable cost. Second, nonpoint pollution is inherently stochastic. For example, weather plays a causal role in the process. These features of nonpoint pollution make the application of the emission-based policy instruments which have been the focus of economic inquiry infeasible.

However, there are means for partially alleviating the problems these aspects of nonpoint pollution pose for the specification of allocatively efficient policies. Models which estimate or predict nonpoint pollutant flows utilizing information on farm management practices, weather, soil characteristics, and other relevant factors have been developed, and research to improve the state of the art continues (Decoursey; EPA 1976, 1979). While such models will never provide error-free predictions and, therefore, a perfect substitute for accurate monitoring of actual flows, they can serve as an important tool for diminishing the uncertainty about nonpoint loadings. Furthermore, predictions obtained from such models offer an alternative to actual flows as a basis for the application of policy instruments (Griffin and Bromley; Harrington, Krupnick, and Peskin).

This paper examines the relative expected efficiency (net benefits) of four general strategies that have been suggested for achieving agricultural nonpoint pollution abatement. Referring to the flow of pollutants from a farm as runoff, the four strategies considered are (*a*) economic incentives applied to estimated runoff (e.g., a tax on estimated soil loss); (*b*) estimated runoff standards (e.g., estimated soil loss standards); (*c*) economic incentives applied to farm management practices (e.g., taxes on nutrient applications); and (*d*) farm management practice standards (e.g., required use of no-till).[1]

Griffin and Bromley evaluate versions of

James S. Shortle is an assistant professor and James W. Dunn is an associate professor, Department of Agricultural Economics and Rural Sociology, Pennsylvania State University.

Pennsylvania State University Agricultural Experiment Station Journal Series No. 7212.

Review was coordinated by Peter Berck, associate editor.

[1] Nonpoint pollution is often referred to as runoff pollution, although it is the constituents of runoff which result in water quality damages. Nevertheless, the analysis and exposition presented in this paper are simplified by focusing on runoff rather than its constituents. The fundamental implications of the analysis would be unaffected by a more general treatment.

these four strategies assuming that (*a*) returns to farmers from alternative management practices are known by public planners and that (*b*) runoff from farms can be observed without error by observing farm management practices and entering the observations in a nonstochastic "nonpoint production function." Adapting the Baumol and Oates framework for examining the characteristics of least-cost pollution control, they conclude that the four approaches are equally efficient when properly specified. Empirical studies of the relative costs of alternative control strategies for agriculture have typically proceeded under similar assumptions (e.g., Miranowski et al., Seitz et al., Taylor and Frohberg).

The analysis presented in this paper incorporates the stochastic nature of runoff and recognizes that models for predicting runoff cannot serve as a perfect substitute for accurate direct monitoring. In addition, it is assumed that farmers have better information about the effects of changes in farm management practices on their profits when choosing how to reallocate resources in response to a nonpoint policy than the public planner has when choosing a policy. The existence of such a differential information structure is often assumed in the literature on choices among policy instruments (Adar and Griffin; Dasgupta; Dasgupta, Hammond, and Maskin; Fishelson; Kerwel; Roberts and Spence; Weitzman 1974, 1978; Yohe 1976, 1978). Moreover, it is worth exploring in the case at hand. Farmers may be expected to acquire specialized information about their farm operations and are notoriously reluctant to reveal their profitability to public agencies. This reluctance may become especially strong if they anticipate that the information will be used to regulate their behavior. Conversely, the public sector actively engages in generating and delivering information of use to farmers. Hence, the flow of information generally promotes the existence of the assumed differential information structure.

The analysis begins with the development of a model of nonpoint pollution for the limiting case of a single farm. Although situations of practical interest generally will involve numerous polluters, the outcome of nonpoint policies depend upon the responses they elicit at the individual farm level. It is initially assumed that the farmer is a price-taking, expected profit maximizer. The implications of multiple sources and risk aversion will be examined subsequently.

The Single-Farm Case

Consider the problem of an environmental agency which must choose a policy for promoting efficient water pollution abatement for a single farm. The agency is unable to observe runoff from the farm at reasonable cost. However, it can form expectations which are conditional upon observations of farm management practices and other relevant data in order to reduce, but not eliminate, the uncertainty about runoff, and thereby enhance the efficiency of the control program. These expectations are viewed as the agency's estimates of runoff under specified circumstances. The agency's choice of a runoff model for computing such estimates, while of interest, is not considered.

The general form of the agency's runoff model is

$$r = g(X, w, \lambda),$$

where r is the true but unobservable flow of runoff from the farm; X is a vector of farm management decisions; w is an index for weather conditions such as rainfall, which have a causal role in the pollution process; and λ is a random variable representing the agency's imperfect knowledge of the runoff function.[2] Particular vectors of farm management decisions, e.g., X_0, define particular farm management practices.

This model essentially represents a stochastic specification of Griffin and Bromley's "nonpoint production function." It incorporates two important sources of uncertainty about runoff faced by the agency when planning a pollution control policy. One is imperfect information about the runoff process itself, represented here by λ, which, along with the unobservability of the flow, precludes the agency from knowing runoff even though it knows the form of $g(\cdot)$. In other words, the model does not offer a perfect substitute for accurate direct monitoring. The second is *ex ante* uncertainty about weather conditions. This uncertainty is captured by treating w as a random variable.

Let $f(w, \lambda)$ be the agency's joint density function for w and λ. Using this function and $g(\cdot)$, two conditional estimates (expectations) of interest may be defined. One is the esti-

[2] As with runoff, replacing w and λ by more descriptive vectors would add more to the complexity than the substance of the analysis.

mated runoff for a given management practice prior to the occurrence of weather:

$$(1) \quad \bar{r}(X) = \int\int g(X, w, \lambda) f(w, \lambda) dw d\lambda.$$

The second is the expected runoff for a given management practice after observing the weather:

$$(2) \quad \bar{\bar{r}}(X, \hat{w}) = \int g(X, \hat{w}, \lambda) f(\hat{w}, \lambda) d\lambda,$$

where $\hat{w}$ denotes the realized value of w.

The water quality damage cost function is written $D(r)$ and assumed to be convex. Although public planners are uncertain about damage cost functions, it is initially assumed that this function is known by the agency. It is shown below that this assumption does not affect the results as long as it is assumed that the agency can evaluate the expected net benefits of its policy choices.

The maximum profit obtainable by the farmer in the absence of public intervention for pollution control is assumed to be given by the function $\pi(X, w, \theta)$, where θ represents specialized knowledge of the operation of the farm. This function is taken to be concave in X for all w and θ. The weather index is included in this function since weather also influences farm profits.

The farm manager is assumed, like the agency, to be uncertain of the weather and to choose a farm management practice prior to its occurrence. In addition, it is assumed that the farmer and the agency have identical information structures for weather. The additional uncertainty faced by the agency about the costs of intervention in farm decision making is introduced by assuming that the farmer knows θ when choosing a management practice, but that the agency does not have this knowledge when choosing a policy.[3] The actual value of θ is denoted $\hat{\theta}$. The results of this analysis would hold under weaker assumptions about the nature of this information differential as long as the farmer's advantage is retained.

An important implication of the differential information about θ is that the agency generally will be uncertain about how the farmer will respond to a policy prior to implementing it and observing the outcome. Specifically, the farmer's response to a policy will depend upon the features of the policy and the farmer's specialized knowledge. Since the agency does not possess this specialized knowledge, it cannot perfectly anticipate the farmer's choice of management practice in response to any policy which does not exhaust the farmer's flexibility.

The realized net benefit of any management practice X may now be written

$$(3) \quad \pi(X, \hat{w}, \hat{\theta}) - D[g(X, \hat{w}, \hat{\lambda})],$$

where $\hat{\lambda}$ denotes the true value of λ. A policy which provides a realized efficient solution will either mandate or otherwise induce the farmer's choice of the practice which maximizes (3). However, the agency cannot choose among policies on the basis of their realized efficiency since it does not know $\hat{w}$, $\hat{\theta}$, and $\hat{\lambda}$ when making its policy decision. As in past work on choices of policy instruments under uncertainty, it is assumed that the agency evaluates policies on the basis of their expected performance with the preferred policy being the one that yields the greatest expected net benefit. Hence, the agency's objective function is generally expressed as

$$(4) \quad E\{\pi(X, w, \theta) - D[g(X, w, \lambda)]\}.$$

Management Practice Incentives

Define a management practice incentive as a tax or subsidy which is directly based upon the management practice chosen by the farmer. The optimal management practice incentive is a tax (positive or negative) of the form $T_1(X) + c_1$, where

$$(5) \quad T_1(X) = \int\int D[g(X, w, \lambda)] f(w, \lambda) dw d\lambda,$$

and c_1 is any constant that does not result in an inefficient decision about whether or not to produce.[4] Note that $T_1(X)$ will be nonlinear if $D(r)$ or $g(\cdot)$ is nonlinear.[5]

The optimality of this incentive can be determined as follows. First, let $X(\hat{\theta})$ be the practice which maximizes

$$(6) \quad \int \pi(X, w, \hat{\theta}) h(w) dw - \int\int D[g(X, w, \lambda)] f(w, \lambda) dw d\lambda,$$

where $h(w)$ is the marginal density function for w. Since (6) is the agency's objective function

[3] This treatment of the differential information structure is consistent with other examinations of the implications of uncertainty regarding costs and benefits for pollution control policy choices.

[4] Dasgupta has discussed an analogous tax for the case in which emissions are nonstochastic and can be accurately monitored at reasonable cost.

[5] It is not necessary that the farmer evaluate the right-hand side of (5) to obtain $T_1(X)$. Rather, the agency can perform this complex task and provide the farmer with a schedule showing the charge or subsidy associated with different choices.

absent its uncertainty about $\hat{\theta}$, $X(\hat{\theta})$ is the practice the agency would choose to maximize (4) if it had the farmer's specialized knowledge. Accordingly, the agency will prefer a policy that attains $X(\hat{\theta})$ to one that does not. This practice may be referred to as the *ex ante* optimum to distinguish it from the practice which maximizes the realized net benefit.

Next, observe that the expected farm profit function after the imposition of the tax $T_1(X) + c_1$ is

$$(7) \quad \int \pi(X, w, \hat{\theta})h(w)dw - [T_1(X) + c_1].$$

Given the definition of $T_1(X)$, this function is identical to (6) up to the constant c_1. Since the maximization of (7) is unaffected by c_1, the farmer will choose $X(\hat{\theta})$ to maximize expected after-tax profit. Hence, the tax induces the farmer to choose the *ex ante* optimal practice.

Management Practice Standards

Define a management practice standard as a set of restrictions on the management practices the farmer may choose and, therefore, on the permissible X. In order to examine the potential performance of this approach, first consider the best management practice the agency can identify to maximize (4) given its prior information. This practice, denoted X_0, maximizes

$$(8) \quad \int\int \pi(X, w, \theta)h(w)v(\theta)dwd\theta - \int\int D[g(X, w, \lambda)]f(w, \lambda)dwd\lambda,$$

where $v(\theta)$ the agency's density function for θ.[6] This alternative specification of (4) differs from (6) in that it utilizes the agency's rather than the farmer's knowledge about θ. As noted above, the assumption that the agency does not know $\hat{\theta}$ implies that it will be uncertain of the farmer's choice in response to any policy which does not exhaust the farmer's flexibility. Hence, the only way to insure that X_0 is adopted by the farmer is to impose the completely restrictive management practice standard $X = X_0$.

It is clear from the foregoing discussion that the agency will prefer $X(\hat{\theta})$ to X_0. Accordingly, the agency will prefer the optimal management practice incentive to the management practice standard $X = X_0$. However, management practice standards need not restrict farmers to a single practice. Consider the general standard $X \epsilon \Omega$, i.e., a standard which constrains the farmer to some set of practices. Such a standard will attain $X(\hat{\theta})$ if two conditions are satisfied: (*a*) $X(\hat{\theta}) \epsilon \Omega$ and (*b*) $E\{\pi[X(\hat{\theta}), w, \hat{\theta}]\} > E[\pi(X, w, \hat{\theta})]$ for any $X \neq X(\hat{\theta})$, $X \epsilon \Omega$. Since the farmer does not choose $X(\hat{\theta})$ in the absence of a nonpoint policy, $X(\hat{\theta})$ will be on the boundary of Ω if both conditions are met. However, to define an Ω that assures that these conditions are satisfied requires that the agency know $\hat{\theta}$. Consequently, the optimal management incentive will be preferred to a general management practice standard unless the agency knows $\hat{\theta}$.

Estimated Runoff Incentives

Define an estimated runoff incentive as a tax or subsidy which is based directly upon either $\bar{r}(X)$ or $\bar{\bar{r}}(X, \hat{w})$ as defined by (1) and (2), respectively. Because both of these quantities are defined with respect to the agency's runoff model, the use of an estimated runoff incentive strategy requires that the agency provide the farmer with two types of schedules. One would show the relationship between the tax or subsidy and the estimated runoff base. The second would show the relationship between the estimated runoff base and management practices. Since the two schedules link the farmer's choice of a practice to the tax or subsidy, an estimated runoff incentive is, effectively, an indirect form of a management practice incentive.

In order to examine whether the agency can specify an estimated runoff incentive that will attain $X(\hat{\theta})$, first consider a tax (positive or negative) on the estimated runoff base $\bar{r}(X)$ having the general form $T_2[\bar{r}(X)] + c_2$, where c_2 is a constant. The function $T_2(\cdot)$ is the schedule showing the relationship between the variable portion of the tax and the estimated runoff base, while $\bar{r}(X)$ is the schedule showing the relationship between the tax base and management practices. The farmer's expected after-tax profit function takes the general form

$$(9) \quad V_1(X, \theta) = \int \pi(X, w, \theta)h(w)dw - \{T_2[\bar{r}(X)] + c_2\}.$$

Since the farmer will choose the practice that maximizes $V_1(X, \hat{\theta})$, a specification of $T_2(\cdot)$ and c_2 which induces the farmer to choose $X(\hat{\theta})$ must be such that $V_1[X(\hat{\theta}), \hat{\theta}] > V_1(X, \hat{\theta})$ for any $X \neq X(\hat{\theta})$. However, since the

[6] It is reasonable to assume that θ and w are independent since the uncertainty about w is due to imperfect knowledge about forthcoming weather, while θ represents the agency's uncertainty about the farmer's specialized knowledge.

agency does not know $\hat{\theta}$, it cannot know $V_1(X, \hat{\theta})$ for any $T_2(\cdot)$ and c_2. Accordingly, the agency cannot know that a given specification will attain $X(\hat{\theta})$ unless the stronger condition, $V_1[X(\hat{\theta}), \hat{\theta}] > V_1(X, \theta)$ for any realizable θ and $X \neq X(\hat{\theta})$, is satisfied. Comparing (5), (6), and (9), it is evident that this stronger condition requires that

$$(10)\quad T_2[\bar{r}(X)] = \int\int D[g(X, w, \lambda)]f(w, \lambda)dwd\lambda.$$

Given the definition of $\bar{r}(X)$, (10) implies that $T_2[E(r)] = E[D(r)]$. This equality can only be satisfied if $E[D(r)] = D[E(r)]$ and $T_2(\cdot)$ is defined as $D(\cdot)$. However, the necessary equality between $E[D(r)]$ and $D[E(r)]$ can hold only when $D(r)$ is linear. Hence, if the agency is uncertain of $\hat{\theta}$, a tax of the form $T_2[\bar{r}(X)] + c_2$ which assures that the farmer chooses $X(\hat{\theta})$ can be defined only if $D(r)$ is linear. Alternatively, since the right-hand side of (10) is also $T_1(X)$, it is evident that an estimated runoff incentive which attains $X(\hat{\theta})$ must be identical to the optimal management practice incentive up to an additive constant when the agency does not know $\hat{\theta}$. This equivalence between the two incentives can be satisfied only if $D(r)$ is linear.

Suppose that the agency knows $\hat{\theta}$. It will still be the case that $D(r)$ must be linear for a tax of the form $T_2[\bar{r}(X)] + c_2$ to be specified to attain $X(\hat{\theta})$. However, the agency has somewhat greater flexibility in defining $T_2(\cdot)$. To demonstrate, suppose that $X(\hat{\theta})$ simultaneously solves the first-order conditions for maximizing (6) and $V_1(X, \hat{\theta})$ for a particular specification of $T_2(\cdot)$. Given the definition of $\bar{r}(X)$, it must then be true that

$$(11)\quad \frac{\partial E[D(r)]}{\partial x_i} = T_2'\{\bar{r}[X(\hat{\theta})]\}\frac{\partial E(r)}{\partial x_i},\quad i = 1, 2, \ldots, n,$$

holds at $X(\hat{\theta})$, where x_i is the ith element of X. Hence, it must be the case that $\partial E[D(r)]/\partial x_i$ and $\partial E(r)/\partial x_i$, $i = 1, 2, \ldots, n$, are equally proportional at $X(\hat{\theta})$ if a tax of the form $T_2[\bar{r}(X)] + c_2$ is to attain $X(\hat{\theta})$. For this to be the case, $E[D(r)]$ must be linear in $E(r)$, which requires that $D(r)$ is linear. However, if this is the case, the agency has some flexibility in specifying the tax schedule $T_2(\cdot)$ since $T_2[\bar{r}(X)]$ must be identical to $D[\bar{r}(X)]$ only at $X(\hat{\theta})$ to satisfy (11).

Now consider a tax on the estimated runoff base $\bar{r}(X, \hat{w})$ having the general form $T_3[\bar{\bar{r}}(X, \hat{w})] + c_3$, where c_3 is a constant. Since the farmer does not know $\hat{w}$ when choosing a management practice, his expected after-tax profit function takes the general form

$$(12)\quad V_2(X, \theta) = \int(\pi(X, w, \theta) - \{T_3[\bar{\bar{r}}(X, w)] + c_3\})h(w)dw.$$

Applying the logic of the preceding case, a specification of the tax which assures the agency that the farmer will choose $X(\hat{\theta})$ when the agency does not know $\hat{\theta}$ must be such that $V_2[X(\hat{\theta}), \hat{\theta}] > V_2(X, \theta)$ for any realizable θ and $X \neq X(\hat{\theta})$. This condition requires that

$$(13)\quad T_3[\bar{\bar{r}}(X, w)] = \int D[g(X, w, \lambda)]\left[\frac{f(w, \lambda)}{h(w)}\right]d\lambda,\ h(w) \neq 0.$$

As with (10), this equality can hold only if $D(r)$ is linear.

Intuitively, an incentive applied to either estimated runoff base cannot fully convey the agency's preferences to the farmer unless the expected damage cost function can be written as a function of expected runoff, which requires a linear damage cost function. With the exception of the linear case, higher moments than the mean of the distribution of runoff influence the expected damage cost. Accordingly, the effects of the farmer's choices on higher moments must be imbedded in an incentive scheme which attains $X(\hat{\theta})$. Alternatively, an indirect management practice incentive cannot convey the same information to the farmer as the optimal management practice incentive unless the two incentives are essentially identical.[7] This can be the case only when $D(r)$ is linear.

Estimated Runoff Standards

Estimated runoff standards are restrictions on the permissible values of either $\bar{r}(X)$ or $\bar{\bar{r}}(X, \hat{w})$. Although the farmer chooses X prior to knowing the weather, a standard imposed upon $\bar{\bar{r}}(X, \hat{w})$ need not put the farmer in the position of making decisions in the face of a stochastic constraint. For example, a probabilistic standard of the form $P[\bar{\bar{r}}(X, \hat{w}) \geq r^*] \leq \alpha$ implies a constraint on moments of $\bar{\bar{r}}(X, w)$ (Beavis and Walker). In any case, since estimated runoff is defined with respect to the agency's runoff model, the use of an estimated runoff standard will require that the agency

[7] While the policies would be mathematically equivalent, they would differ administratively.

provide the farmer with a schedule showing the practices which satisfy the standard. This implies that estimated runoff standards are indirect forms of general management practice standards. For example, if a particular runoff standard is satisfied by any $X\epsilon\beta$, the agency is in effect imposing a general management practice standard of this form. For the estimated runoff standard to attain $X(\hat{\theta})$, it must be the case that $X(\hat{\theta})\epsilon\beta$ and $E\{\pi[X(\hat{\theta}), w, \hat{\theta}]\} > E[\pi(X, w, \hat{\theta})]$ for any $X \neq X(\hat{\theta})$, $X\epsilon\beta$. Given the agency's inability to specify a general management practice standard that assures the farmer's choice of $X(\hat{\theta})$ when it is uncertain of $\hat{\theta}$, it cannot specify an estimated runoff standard to attain this result.

As in the case of estimated runoff incentives, the ability of the agency to specify estimated runoff standards that attain $X(\hat{\theta})$ when the agency knows $\hat{\theta}$ depends upon the form of $D(r)$. To illustrate, consider a standard of the form $\bar{r}(X) \leq r^*$ and suppose that $X(\hat{\theta})$ simultaneously solves the first-order conditions for maximizing (6) and $E[\pi(X, w, \hat{\theta})]$ subject to $\bar{r}(X) \leq r^*$. It must then be true that

$$(14) \quad \frac{\partial E[D(r)]}{\partial x_i} = \rho\{\bar{r}[X(\hat{\theta})]\}\frac{\partial E(r)}{\partial x_i}, \quad i = 1, 2, \ldots, n,$$

holds at $X(\hat{\theta})$ where $\rho\{\bar{r}[X(\hat{\theta})]\}$ is the shadow price of the constraint $\bar{r}(X) \leq r^*$ at the optimum. Since $X(\hat{\theta})$ must be on the boundary of the set defined by this constraint, it must be the case that $\bar{r}[X(\hat{\theta})] = r^*$. Hence, setting an estimated runoff standard of the form $\bar{r}(X) \leq \bar{r}[X(\hat{\theta})]$ will attain $X(\hat{\theta})$ if (14) holds at the optimum. However, from the previous discussion of estimated runoff incentives, it is evident that (14) can hold at $X(\hat{\theta})$ only if $D(r)$ is linear. Accordingly, even if the agency knows $\hat{\theta}$, it cannot specify an estimated runoff incentive of the form $\bar{r}(X) \leq r^*$ to attain $X(\hat{\theta})$ unless $D(r)$ is linear.

Although the agency may specify more complex forms of estimated runoff standards, such as the probabilistic constraint mentioned above, the capacity of the approach to attain $X(\hat{\theta})$ even when the agency knows $\hat{\theta}$ is inherently constrained by focus on one moment of the distribution of runoff. While the agency can define a general management practice standard that attains $X(\hat{\theta})$ when it knows $\hat{\theta}$, it may not be able to define an estimated runoff standard that implies an optimal management practice standard.

Damage Cost Uncertainty

Damage cost uncertainty has been present in the foregoing analysis. However, it has been entirely due to uncertainty about runoff. It is useful to pause at this point to consider the implications of damage cost uncertainty for reasons other than runoff uncertainty.

Suppose that the agency is uncertain of the damage cost resulting from a given level of runoff but can formulate a conditional expectation of this cost. Examination of this case requires some modifications in the analysis but the results are unaffected. To demonstrate, let $D(r, \epsilon)$ denote the damage cost when r is the runoff level and ϵ is a random variable representing the agency's uncertainty about the cost of the runoff level. It is reasonable to assume that ϵ is independently distributed with respect to θ since ϵ represents uncertainty about the effects of runoff, while θ represents uncertainty about the farmer's specialized knowledge. However, ϵ and w may not be independent since the water quality effects of a given runoff level will be affected by stream conditions, which are in turn influenced by the weather. Hence, in the interest of generality, let $k(w, \lambda, \epsilon)$ be the joint density function of w, λ, and ϵ. The marginal density function for w is still written $h(w)$.

The *ex ante* optimal management practice is now defined as the practice which maximizes

$$(15) \quad [\int\pi(X, w, \hat{\theta})h(w)dw] - \{\int\int\int D[g(X, w, \lambda), \epsilon]k(w, \lambda, \epsilon)dwd\lambda d\epsilon\}.$$

The farmer's choice of this practice can again be obtained by imposing a management practice tax which is the second bracketed term of (15). However, because the practice is contingent upon the farmer's specialized knowledge, it remains the case that management practice standards cannot be defined to induce the farmer to choose the practice unless the agency has acquired this knowledge. It follows from this that estimated runoff standards cannot obtain the *ex ante* optimal choice since such standards imply general management practice standards. Whether an estimated runoff incentive can induce the *ex ante* optimal choice will again depend upon whether the damage cost function is linear in runoff.

To go a step further, many economists argue that too little is known about damage costs to use them in specifying pollution control incentives. If this is the case, it must also be the case that too little is known to choose between policies on the basis of their expected net benefits. Accordingly, an alternative criterion must be used to choose between policies. Identifying the range of feasible alternatives and evaluating the relative performance of the policies under consideration with respect to them is beyond the scope of this paper. However, it can be said that the results derived above will hold for a criterion which involves substituting a runoff preference function provided by public decision makers for the unknown damage cost function based on the preferences of individuals. Support for such an approach can be found in the considerable literature on evaluating public projects with benefits which cannot be measured in dollar terms (e.g., Eckstein, Freeman).

Multiple Sources and Risk Aversion

The results of the analysis to this point may be summarized as follows: First, setting aside policy transaction costs, Griffin and Bromley's conclusion that the four approaches can be equally efficient when appropriately specified holds for the special case of a single polluting farm which is operated by an expected profit-maximizing farmer given that the farmer has no informational advantage over the agency and that the damage cost function is linear. Under these conditions, each of the four strategies can provide the maximum of (4). Second, allowing for a differential information structure results in a situation in which incentive schemes can be specified to provide a greater expected net benefit than quantity control schemes. The logic behind this is straightforward. An appropriately specified incentive scheme transforms the farmer's objective function in such a manner that the farmer maximizes the agency's objective function while utilizing his specialized knowledge about the farm operation to obtain a better outcome than the agency is able to achieve by quantity controls. Finally, allowing for both a differential information structure and a nonlinear damage cost function results in a situation in which only an appropriately specified management practice incentive is optimal. Estimated runoff incentives fail in this case because the incentive cannot be specified to fully internalize the expected damage cost when the damage cost function is nonlinear.

The remainder of this section examines the implications of relaxing two important assumptions which were used to simplify the preceding analysis: first, that there is a single polluting farm and, second, that the farmer is an expected profit maximizer. Together, these assumptions imply that the farmer's objective function can be made equivalent to the agency's by the imposition of a tax which is the expected damage cost function for runoff from the farm. This tax defines the optimal management practice incentive up to an additive constant. Moreover, if the damage cost function is linear, this tax also defines an optimal estimated runoff incentive. However, the specification of an optimal incentive becomes considerably more complex if the farmer is risk averse and/or there are multiple sources.

Multiple Sources

Water pollution problems of practical interest generally will involve a number of polluters and, often, a mix of point and nonpoint sources. In order to examine the issues which emerge when there are multiple sources, consider a situation in which a water quality problem is caused by the runoff from n farms. Each farmer is an expected profit maximizer and has specialized information about the operation of his particular farm. The profit function for the ith farm is $\pi_i(X_i, w_i, \theta_i)$, $i = 1, 2, \ldots, n$. The runoff from each farm is unobservable but can be estimated in a manner analogous to that outlined previously. The damage cost function is $D(r_1, r_2, \ldots, r_n)$, where r_i is the runoff from the ith farm.

Define the first-best solution as the set of farm management practice vectors which maximize the sum of expected farm profits less the expected damage cost of runoff from the n farms. There are two important points to be made about these vectors. First, as long as the damage cost function is not linearly separable in the runoff from each of the n farms, the damage cost due to runoff from one farm is contingent upon the runoff from the remaining farms. This implies that the vectors which provide the first-best solution must be jointly determined. In other words, the first-best vector for one farm is contingent upon the first-best vectors for the remaining farms. Second, it is evident from the foregoing analysis that a solution which maximizes expected profits less expected damage costs must utilize farmers' specialized knowledge. Together, these considerations imply that the first-best vector for any given farm generally will be contingent on the specialized knowledge of all n farmers. Accordingly, the first-best vector for the ith farm is denoted $X_i(\hat{\theta}_1, \hat{\theta}_2, \ldots, \hat{\theta}_n)$.

Can any of the four strategies under consideration be specified to attain the first-best solution? It is clear from the analysis of the preceding section that management practice standards cannot attain the first-best solution given the environmental agency's uncertainty about $\hat{\theta}_1, \hat{\theta}_2, \ldots, \hat{\theta}_n$. Furthermore, because estimated runoff standards imply management practice standards, it is also clear that estimated runoff standards cannot attain the first-best solution with this type of uncertainty on the part of the agency. Hence, neither of the quantity control schemes can be first-best policies. Consider management practice taxes of the form $T_{im}(X_i) + c_{im}$ $i = 1, 2, \ldots, n$, where the subscript m is to distinguish these taxes from those considered previously. For the ith farmer to choose $X_i(\hat{\theta}_1, \hat{\theta}_2, \ldots, \hat{\theta}_n)$ to maximize $E[\pi_i(X, w, \hat{\theta}_i)]$ less the quantity $T_{im}(X_i) + c_{im}$, the function $T_{im}(X_i)$ must incorporate the information $\hat{\theta}_j$, $i \neq j, j = 1, 2, \ldots, n$. Clearly, the agency cannot provide such a specification if it does hold this information. Hence, management practice taxes of this form cannot be designed to achieve the first-best optimum. Similar reasoning implies that estimated runoff taxes cannot be specified to attain the first-best optimum.

From this discussion, it is evident that the agency's relative uncertainty about profits poses a significant barrier to the attainment of the first-best optimum when there are multiple sources. This problem has been previously recognized and has motivated an interest in pollution control schemes which elicit truthful transfers of specialized knowledge to environmental regulators to achieve a first-best solution. Incentive schemes which can achieve such transfers in principle have been identified for the case of nonstochastic emissions which can be readily monitored by source (Dasgupta, Hammond, and Maskin). Undoubtedly, such mechanisms could be identified to provide such transfers from nonpoint sources. However, the administrative complexity of incentive schemes for inducing individuals to reveal truthfully information to policy makers when such transfers are opposed to economic self-interest severely limit their practical applicability (Boadway, Dasgupta). Accordingly, an appropriate policy may pursue a second-best solution, which can be achieved with information more easily obtained by the policy maker. In light of this possibility and the expressed interest in the four strategies examined here, the relative performance of these strategies remains an issue of importance.

The advantage of management practice incentives relative to the other strategies in the single-farm case is due to two factors. First, they permit farmers to utilize fully specialized knowledge of their own farm operations. Second, such incentives can convey at least as much, and generally more, information to farmers about the expected external costs of their management decisions than the quantity control schemes and estimated runoff incentives. These features do not vanish when multiple sources are allowed for. Hence, an appropriately specified management practice incentive should generally yield a greater expected net benefit than management practice standards, expected runoff standards, and estimated runoff incentives.

Risk Aversion

The issues which emerge with risk aversion on the part of the farmer may be illustrated by returning to the single-farm case and assuming that the farmer maximizes expected profit less a weighted variance of profit where the weight is a coefficient of risk aversion. Although the mean-variance structure is consistent with expected utility theory only under certain restrictive circumstances, it is sufficient for demonstrating the implications of risk aversion. With this assumption, the farmer's objective function after the imposition of the tax $T_1(X) + c_1$ is

$$(16) \quad \int \pi(X, w, \hat{\theta})h(w)dw - \alpha \mathrm{Var}[\pi(X, w, \hat{\theta})] - [T_1(X) + c_1]$$

where α is the coefficient of risk aversion. If the farmer is not risk neutral ($\alpha \neq 0$), the gain from exploiting the farmer's specialized knowledge via the tax $T_1(X) + c_1$ will be offset to some extent by the fact that the farmer is no longer maximizing the agency's objective function. The extent of the expected loss depends upon the degree of the risk aversion.

If the agency believes that the farmer is strongly influenced by risk, it will wish to choose a policy that accounts for risk preferences. It can be shown that whether the farmer is risk neutral or not, a management practice tax $T_4(X) + c_4$ will attain $X(\hat{\theta})$ where

$$(17) \quad T_4(X) = -\alpha \mathrm{Var}[\pi(X, w, \hat{\theta})] + T_1(X)$$

and c_4 is some constant. There is, however, a severe practical difficulty with this first-best management practice incentive which will also characterize other policy schemes that attempt to attain $X(\hat{\theta})$. This difficulty is that the tax depends upon perfect knowledge of the farmer's risk preferences and specialized knowledge of the farm operation. Hence, the problem of eliciting truthful revelation of the specialized knowledge of the farm operation which emerges with multiple sources is compounded by the need for truthful revelation of risk preferences.

Nevertheless, management practice incentives still appear to be relatively advantageous when compared to the other three strategies considered in this analysis. As in the case of multiple sources, the two features of management practice incentives which provide them with a comparative advantage over the other three strategies do not vanish because there is risk aversion. It remains that such incentives can reveal the agency's evaluation of the expected external cost of farmer's decisions to a greater degree than the other three strategies while still permitting the farmer to utilize fully his informational advantage about the returns from alternative management practices to maximize his welfare.

It is worth noting that estimated runoff incentives and some forms of estimated runoff standards based upon $\bar{r}(X, \hat{w})$ as opposed to $\bar{r}(X)$ will impose additional uncertainty on the farmer because the farmer chooses a practice prior to knowing the weather. When the farmer is risk averse, this additional uncertainty represents a policy-induced cost. Yet, since the farmer cannot act upon the additional information that becomes available when the weather is realized to increase profits or reduce damage costs, there is no offsetting gain. Hence, such policy-induced uncertainty should be avoided.

Conclusions

The ineffectiveness of existing programs for nonpoint source water pollution abatement along with the leading role of nonpoint sources in the nation's remaining water quality problems has led to a renewed interest in policy options for nonpoint control (EPA 1984; GAO; Harrington, Krupnick, and Peskin; Libby; Savage; Thomas). This interest is focused on agriculture as the most pervasive nonpoint source.

This paper has examined the relative expected efficiency of four general strategies which have been proposed for achieving agricultural nonpoint pollution abatement. Emphasis was placed on the implications of differential information about the costs of changes in farm management practices, the impracticality of accurate direct monitoring, and the stochastic nature of nonpoint pollution. The possibility of using hydrological models to reduce, but not eliminate, the uncertainty about the magnitude of nonpoint loadings was incorporated into the analysis.

Setting aside policy transactions costs, the principal result of this analysis is that an appropriately specified management practice incentive should generally outperform estimated runoff standards, estimated runoff incentives, and management practice standards for reducing agricultural nonpoint pollution. The logic behind this result is that an appropriate management practice incentive has a greater capacity to induce a farmer to choose management practices to maximize the expected social net benefits of his decisions than the other three strategies. It is worth noting that it is the quality rather than the quantity of information conveyed to the farmer by management practice incentives that is important. For example, where an estimated runoff incentive requires that the farmer be provided with an estimated runoff schedule in addition to a tax or subsidy schedule, a management practice incentive requires only a tax or subsidy schedule. It is emphasized, however, that none of the four strategies can be defined to attain a first-best optimum, as defined in this analysis, where there are multiple sources, and/or, farmers are risk averse.

The relative performance of management practice incentives is of particular interest under present political conditions. If is often the case that policy approaches which fare well on economic grounds are politically unacceptable. Yet, in the case of pollution control for agricultural sources, prevailing political preferences appear to favor management practice incentive schemes. Consequently, it would seem that a well-specified management practice incentives approach may be politically acceptable as well as economically advantageous.

[*Received July 1984; final revision received June 1985.*]

References

Adar, Zvi, and James M. Griffin. "Uncertainty and the Choice of Pollution Control Instruments." *J. Environ. Econ. Manage.* 3(1976):178–88.

Baumol, William J., and Wallace E. Oates. *The Theory of Environmental Policy.* Englewood Cliffs NJ: Prentice-Hall, 1975.

Beavis, Brian, and Martin Walker. "Achieving Environmental Standards with Stochastic Discharges." *J. Environ. Econ. Manage.* 10(1983):103–11.

Boadway, Robin W. *Public Sector Economics.* Cambridge MA: Winthrop Publishers, 1979.

Dasgupta, Partha. "Environmental Management Under Uncertainty." *Explorations in Natural Resource Economics,* ed. V. Kerry Smith and J. V. Krutilla, pp. 109–39. Baltimore MD: Johns Hopkins University Press for Resources for the Future, 1982.

Dasgupta, Partha, Peter J. Hammond, and Eric Maskin. "A Note on Imperfect Information and Optimal Pollution Control." *Rev. Econ. Stud.* 47(1980):857–60.

Decoursey, D. G. "ARS Small Watershed Model." Collaborative Paper CP-82-89, International Institute of Applied Systems Analysis, Laxenburg, Austria, 1983.

Eckstein, Otto. "A Survey of Public Expenditure Criteria." *Public Finance: Needs, Sources, and Utilization,* pp. 439–504. Universities–National Bureau Committee on Economic Research. Princeton NJ: Princeton University Press, 1961.

Fishelson, Gideon. "Emissions Control Policies Under Uncertainty." *J. Environ. Econ. Manage.* 3(1976): 189–97.

Freeman, A. Myrick, III. "Project Design and Evaluation with Multiple Objectives." *Public Expenditure and Policy Analysis,* 2nd ed., ed. Robert H. Haveman and Julius Margolis, pp. 239–56. Chicago: Rand-McNally & Co., 1977.

Griffin, Ronald, and Daniel Bromley. "Agricultural Runoff as a Nonpoint Externality." *Amer. J. Agr. Econ.* 64(1982):547–52.

Harrington, Whinston, Alan J. Krupnick, and Henry M. Peskin. "Policies for Nonpoint-Source Water Pollution Control." *J. Soil and Water Conserv.* 40 (1985):27–32.

Kerwel, Evan. "To Tell the Truth: Imperfect Information and Optimal Pollution Control." *Rev. Econ. Stud.* 44(1977):595–601.

Libby, Lawrence W. "Paying the Nonpoint Pollution Control Bill." *J. Soil and Water Conserv.* 40 (1985):33–36.

Miranowski, John, Michael J. Monson, James S. Shortle, and Lee D. Zinser. *Effect of Agricultural Land Use on Stream Water Quality: Economic Analysis.* Final Report to the U.S. Environmental Protection Agency. Springfield VA: National Technical Information Service, 1983.

Roberts, Marc J., and Michael Spence. "Effluent Charges and Licenses under Uncertainty." *J. Public Econ.* 5(1976):193–208.

Savage, Roberta. "State Initiatives in Nonpoint Source Pollution Control." *J. Soil and Water Conserv.* 40(1985):53–54.

Seitz, W. D., P. M. Gardner, S. K. Grove, K. L. Guntermann, J. R. Karr, R. G. F. Spitze, E. R. Swanson, C. R. Taylor, D. L. Uchtmann, and J. C. Van Es. *Alternative Policies for Controlling Nonpoint Agricultural Sources of Water Pollution.* Washington DC: U.S. Environmental Protection Agency, Environmental Research Laboratory, April 1978.

Taylor, Robert C., and Klaus K. Frohberg. "The Welfare Effects of Erosion Controls, Banning Pesticides, and Limiting Fertilizer Application in the Corn Belt." *Amer. J. Agr. Econ.* 59(1977):25–36.

Thomas, Lee M. "Viewpoint: Management of Nonpoint Source Pollution: What Priority? *J. Soil and Water Conserv.* 40(1985):8.

U.S. Environmental Protection Agency, Environmental Research Laboratory. *Effectiveness of Soil and Water Conservation Practices for Pollution Control.* Washington DC, Oct. 1979.

———, Office of Research and Development. *Loading Functions for Assessment of Water Pollution from Nonpoint Sources.* Washington DC, May 1976.

———, Office of Water Program Operations. *Report to Congress: Nonpoint Pollution in the U.S.* Washington DC, Jan. 1984.

U.S. General Accountinug Office. *Cleaning Up the Environment: Progress Achieved but Major Unresolved Issues Remain.* Washington DC, July 1982.

Weitzman, Martin. "Optimal Rewards for Economic Regulation." *Amer. Econ. Rev.* 68(1978):683–91.

———. "Prices vs. Quantities." *Rev. Econ. Stud.* 41(1974):477–91.

Yohe, Gary. "Substitution and the Control of Pollution. A Comparison of Effluent Charges and Quantity Standards under Uncertainty." *J. Environ. Econ. and Manage.* 3(1976):312–24.

———. "Towards a General Comparison of Price Controls and Quantity Controls under Uncertainty." *Rev. Econ. Stud.* 45(1978):229–38.

[3]

JOURNAL OF ENVIRONMENTAL ECONOMICS AND MANAGEMENT 15, 87–98 (1988)

Uncertainty and Incentives for Nonpoint Pollution Control

KATHLEEN SEGERSON*

Department of Economics, University of Connecticut, Storrs, Connecticut 06268

Received May 14, 1985; revised June 1986

In dispersed or nonpoint pollution problems, monitoring of individual polluting actions is difficult and those actions cannot generally be inferred from observed ambient pollution because (i) ambient pollutant levels have a random distribution that is contingent on the level of abatement undertaken and/or (ii) the actions of several polluters contribute to the ambient levels and only combined effects are observable. This paper describes a general incentive scheme for controlling nonpoint pollution. Rewards for environmental quality above a given standard are combined with penalties for substandard quality. The mechanism is discussed in the context of both a single suspected polluter and multiple suspected polluters where free riding must be avoided. © 1988 Academic Press, Inc.

At least theoretically, appropriate reductions in pollution from point sources can be achieved by direct regulation or by a system of effluent charges, with transferable discharge permits offering a promising compromise to the practical problems of each. However, the appropriate economic incentives for control of nonpoint pollution (NPP) have not yet been addressed adequately at either a theoretical or a practical level. For example, the suggestion that "best management practices" (BMPs) be required to reduce nonpoint surface pollution does not allow for flexibility and cost-minimizing abatement strategies unless applied on a site-specific basis, which is generally impractical. Likewise, the suggested use of a soil loss tax to reduce agricultural NPP ignores the important distinction between "discharges" and the resulting pollutant levels that determine damages, since lands with high erosion rates are not necessarily those causing significant NPP problems and vice versa.

The standard solutions that have been successful in controlling point source problems are unworkable for NPP partly because it is generally not possible to observe (without excessive cost) the level of abatement or discharge of any individual suspected polluter or to infer those levels from observable ambient pollutant levels. There are two possible reasons for the inability to infer behavior from observed outcomes: (1) given any level of abatement, the effects on environmental quality are uncertain due to stochastic variables,[1] i.e., there is not a one-to-one relationship between discharge and ambient levels, or (2) the emissions of several

*Most of the work on this paper was done while the author was an Assistant Professor of Agricultural Economics, at the University of Wisconsin. The author thanks Michael Carter, Jean-Paul Chavas, Daniel Bromley, and two anonymous reviewers for helpful comments on earlier drafts, without implicating them for any remaining errors. This research was funded in part by the Water Resources Center and the College of Agricultural and Life Sciences at the University of Wisconsin.

[1] This may also be true for point sources of pollution. Many have studied the role of uncertainty in the control of point sources, including [1, 5, 6, 9, 12, 15, 16]. However, their discussion of policies is limited to those that apply directly to emissions. In addition, the analyses do not consider the problem of multiple polluters where only the joint impact of their behavior is observable. Thus, these analyses do not adequately address the policy questions that are relevant in the control of stochastic nonpoint pollution.

0095-0696/88 $3.00

polluters contribute to the ambient levels and only combined effects are observable. It is these characteristics that have made control of NPP so elusive, and policy instruments designed to address NPP must recognize them.

This paper describes an economic incentive scheme that could be used to control NPP even in the presence of uncertainty and monitoring difficulties. The general mechanism combines rewards for water quality above a given standard with penalties for substandard water, although a special case includes only penalties. It can be applied either when there is a single suspected polluter or when there are several suspected polluters. In the latter case it can be designed to eliminate problems of free riding.[2]

It should be noted that, although the discussion of economic incentives here is in the context of nonpoint surface water pollution, the results are applicable to other dispersed pollution problems characterized by uncertainty and monitoring difficulties, such as many cases of groundwater contamination and acid rain.

1. UNCERTAINTY AND INCENTIVES

The physical uncertainty feature of nonpoint pollution problems—that the ambient pollutant levels resulting from any given operating practice depend on a number of climatic and topographic conditions in a manner that cannot be predicted with certainty—implies that there will be a range of possible ambient levels associated with any given abatement practice or discharge level at any given time. More generally, there is a range of possible damages in terms of the impacts on human health and welfare that depend not only on pollutant levels but also on factors such as stream flow and exposure risks. The analysis could be applied to this broad range of impacts, but for simplicity we focus here only on the range of possible ambient levels. This range can be represented by a probability density function (pdf) that is conditional on the abatement practice. The pdf gives the probability that ambient pollutant levels of a given magnitude will occur at the specified time, where the probability depends on the abatement practices being used. The objective of pollution control policies is then to increase the probability that ambient levels will fall below some tolerance level, i.e., to shift the pdf to the left, as illustrated in Fig. 1, so that the new distribution dominates the old one in the sense of first-order stochastic dominance.[3]

If direct monitoring of all firm operations were economically feasible or voluntary compliance with regulations were guaranteed, then the distribution could be shifted through site-specific mandatory abatement practices. Alternatively, even in the absence of direct monitoring, direct regulation can be used if it is possible to infer the actions of an individual polluter (and thus detect noncompliance) from an observation of ambient pollutant levels. This would be possible, for example, if there were a single polluter whose emissions entered a given body of water and if the relationship between his discharge and ambient pollutant levels were determinis-

[2]Free rider problems in the context of pollution control have been discussed by Dasgupta [5]. In his model the need for incentives arises from an information gap rather than a monitoring problem. He devises an incentive compatible scheme to ensure correct revelation of preferences for improvements in environmental quality.

[3]In the absence of uncertainty, the distribution simply collapses to its mean and the objective is then merely to reduce the non random ambient pollutant levels.

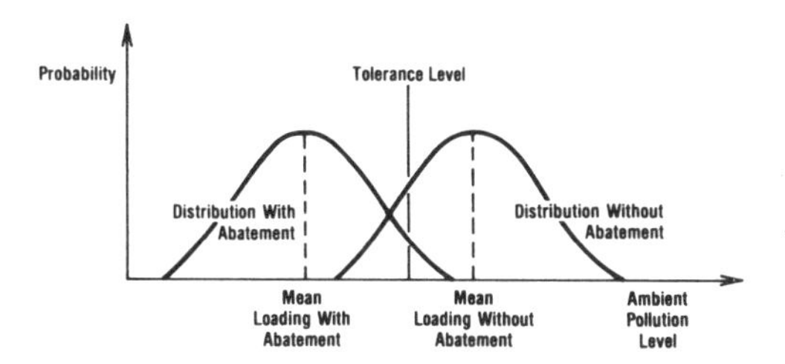

FIGURE 1

tic. However, when that relationship is stochastic, then the actions or discharge of the single polluter cannot be inferred from observed ambient levels. In this case, a mechanism that provides an incentive (either positive or negative) for compliance must be used instead of direct regulation of the polluter's discharge.

A similar problem arises when many polluters jointly contribute to ambient pollutant levels and the actions or discharge levels of the individual polluters cannot be directly observed. Since ambient pollutant levels depend on the behavior of all polluters, it is not possible to infer the actions of individual polluters from observations on ambient levels. Direct regulation of individual actions is not possible in this case either, regardless of whether the joint effect of individual actions on ambient levels is stochastic or deterministic. Because a deterministic relationship between discharges and ambient pollutant levels is a special case of the more likely stochastic relationship, in Section 3, we discuss the multiple polluter problem in the context of uncertainty even though this is not a necessary condition for incentive problems to arise. This discussion highlights the similarity between the single and multiple polluter cases: For both, it is possible to rely on an incentive mechanism based on the observable variable (ambient pollutant levels) to induce certain unobservable actions.

2. SINGLE POLLUTER PROBLEM

Consider first the problem when there is only one suspected polluter. Let x be the ambient level of a given pollutant in the stream, and let $\bar{x}$ be a specified cutoff level, which is set by authorities.[4] The ambient level x will depend upon both the abatement actions taken by the polluter and the random variables reflecting unpredictable weather and stream conditions, as illustrated in Fig. 1.

A general incentive scheme designed to shift the distribution of ambient levels could take the form of automatic, required payments $T(x)$ that depend upon the

[4] In the short run the choice of $\bar{x}$ is somewhat arbitrary because it does not affect the socially optimal level of abatement undertaken. However, it does affect the values for t and k (the parameters of the incentive scheme) that are necessary to ensure that polluters choose that optimal level. The choice may also be important in terms of political acceptability and certainly determines the financial impacts of the incentive scheme, since it determines the cutoff for taxes/subsidies and penalties. Thus, it affects the long run market equilibrium position. This is discussed more fully below.

ambient level of pollutants as compared to the cutoff level $\bar{x}$ and are given by

$$T(x) = \begin{cases} t(x - \bar{x}) + k & \text{if } x > \bar{x} \\ t(x - \bar{x}) & \text{if } x \leqq \bar{x}. \end{cases}$$

The regulating authority sets the constants t and k so that the payment scheme provides the incentive necessary to induce the polluter to undertake the level of abatement that is deemed socially desirable.

The payment scheme is composed of two parts. The first, reflected in t, is a tax/subsidy payment that depends upon the extent to which x differs from $\bar{x}$. If ambient levels exceed the cutoff level, the suspected polluter pays a tax proportional to that excess, while ambient levels below the cutoff result in a subsidy or credit to the polluter. Note that ambient levels may differ from cutoff levels because of either the abatement actions of the polluter or the influence of the random variables. Thus, the polluter may be liable for tax payments that result from influences outside his control.[5] Likewise, his liability may be reduced (and he will even receive subsidies if x falls below $\bar{x}$) due to favorable environmental conditions even if he has taken no action to control pollution. In choosing his level of abatement, he gambles on what his tax liability will be and weighs the additional cost of pollution abatement against the decrease in expected payments that results from increased abatement (see further discussion below). However, once $\bar{x}$ is fixed, this feature will also allow him to take advantage of the naturally fluctuating assimilative capacity of the waterway. During periods of high stream flow when the waterway has a large assimilative capacity, the level of abatement can be reduced; and during periods of low assimilative capacity, firms will have an incentive to curtail polluting activities.

The same type of incentive is provided by the second component of the payment scheme, reflected in k, which is a fixed penalty imposed whenever ambient levels exceed the cutoff.[6] The amount of the penalty is independent of the amount by which the cutoff is exceeded. When deciding on additional abatement, the suspected polluter can again weigh the cost of that abatement against the resulting decrease in the probability that x will exceed $\bar{x}$, i.e., that he will incur the penalty. Note that the effect of this penalty scheme differs from that of penalties applied to actions (or inactions) that are directly under the control of the polluter (e.g., penalties for point emissions in violation of standards). In the stochastic case where penalties depend upon ambient levels rather than emissions, there will always be an incentive for additional abatement since it will decrease the expected penalty by decreasing the probability that x will exceed $\bar{x}$. In contrast, when the penalties are for emissions in

[5]Holmstrom [10] has shown that an incentive scheme can be improved by releasing an agent from liability for outcomes that are clearly outside of his control and could not have been influenced by the agent's actions, e.g., natural disasters. Thus, for example, farmers should not be held liable for loadings clearly attributable to an external cause such as a chemical spill from an upstream manufacturing plant. Holmstrom [10] and Shavell [14] investigate the conditions under which additional variables or signals that provide some generally imperfect information about the agent's actions can be used to improve incentive contracts. Although the use of additional information is not discussed here explicitly, appropriate modifications to the analysis could be made.

[6]An alternative formulation of the incentive scheme would be to have $T(x) = t(x - \bar{x})$ if $x \geqq \bar{x}$ and $T(x) = t(x - \bar{x}) - k$ if $x < \bar{x}$. In this case, there would be no fixed penalty for ambient levels above $\bar{x}$. Instead, an additional fixed bonus would be given when those levels were below $\bar{x}$.

excess of standards, incentives exist to reduce emissions to the standard level but not below.

2.1. Short Run Analysis

In the short run where output price is fixed, either component of the incentive mechanism can be used by itself to induce a desired level of abatement, or they can be used in combination. To see this, let a denote the level of abatement and write the ambient pollution level as $x(a, e)$, where e is a random variable and $\partial x/\partial a \leqq 0$. Let y be the level of the good produced by the polluting firm, let $C(y, a)$ be the cost of producing y while abating to level a,[7] and let $F(\bar{x}, a)$ be the probability that x is less than the cutoff level $\bar{x}$, given a, where $\partial F/\partial a \geqq 0$. If the benefit of increasing abatement from zero to a, denoted $B(x(0, e) - x(a, e))$, is known,[8] then the social planner seeks the levels of output and abatement that maximize[9]

$$py + E[B(x(0, e) - x(a, e))] - C(y, a), \tag{1}$$

where p is the output price (reflecting the marginal utility of the good) and E is the conditional expectation operator over the random variable e. The optimal levels of output and abatement, denoted y^* and a^*, are implicitly defined by the first-order condition

$$p - C_y = 0 \tag{2a}$$

$$E[B' \cdot x_a] + C_a = 0, \tag{2b}$$

where the subscripts denote partial derivatives. This is a necessary and sufficient condition for global optimality if the objective function is concave. Note that (2a) can be used to define y^* as a function of a. If (2b) is evaluated at $y = y^*(a)$, then (2b) alone can be used to define the optimal abatement level a^*.

Given the socially optimal abatement level, the incentive scheme can be designed to induce a competitive firm to abate to that level. The firm is assumed to choose the levels of output and abatement that maximize[10]

$$py - C(y, a) - E(T(x(a, e))). \tag{3}$$

Since $E[T(x(a, e))] = t \cdot E[x(a, e)] - t\bar{x} + k(1 - F(\bar{x}, a))$, his choices, denoted $\hat{y}$

[7]These costs include all social opportunity costs of pollution abatement. We assume that the firm's cost function for output and abatement is identical to the social cost function.

[8]If the benefits of abatement are not known, then the social planner could simply choose the level of abatement that would on average meet an exogenous target level of ambient pollution $\hat{x}$. Then the optimal level of abatement a^* would be implicitly defined by $E[x(a, e)] = \hat{x}$, and y^* would be chosen to maximize $py - C(y, a^*)$.

[9]For simplicity, it is assumed that society is risk neutral. An optimal abatement level could also be chosen under a more general expected utility framework. For a discussion of the appropriate attitude of the public sector toward risk, see [2 and 7].

[10]This assumes that polluters are risk neutral. If they are not but the polluter's utility function is known, the values of t and k can still be set to ensure that the socially optimal level of abatement is undertaken.

and $\hat{a}$, are implicitly defined by

$$p - C_y = 0 \tag{4a}$$

$$t \cdot E[x_a] - k(F_a) + C_a = 0. \tag{4b}$$

If it is assumed that the objective function is concave, then (4) is necessary and sufficient for a global maximum. Again (4a) can be used to define $\hat{y}$ as a function of a, which is identical to the function defined by (2a). Then (4b) evaluated at $y = y^*(a)$ alone defines $\hat{a}$.

Condition (4b) implies that the polluter will be induced to choose the socially optimal level of abatement (i.e., $\hat{a} = a^*$) if, given $\bar{x}$, t and k are set in one of the following ways:

(a) $k = 0$ and $t = E[B' \cdot x_a]/E[x_a]$,[11] (5a)

(b) $t = 0$ and $k = -E[B' \cdot x_a]/F_a$, (5b)

or

(c) t is arbitrary and $k = (-E[B' \cdot x_a] + tE[x_a])/F_a$, (5c)

where in each case the derivatives are evaluated at a^* and $y^*(a^*)$.[12] Thus, in the short run a pure tax/subsidy scheme, a pure penalty scheme, or a combined scheme can be used to ensure optimal abatement. However, the implications of these alternatives in terms of total polluter or government payments are clearly different. Because they imply different total costs for the polluters, they imply different industry sizes in the long run (see, e.g., [3]). The following subsection discusses the appropriate design of the incentive scheme in the long run where output price adjusts endogenously to the entry and exit of firms.

2.2. *Long Run Analysis*

In the above partial equilibrium analysis where output price is assumed to be fixed, short run efficiency could be achieved by an infinite number of combinations of $\bar{x}$, t, and k given in (5). However, this apparent indeterminacy in the short run actually provides the flexibility necessary to ensure efficiency in the long run, where long run efficiency is defined not only in terms of the optimal abatement of the firm but also in terms of its output level and the industry size.

To see this, let N be the number of firms in the industry and let $p(Ny)$ be the inverse demand curve for the output. Then the long run efficiency conditions

[11] In this case, if benefits are known, then the optimal tax rate is equal to marginal benefits B' if B' is constant. Under a nonlinear benefit function $t \neq E(B')$. However, $E(B')$ may be a sufficient local approximation to the optimal t, or serve as a guide in setting t. The case of a linear benefit function is discussed more fully in the context of the multiple polluter problem.

[12] Thus, setting the optimal levels of t and k requires that the regulating authority know the effect of abatement on the distribution of ambient pollutant levels. It does not require that it know a one-to-one relationship between abatement and ambient conditions. The premise of the problem discussed here is that no such relationship exists. If it did, there would be no need for a control scheme based on ambient levels since abatement could be inferred and thus controlled directly.

become

$$p(Ny) - C_y = 0, \tag{6a}$$

$$E[B' \cdot x_a] + C_a = 0, \tag{6b}$$

and

$$p(Ny)y + E[B(x(0,e) - x(a,e))] - C(y,a) = 0, \tag{6c}$$

where the first two conditions correspond to (2a) and (2b) and the third condition requires that the expected benefits from operation of the firm equal the costs of that operation. These three conditions define the efficient levels of abatement and output per firm and the efficient industry size (a^*, y^*, and N^*).

Under the incentive scheme given above, the long run equilibrium conditions of a competitive market are given by

$$p(Ny) - C_y = 0, \tag{7a}$$

$$t \cdot E(x_a) - kF_a + C_a = 0, \tag{7b}$$

and

$$p(Ny)y - t(E[x - \bar{x}]) - k[1 - F(\bar{x}, a)] - C(y,a) = 0, \tag{7c}$$

where the first two conditions are from profit maximization and the third condition states that in equilibrium profits must be zero. These three conditions simultaneously define the equilibrium levels of a, y, and N as functions of the parameters of the incentive scheme (t, k, and $\bar{x}$).

The planner can ensure long run efficiency by choosing t, k, and $\bar{x}$ such that

$$\hat{a}(t,k,\bar{x}) = a^*, \tag{8a}$$

$$\hat{y}(t,k,\bar{x}) = y^*, \tag{8b}$$

and

$$N(t,k,\bar{x}) = N^*. \tag{8c}$$

The unique combination of t, k, and $\bar{x}$ that solves these three equations will ensure long run efficiency. Thus, of the infinite number of combinations of t, k, and $\bar{x}$ that yield short run efficiency only one also yields long run efficiency.[13] For the special case where the benefits of abatement are linear (B' is constant) the unique combination that ensures long run efficiency is $k = 0$, $t = B'$, and $\bar{x} = E[x(0,e)]$. This is analogous to the pure tax policy given in (5a) with an appropriate choice of $\bar{x}$.

[13] The reader may wonder why adding only one additional equilibrium and efficiency condition in moving from the short run to the long run reduces the degrees of freedom in the choice of parameters by two. In the short run, choosing the parameters to ensure $\hat{a} = a^*$ is sufficient to also guarantee that $\hat{y} = y^*$ since $\hat{y}$ does not depend on t, k, or $\bar{x}$ directly but only indirectly through their effect on a. Thus, in the short run only one degree of freedom is necessary to meet two goals. However, in the long run $\hat{a} = a^*$ is not sufficient to guarantee $\hat{y} = y^*$ since $\hat{y}$ depends on the parameters directly as well as indirectly through a. In other words, it depends on the expected total payment under the incentive scheme (since this affects average costs) and not just on the scheme's marginal effects.

3. MULTIPLE POLLUTERS PROBLEM

In most NPP cases, it is likely that several polluters will be possible contributors to the ambient pollutant levels of a given waterway. An incentive scheme similar to the one introduced above can still be used, if t and k are allowed to vary across polluters, i.e., if the payments of polluter i are given by

$$T_i(x) = \begin{cases} t_i(x - \bar{x}) + k_i & \text{if } x > \bar{x} \\ t_i(x - \bar{x}) & \text{if } x \leq \bar{x}. \end{cases}$$

This mechanism is similar to one described by Holmstrom [11] as a solution to free riding in the context of organizational structure. Note that each polluter's liability depends on ambient levels that are determined by emissions from the whole group, not just his individual contribution, since at any given time individual contributions are not known or observable. This is equivalent to putting a "bubble" over the entire group of suspected polluters and setting standards for the whole bubble rather than for each source within the bubble. It is also similar to imposition of the legal doctrine of strict (no-fault) joint liability.

Again, t_i and k_i can be set to ensure optimal levels of abatement by each source. To see this for the short run where output price is fixed, let a_i be the abatement level of polluter i, let $C_i(y_i, a_i)$ be i's cost function, and interpret a in $x(a, e)$ and $F(\bar{x}, a)$ as the vector $a = (a_1, \ldots, a_n)$, where n is the number of suspected polluters. If individual polluters are risk neutral and competitive in their output markets, they will choose y_i and a_i to maximize

$$py_i - E\left[T_i(x(a, e))\right] - C_i(y_i, a_i) \tag{9}$$

given a set of expectations about the actions of all other polluters. For simplicity we assume that (i) each polluter is a Cournot firm and takes the abatement levels of all other polluters as given when deciding on his own abatement level and (ii) the regulatory agency setting the incentive parameters knows that this is how individual expectations are held.[14] A Cournot–Nash equilibrium where all expectations are

[14]Alternative assumptions regarding expectations are possible. For example, one could specify a set of conjecture functions $a_j^i(a_i)$ that indicate firm i's expectation about the reaction of firm j to i's choice of a_i. In this case, firm i would seek to maximize

$$py_i - C_i(Y_i, a_i) - E\left[T\left(x\left(a_1^i(a_i), \ldots, a_i, \ldots, a_n^i(a_i), e\right)\right)\right].$$

If mistaken expectations are not detectable so that consistency of expectations with actual outcomes is not necessary for equilibrium, then the results discussed under the simpler Cournot assumption still will follow for arbitrary a_j^i functions if partial derivatives with respect to a_i are replaced with total derivatives that reflect both direct effects and anticipated indirect effects through the behavior of other firms. If consistency is required for equilibrium, then the allowable forms for $a_j^i(\cdot)$ must be restricted to ensure the existence of an equilibrium set of abatement actions (see, e.g., [4, 8, 13]). If the consistent conjectures are independent of the parameters of the incentive scheme, then the results in the text would again hold by using total derivatives. Alternatively, if the consistent conjectures depend on those parameters or if the polluters are assumed to collude under an enforceable agreement, then the regulator could still be able to set the t_i's and k_i's to ensure optimal abatement by all firms. However, the optimal values for t_i and k_i will not take the forms discussed below. Of course, in any of these cases, the ability to induce optimal behavior requires that the regulatory agency know (or be able to deduce) the conjecture functions of the individual firms.

realized and each polluter is induced to choose its socially optimal level a_i^* would be possible under any one of the following incentive schemes:

(a) $k_i = 0$ and $t_i = E[B' \cdot \partial x/\partial a_i]/E[\partial x/\partial a_i]$, (10a)

(b) $t_i = 0$ and $k_i = -E[B' \cdot \partial x/\partial a_i]/E[\partial F/\partial a_i]$, (10b)

or

(c) t_i is arbitrary and $k_i = (-E[B' \cdot \partial x/\partial a_i] + t_i E[\partial x/\partial a_i])/(\partial F/\partial a_i)$, (10c)

where in each case the derivatives are evaluated at a_i^* and $y_i^*(a_i^*)$ for all i.

The free rider problem is eliminated under this scheme since the costs of additional pollution are borne by polluters in a way that does not distort marginal incentives.[15] To see why this eliminates free riding, assume that the benefits of abatement are known and consider the simple case of a linear benefit function, i.e., constant B', and the pure tax/subsidy form of the incentive scheme where $k_i = 0$ for all i. In this case, the optimal tax/subsidy rates are given by

$$t_i = \frac{E[B' \cdot \partial x/\partial a_i]}{E[\partial x/\partial a_i]} = B'.$$

Thus, each polluter pays the full marginal benefit of reduced ambient pollutant levels, rather than just paying a share equal to B'/n. For example, if marginal damages are valued at \$100, the regulatory agency will collect \$100 from each polluter for the marginal unit of ambient pollution, for a total collection of \$(100n). Although the total collection for the marginal unit exceeds the marginal damages, in deciding on a marginal unit of pollution each polluter faces the correct marginal incentives since each will compare his potential abatement cost savings to the full marginal damage (rather than just $1/n^{\text{th}}$ of it) times the likely effect of his reduced abatement on ambient levels, given by $E[\partial x/\partial a_i]$. In this case all polluters face the same tax/subsidy rate B' per unit of ambient pollution regardless of whether they are likely to contribute heavily to marginal ambient levels, i.e., regardless of the magnitude of $E[\partial x/\partial a_i]$. This is necessary because, if their pollution does contribute to those levels, the damages associated with that contribution are assumed to be the same regardless of the source. Note that although they pay the same marginal rate *per unit of additional ambient pollution*, they do not pay the same expected rate *per unit of abatement*. (This latter rate depends upon each polluter's expected contribution to marginal ambient levels.) Thus, despite the constant tax rate in this case, the correct marginal incentives are maintained; polluters weigh the expected marginal benefits of abatement against their marginal abatement costs.

As in the case of the single polluter problem, the alternative forms of the incentive scheme imply different total costs, since the expected values of incentive payments by polluters are different. Thus, in the short run, the planner can alter the financial impact of the plan on any individual firm by appropriately changing t_i, k_i,

[15] This is analogous to the result obtained by Holmstrom [11], where free riding in an organization can be avoided by breaking the balanced-budget constraint, i.e., by allowing total payments to contributors to be less than total output.

and $\bar{x}$ without altering marginal incentives. However, in the long run this flexibility is again lost. Because the alternative forms in (10) differ in terms of total costs, they will result in different long run equilibrium positions. If all firms are identical with respect to output costs, abatement costs, and polluting characteristics, then the extension of the above short run analysis to incorporate long run exit and entry is the same as it was for the single polluter case. However, if polluting characteristics differ across firms, then the zero net benefit/profit conditions no longer can hold for all firms. Those firms with low contributions to ambient pollution levels (for example, farms with flat land located far from waterways) would in general be expected to earn positive profit under an efficient abatement policy,[16] while firms with higher contributions (and thus higher efficient abatement levels) would have lower profits. The marginal firm would earn zero profit. Although the efficiency and equilibrium conditions are more complicated in this case, the planner still can choose the parameters of the incentive scheme to achieve long run efficiency. In the special case of a linear benefit function, a pure tax scheme with $t_i = B'$ for all i and $\bar{x} = E[x(0, e)]$ is the form of the incentive scheme that guarantees long run efficiency.

4. ADVANTAGES AND DISADVANTAGES

The use of the incentive mechanism described above has several advantages for controlling dispersed sources of pollution. First, it involves a minimum amount of government interference in daily firm operations, and firms are free to choose the least cost pollution abatement techniques. Since individual firms are in a better position to determine the abatement practices that will be most effective for them (and will have an incentive to do so), their freedom to choose the techniques used provides the flexibility necessary to ensure that any given level of abatement is achieved at the lowest possible cost.

Second, the incentive mechanism does not require continual monitoring of firm practices or metering of "emissions." It does, however, require that the regulatory authority monitor ambient pollutant levels. The difficulty and expense of this form of monitoring would depend upon the specific pollutant of concern.[17] This might be reduced by identifying a small number of "hot spots" and crucial time periods that could be targeted for monitoring. Once ambient pollutant levels are recorded, the necessary tax or subsidy payment could be calculated easily. Accounts can be cumulated over time with payments made periodically. If, over the time period, tax liability exceeds subsidy payments, then no government outlays would be necessary under the pure tax/subsidy or combined approaches. The subsidies would simply act as credits against tax liability.

Third, if desired, in the short run cost-sharing mechanisms could be used to prevent placing excessive burdens on the polluting sector, and other considerations regarding an appropriate distribution of costs could be accommodated, as long as

[16] This assumes that there is a limit on the number of locations with low polluting characteristics. If there were no limit, all firms would eventually locate in low polluting areas (ceteris paribus), and thus in equilibrium all firms would be identical.

[17] As with many policies that are theoretically appealing because of their efficiency properties, the practical difficulties and administrative costs of implementation may be sufficiently high to offset any efficiency gains. This would have to be judged on a case-by-case basis.

the parameters of the payment scheme are adjusted accordingly to maintain proper incentives. In the long run, the parameters of the scheme can be chosen to ensure long run efficiency.

Finally, the incentive scheme focuses on environmental quality rather than emissions or erosion, which is more appropriate for controlling many forms of stochastic pollution. To the extent that some of the fluctuations in ambient pollutant levels can be anticipated, there would be an incentive for polluters to try to offset peaks by, for example, avoiding heavy pesticide or fertilizer applications prior to anticipated rain or wind storms.

The disadvantages of this incentive scheme include the information requirements that are necessary to set the levels of the t_i and k_i parameters initially to provide the correct incentive. (In general, this is a problem with any regulatory device seeking to achieve socially optimal outcomes.) The necessary information includes abatement cost estimates, estimates of damages from ambient pollution, and estimates of how each polluter's abatement affects the distribution of those ambient levels.

A second possible disadvantage of the mechanism is its implications with regard to discriminatory taxation. It would have to be structured so that allowing the t_i and k_i parameters to vary across sources would not be considered to be discriminatory taxation, since discriminatroy taxation is illegal. Of course, in the special case of a linear benefit function, the pure tax/subsidy approach would have t_i be the same for all firms, thereby eliminating this potential problem.

5. SUMMARY

The standard pollution control devices such as direct regulation or the use of emission taxes are inappropriate for nonpoint pollution problems characterized by physical uncertainty and monitoring difficulties. In these cases, we cannot identify with certainty the source of an observed pollutant or infer a firm's level of abatement from observations of ambient pollution levels, especially when there are many suspected polluters contributing pollutants to a common waterway. Thus, mechanisms that focus on ambient pollutant levels rather than emissions are needed in order to control environmental quality efficiently. However, these mechanisms must be designed to ensure socially optimal abatement levels. In the context of multiple polluters, this requires that the mechanism eliminate free riding. This paper has suggested a possible incentive scheme that could be used to induce optimal abatement for single or multiple suspected polluters. The mechanism has several advantages, including an emphasis on environmental quality, flexibility regarding choice of abatement technique, elimination of the need to monitor individual polluting activities, and the ability to alter financial impacts in the short run and/or ensure efficiency in the long run.

REFERENCES

1. Z. Adar and J. M. Griffin, Uncertainty and the choice of pollution control instruments, *J. Environ. Econom. Management* **3**, 178–188 (1976).
2. Kenneth Arrow and Robert C. Lind, Uncertainty and the evaluation of public investment decision, *Amer. Econom. Rev.* **60**, 364–378 (1970).

3. William J. Baumol and Wallace E. Oates, "The Theory of Environmental Policy: Externalities, Public Outlays and the Quality of Life," Prentice–Hall, Englewood Cliffs, NJ (1975).
4. Timothy F. Bresnahan, Duopoly models with consistent conjectures, *Amer. Econom. Rev.* **71**(5), 934–945 (1981).
5. P. Dasgupta, Environmental management under uncertainty, *in* "Explorations in Natural Resource Economics" (Smith and Krutilla, Eds.), Johns Hopkins Univ. Press, Baltimore (1982).
6. G. Fishelson, Emission control policies under uncertainty, *J. Environ. Econom. Management* **3**, 189–197 (1976).
7. Anthony C. Fisher, Environmental externalities and the Arrow–Lind public investment theorem, *Amer. Econom. Rev.* **63**(4), 722–725 (1973).
8. Joel M. Guttman and Michael Miller, Endogenous conjectural variations in oligopoly, *J. Econom. Behav. Organiz.* **4**, 249–264 (1983).
9. Jon D. Harford, Firm behavior under imperfectly enforceable pollution standards and taxes, *J. Environ. Econom. Management* **5**, 26–43 (1978).
10. B. Holmstrom, Moral hazard and observability, *Bell J. Econom.* **10**(1), 74–91 (1979).
11. B. Holmstrom, Moral hazard in teams, *Bell J. Econom.* **13**(2), 324–40 (1982).
12. R. E. Just and D. Zilberman, Asymmetry of taxes and subsidies in regulating stochastic mishap, *Quart. J. Econom.* **43**(1), 139–148 (1979).
13. Martin K. Perry, Oligopoly and consistent conjectural variation, *Bell J. Econom.* **13**(1), 197–205 (1982).
14. S. Shavell, Risk sharing and incentives in the principal and agent relationship, *Bell J. Econom.* **10**(1), 55–73 (1979).
15. W. D. Watson and R. G. Ridker, Losses from effluent taxes and quotas under uncertainty, *J. Environ. Econom. Management* **11**(4), 310–326 (1984).
16. M. Weitzman, Prices vs. quantities, *Rev. Econom. Stud.* **41**(4), 477–491 (1974).

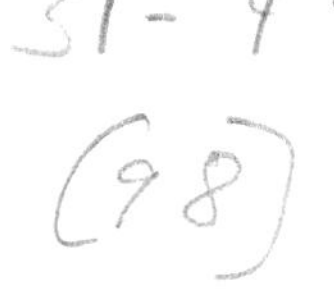

[4]

JOURNAL OF ENVIRONMENTAL ECONOMICS AND MANAGEMENT **36**, 186–199 (1998)
ARTICLE NO. EE981041

Ambient Taxes When Polluters Have Multiple Choices[1]

Richard D. Horan

Agricultural Economist, Resource and Environmental Policy Branch, Resource Economics Division, Economic Research Service, U.S. Department of Agriculture, Washington, D.C., 20250

James S. Shortle[2]

Professor of Agricultural Economics, Department of Agricultural Economics and Rural Sociology, The Pennsylvania State University, University Park, Pennsylvania 16802

and

David G. Abler

Associate Professor of Agricultural Economics, Department of Agricultural Economics and Rural Sociology, The Pennsylvania State University, University Park, Pennsylvania 16802

Received November 13, 1997; revised May 1998

Ambient-based tax–subsidy schemes have drawn considerable interest in nonpoint pollution literature as alternatives to emissions-based instruments. Expanding especially on Segerson's [*J. Environ. Econom. Management* **15**, 88–98 (1988)] seminal article, this article examines the optimal design and budget-balancing properties of ambient tax–subsidy schemes under more realistic assumptions about the dimensions of firms' choice sets than prior research. © 1998 Academic Press

INTRODUCTION

Economic research on environmental policy design has largely been concerned with the merits of emissions-based economic incentives (e.g., emissions charges, emissions reduction subsidies, transferable discharge permits). This literature has limited relevance to nonpoint pollution because emissions in this case are for all practical purposes unobservable and typically stochastic [1, 12, 14, 15, 17]. Other constructs must therefore be used to monitor performance and as a basis of the application of pollution control instruments. Instruments that are used in practice or that have been proposed include taxes on polluting inputs (e.g., fertilizers, pesticides), subsidies for purchases of pollution control equipment, liability for damages, environmental bonds, tax–subsidy schemes applied to ambient concentra-

[1]The views expressed here are the authors' and do not necessarily reflect those of the Economic Research Service or the USDA. We appreciate the helpful suggestions of the anonymous reviewers and the associate aditor for helpful comments. All remaining errors are our own.

[2]E-mail address: jshortle@psu.edu.

0095-0696/98 $25.00

tions, and tradeable permits in polluting products [14, 10]. Ambient-based instruments have drawn particular interest in economic literature and are the focus of this article.

Ambient-based instruments shift the location of monitoring from the choices of firms (or households) that are suspected of contributing to environmental degradation, to the environmental media. Segerson [12] first suggested the use of ambient-based instruments. Subsequent contributions include Cabe and Herriges [2], Herriges, Govindasamy, and Shogren [5], and Xepapadeas [19, 20, 21]. This literature focuses on moral hazard in nonpoint pollution. Unobservable emissions mean that polluters' performance cannot be observed directly. To the extent that input choices are costly to monitor and their relationship to ambient concentrations is uncertain, there will be uncertainty about polluters' effort and a moral hazard problem. However, much of the research on ambient-based instruments has thus far assumed either that polluters can observe their own emissions and control them deterministically (e.g., [19, 20, 21]), or that the distribution of emissions is determined by a single variable, referred to generally as abatement effort (e.g., [2, 12]). In either case, the nonpoint pollution control problem is overly simplified [1, 14].[3]

We expand on Segerson's [12] analysis by exploring the design of ambient taxes when each firm's choice set is multidimensional. Specifically, the distribution of each source's emissions, which are stochastic and not directly observable by firms or the regulator, depends on the polluter's choices over a set of variables. After developing the model, efficient input use as well as efficient entry and exit are characterized. Segerson's linear ambient tax–subsidy scheme is then shown to be efficient only under restricted conditions. These restrictions relate to firms' opportunity sets, the nature of the stochastic relationships, and whether the tax rate is state-independent or state-dependent. Two alternative ambient schemes are developed. One is a modified version of Segerson's linear scheme in which the tax rate is determined ex post along with the ambient tax base. The second is a nonlinear ambient tax. Both structures require a lump sum tax–subsidy scheme to induce optimal entry and exit.

The optimal taxes are shown to require (potentially) large transfers between polluters and the government in the form of taxes and/or subsidies. However, a budget-balancing solution may be a desirable property of the tax system. Necessary conditions for budget-balancing outcomes are derived, and both state-dependent and state-independent budget-balancing solution schemes are presented.

[3]To illustrate, nitrates from fertilizer and animal manures are a major water pollutant from agricultural nonpoint sources [18]. Nitrates that are not used by the target crop may enter surface or ground waters, the atmosphere, or remain in the soil. In any case, the fate cannot be routinely measured. The outcome will depend on production choices, weather events that affect plant growth, soil denitrification, and leaching, and surface and subsurface runoff, and soil characteristics. Production choices that influence the volume of nitrates that enter surface or ground waters from a farm field include the timing, amount and method of fertilizer applications, the crop, the particular crop cultivar, the application of other inputs and field management practices that influence crop growth (and nitrogen consumption by the plants), and the movement of water across fields, and the use of pollution abatement techniques such as grassed waterways [9].

A NONPOINT POLLUTION MODEL

Consider a model in which a particular resource (e.g., a lake) is damaged by a single residual (e.g., nitrogen) from nonpoint sources of emissions. The ambient concentration of the pollutant is given by

$$a = a(r_1, \ldots, r_i, \ldots, r_n, b, w, \lambda),$$

where a is the ambient concentration, r_i $(i = 1, 2, \ldots, n)$ is emissions from nonpoint source i $(\partial a/\partial r_i > 0, \forall i)$, b is the natural generation of the pollutant, w represents stochastic environmental variables that influence transport and fate, and λ is a vector of watershed characteristics and parameters.

Emissions cannot be observed directly (at least not at an acceptable cost) and, via stochastic variations in environmental drivers (e.g., weather), are stochastic. Accordingly, each polluter can only influence the distribution of its emissions, which depends on management decisions, environmental variables, and site characteristics. Specifically, firm i's emissions are given by the emissions function $r_i(x_i, \nu_i, \gamma_i)$, where x_i is an $(m \times 1)$ vector of inputs chosen by the firm, ν_i represents environmental drivers at the firm's site, and γ_i is a vector of site characteristics such as soil type and topography.

Polluters are assumed to be risk-neutral. The ith firm's expected profit for any choice of inputs is given by $\pi_i(x_i)$. Larger values of i are assumed to correspond to firms that are either less productive or that have locations that are more conducive to nonpoint pollution (e.g., pervious vs. impervious surfaces, urban vs. rural, soil types, slope of land, etc. See the Appendix for the derivation of the index). Stated another way, larger values of i represent firms with greater (incremental) social economic value. For simplicity, the polluters of this particular resource are assumed not to have any collective influence on the prices of inputs or outputs, and that input and output markets are free from distortions. The economic cost of damages caused by pollution is given by $D(a, \eta)$, where η is a random variable introducing damage cost uncertainty $(\partial D/\partial a > 0,\ \partial^2 D/\partial a^2 \geq 0)$. This uncertainty may result from stochastic processes influencing the economic consequences of ambient levels. All random components in the model are assumed to be jointly distributed according to the function $g(\upsilon, w, \eta)$, where υ is an $(n \times 1)$ vector with ith element ν_i.

An ex ante efficient allocation maximizes expected net surplus (quasi-rents, less environmental damage costs) to society [3, 7],[4]

$$\operatorname*{Max}_{x_{ij}} V = \sum_{i=1}^{n} \pi_i(x_i) - E\{D(a)\}. \tag{1}$$

The necessary conditions for input choices for an interior solution are

$$\frac{\partial V}{\partial x_{ij}} = \frac{\partial \pi_i}{\partial x_{ij}} - E\left\{\frac{\partial D}{\partial a}\frac{\partial a}{\partial r_i}\frac{\partial r_i}{\partial x_{ij}}\right\} = 0, \qquad \forall i, j. \tag{2}$$

[4] However, as with prior reserach on ambient and other nonpoint instruments, we assume that society is risk neutral.

Condition (2) requires that the marginal net benefit from the use of each input equal the expected marginal damage from the use of the input.[5] In addition to the first-order condition (2), the following condition is necessary to define the optimal number of firms in the region (see the Appendix, Eq. (A8)),

$$\pi_n - \Delta_n E\{D(a)\} \geq 0, \tag{3}$$

where $\Delta_n E\{D(a)\} = E\{D(a(r_1, \ldots, r_n, b, w, \lambda), \eta)\} - E\{D(a(r_1, \ldots, r_{n-1}, b, w, \lambda), \eta)\}$ (as defined in the Appendix) is the difference in expected damages from when firm n produces and when it does not (i.e., the incremental effect of firm n on expected damages). Condition (3) describes the incremental impact of firm n on expected net benefits. If the nth firm is defined optimally, then the addition of any other firm has a negative incremental impact. Together, conditions (2) and (3) define the efficient scale of production for the marginal firm.

THE SEGERSON LINEAR AMBIENT TAX

A firm-specific ambient tax can be defined in general terms by the function $T_i(a) + b_i$, where b_i denotes a lump sum tax or subsidy. When subject to such a tax, a firm will choose inputs to maximize expected after-tax profit,

$$\underset{x_{ij}}{\text{Max}}\, U_i = \pi_i(x_i) - E\{T_i(a)\} - b_i. \tag{4}$$

The first-order necessary conditions for an interior solution are

$$\frac{\partial U_i}{\partial x_{ij}} = \frac{\partial \pi_i}{\partial x_{ij}} - E\left\{T_i'(a)\frac{\partial a}{\partial r_i}\frac{\partial r_i}{\partial x_{ij}}\right\} = 0, \qquad \forall i, j. \tag{5}$$

In the seminal article on ambient taxes, Segerson [12] proposed taxing nonpoint polluters based on the observed deviation of the ambient pollution concentration from a target. Specifically, Segerson's ambient tax scheme takes the firm-specific form,

$$T_i(a) = \begin{cases} t_i(a - a_0) + k_i, & a > a_0, \\ t_i(a - a_0), & a \leq a_0, \end{cases}$$

where a_0 is the ambient target, t_i is the marginal tax (subsidy) rate, and k_i is a lump sum penalty or subsidy. The specific structure of the expected tax payment depends on each firm's expectations about how its choices and the choices of others affect the tax base, and how it expects other suspected polluters to behave. Assuming firms adopt the theory of ambient concentrations outlined previously and that the choices of others are observable and taken as given, firm i's expected tax function can be written

$$E\{T_i(a)\} = t_i[E\{a\} - a_0] + k_i(1 - F(a_0, a)), \tag{7}$$

[5] Note that the marginal damage cost will be negative for those inputs that reduce emissions.

where $F(a_0, a) = \text{Prob}(a \leq a_0)$. In this case, (5) becomes

$$\frac{\partial \pi_i}{\partial x_{ij}} = t_i E\left\{\frac{\partial a}{\partial r_i}\frac{\partial r_i}{\partial x_{ij}}\right\} - k_i \frac{\partial F}{\partial x_{ij}}, \qquad \forall i, j. \tag{8}$$

For a Cournot–Nash equilibrium to be efficient, t_i and k_i and must be chosen such that

$$t_i E\left\{\frac{\partial a^*}{\partial r_i}\frac{\partial r_i^*}{\partial x_{ij}}\right\} - k_i \frac{\partial F^*}{\partial x_{ij}} = E\left\{\frac{\partial D^*}{\partial a}\frac{\partial a^*}{\partial r_i}\frac{\partial r_i^*}{\partial x_{ij}}\right\}, \qquad \forall i, j, \tag{9}$$

where $D^* = D(a^*, \eta)$, $F^* = F(a_0, a^*)$, $a^* = a(r_1^*, \ldots, r_n^*, b, w, \lambda)$, $r_i^* = r_i(x_i^*, \nu_i, \gamma_i)$, and x_i^* is the optimal value of x_i. However, condition (9) will not be satisfied for all i and j unless (i) each firm is faced with only one or two production choices, or (ii) the covariance between marginal damages and marginal ambient pollution is zero for all firms and inputs. To demonstrate, Segerson proposes that the optimal linear incentive structure must take one of the following three forms,

$$k_i = 0, \qquad t_i = \frac{E\{(\partial D^*/\partial a)(\partial a^*/\partial r_i)(\partial r_i^*/\partial x_{ij})\}}{E\{(\partial a^*/\partial r_i)(\partial r_i^*/\partial x_{ij})\}}, \qquad \forall i, \tag{10a}$$

$$t_i = 0, \qquad k_i = -\frac{E\{(\partial D^*/\partial a)(\partial a^*/\partial r_i)(\partial r_i^*/\partial x_{ij})\}}{\partial F^*/\partial x_{ij}}, \qquad \forall i, \tag{10b}$$

$$t_i = \text{arbitrary},$$

$$k_i = \frac{-E\{(\partial D^*/\partial a)(\partial a^*/\partial r_i)(\partial r_i^*/\partial x_{ij})\} + t_i E\{(\partial a^*/\partial r_i)(\partial r_i^*/\partial x_{ij})\}}{\partial F^*/\partial x_{ij}}, \qquad \forall i, \tag{10c}$$

for all $j = l, \ldots, m$.

In general, each of these forms is overdetermined with m equations and one unknown. If $m > 2$, there will not, in general, exist a t_i and k_i that will satisfy all m equations in (9). The linear tax does not lead to an overdetermined system in Segerson's model because "abatement" is the only choice variable in that model.

The intuition here is straightforward. A linear ambient tax as in (6) only provides firms with an incentive to choose input levels to control expected ambient levels. However, reductions in $E\{a\}$ do not necessarily correspond to reductions in $E\{D\}$ when damages are convex because higher distributional moments of a may be influenced by the actions firms take to reduce $E\{a\}$ [13]. If var$\{a\}$ increases as $E\{a\}$ is reduced, for example, then $E\{D\}$ may increase. When only a single input influences emissions, the linear tax scheme optimally manages such risk effects. To illustrate, consider the special case where $m = 1$. The optimal linear tax rate is

$$t_i = E\left\{\frac{\partial D^*}{\partial a}\right\} + \frac{\text{cov}\{\partial D^*/\partial a, (\partial a^*/\partial r_i)(\partial r_i^*/\partial x_i)\}}{E\{(\partial a^*/\partial r_i)(\partial r_i^*/\partial x_i)\}}, \qquad \forall i. \tag{11}$$

The first term on the right-hand side (RHS) is the expected marginal damages from pollution. The second term adjusts for the risk effects that result from using the one input, x_i, to control expected ambient levels as opposed to expected damages. With more than one input, the linear tax, to be optimal, must adjust for the differential impacts that each input used to control expected ambient levels has on expected damages. This is not possible except for the special case in which the covariance between marginal damages and marginal ambient pollution, $\text{cov}\{\partial D^*/\partial a, (\partial a^*/\partial r_i)(\partial r_i^*/\partial x_{ij})\}$, is zero for each input for each firm. For this special case, (10a) simplifies to Segerson's result where each firm pays an amount equal to total expected marginal damages [12, p. 95][6]

$$k_i = 0, \qquad t_i = E\left\{\frac{\partial D^*}{\partial a}\right\}, \qquad \forall i.$$

The linear ambient tax defined by (6) can also be optimally designed for the special case where $m = 2$ because of the additional lump sum instrument k_i. Although k_i is a lump sum instrument ex post, it has the ex ante effect of being a linear tax based on a nonlinear relationship of ambient levels (see Eq. (7)). The optimal form of the tax scheme in this case is (10c) because the two instruments must be used conjunctively to adjust for the differential impacts that the two inputs have on expected damages.

The linear ambient tax defined by (6) can therefore be optimally designed only in the special cases where less than three production decisions influence emissions or when there are no risk effects associated with the use of different inputs. However, even in these special cases, additional instruments may be needed to ensure that the long-run efficiency condition (3) is satisfied.

STATE-DEPENDENT LINEAR AND NONLINEAR AMBIENT TAXES

In this section, we describe two alternative ambient taxes that can satisfy the efficiency conditions as a Cournot–Nash equilibrium regardless of the dimensionality of the firms' choice sets. The first tax is linear in the ambient concentration. However, unlike the tax defined by (6), the tax rate is determined ex post. Specifically, consider defining a_0, t_i, and k_i in (6) according to

$$a_0 = k_i = 0, \qquad t_i = \frac{\partial D(a^*, \eta)}{\partial a}, \qquad \forall i. \tag{13}$$

The tax rate is the marginal damage of pollution, evaluated at the ex ante efficient level of input use and conditional on the realization of all random variables. Thus, the tax rate as well as the tax base (i.e., ambient pollution levels) are state-dependent. However, as with (12), this tax rate is applied uniformly to all firms.

Ex ante, the expected tax faced by firm i is

$$E\{T_i(a)\} = E\{ta\} = E\left\{\frac{\partial D(a^*, \eta)}{\partial a} a\right\}, \qquad \forall i, \tag{14}$$

and the firm's first-order conditions become

$$\frac{\partial \pi_i}{\partial x_{ij}} - E\left\{\frac{\partial D(a^*, \eta)}{\partial a}\frac{\partial a}{\partial r_i}\frac{\partial r_i}{\partial x_{ij}}\right\} = 0, \qquad \forall i.$$

[6] Segerson [12] derived (12) under the assumption of a damage function that is linear in the ambient concentration. This assumption is a special case of our model.

The firms' choices will therefore satisfy the efficiency conditions given by (2). This linear, state-dependent tax rate is optimal because, unlike its state-independent counterpart in (6), polluters have an incentive, at the margin, to consider the impact of each input on expected damages.

The second tax is nonlinear in the ambient concentration. Comparison of (5) with (2) implies that a nonlinear ambient tax of the form[7]

$$T_i(a) = D(a, \eta), \qquad \forall i \tag{15}$$

can also be used to achieve an efficient outcome as a Cournot–Nash equilibrium. In contrast to the linear ambient taxes in (7) or (14), each firm pays an amount equal to *total* damages. Consequently, each firm considers the impact of each input on expected damages. The tax is again applied uniformly to each firm.

Before concluding this section, we offer some comments on implementation of these tax schemes. The implementation of state-dependent ambient taxes is not likely to be significantly more demanding than that of state-independent tax schemes. Implementation of the linear tax would involve an announcement to producers that their tax will be levied based on the realized ambient concentration, with a tax rate to be determined according to specific rule (i.e., Eq. (13)). After the realization, they would owe the tax. Implementation is not much different for the nonlinear ambient tax. That case would involve an announcement to producers that their tax will be levied based on the realized ambient concentration, with the tax bill being a known nonlinear function of the realized ambient concentration. Because the realized ambient concentration is unknown when decisions are made, the marginal tax rate is also uncertain. To aid producers in their decisions, each firm could be provided with a schedule of tax rates (for the linear case) or tax bills (for the nonlinear case) corresponding to different realizations of the random variables. Implementation is not greatly different from that of graduated income taxes. Income taxpayers know the rules that are used to determine their taxes (or at least they are legally presumed to know them), but the actual base and rate are uncertain. Except for methods of administration and corresponding transactions costs, both taxes are analogous to well-defined joint-tort liability rules.[8]

LONG RUN EFFICIENCY AND BUDGET-BALANCING PROPERTIES

Optimal entry and exit (i.e., $n = n^*$, where n^* is the solution to (2) and (3)) is a critical aspect in the evaluation of ambient-based instruments because the tax base is a function of group performance. The marginal incentives faced by each firm will be incorrect in any Nash equilibrium involving suboptimal entry and exit.

[7]The damage-based tax (15) is not new to the nonpoint literature. A state-dependent, nonlinear tax was suggested for use in nonpoint pollution control by Shortle and Dunn [15] for the special case of a single firm. It was analyzed in the context of multiple firms as a liability rule by Miceli and Segerson [8], and more recently as a tax by Hansen [4].

[8]It is not immediately apparent that transactions costs for the state-dependent taxes would differ significantly from those of other (state-independent) ambient taxes that are second best. This is because the state-dependent taxes share the same tax base and utilize the same information for (optimal) rate design as other ambient taxes. How the taxes would perform in a second best world is an interesting question, but inherently empirical.

A firm will not operate unless it expects its after-tax profits to be nonnegative (i.e., $U_i \geq 0$, where U_i is defined by (4)). With $b_i = 0$, the nth firm will produce as long as the following condition holds,

$$U_n = \pi_n - T(a) \geq 0, \tag{16}$$

which generally differs from (3). Thus, the ambient taxes defined by (14) and (15) do not provide incentives to ensure long-run efficiency. It is therefore necessary to make use of another instrument to induce the correct market structure. Specifically, consider the use of a firm-specific, lump sum tax–subsidy, b_i.

First, consider the extramarginal firms denoted $i > n^*$. These firms do not expect to earn nonnegative after-tax profits if either $E\{T(a)\}$ or b_i are sufficiently large. The expected variable tax, $E\{T(a)\}$, is at least as great as expected damages if damages are convex. A lump sum tax on extramarginal firms may therefore not be needed if expected damages are significant. Alternatively, extramarginal firms expecting their after-tax profits to be nonnegative when $b_i = 0$ must be charged a lump sum tax if they produce (i.e., the lump sum tax applied to extramarginal firms is conditional on production).[9] Moreover, because each firm's expected variable tax payment depends on the total number of producing firms, the lump sum tax must be sufficiently large to ensure that extramarginal firms do not find it profitable to enter for any combination of firms in the region. For example, setting b_i equal to the profit level that extramarginal firm i would make under competition (i.e., when $T(a) = b_i = 0$) would ensure negative expected after-tax profits (because regulated profits are less than competitive levels), and firms $i > n^*$ will not produce.

It is not necessary to impose lump sum taxes on the marginal and inframarginal firms to satisfy the efficiency conditions unless their decision to operate is influenced by the magnitude of the expected variable part of the tax. If expected damages are large enough to prevent the marginal and inframarginal firms from producing, then subsidies are necessary. However, even if subsidies were not necessary, the resulting monetary transactions and the associated costs could be substantial.[10] These considerations lead us to investigate the properties of a budget-balancing tax scheme.

An environmental tax is said to be ex post budget balancing when the total tax paid equals total damages and ex ante budget balancing when the expected total tax equals total expected damages.[11] The budget-balancing properties of ambient tax schemes has been raised as an issue in prior literature [5, 8, 19].

[9] The lump sum tax would only be applied to extramarginal firms that produce. Extramarginal firms that do not produce pay no tax. Alternatively, efficient entry–exit could be accomplished by subsidizing extramarginal firms to not produce. Essentially, this is how the U.S. Conservation Reserve Program (CRP) operates, which pays farmers to retire land with greater environmental risks.

[10] A characteristic of nonpoint problems is large numbers of sources. For example, tens of thousands of farms contribute to nonpoint pollution in the Chesapeake Bay watershed. For such large numbers, even modest damages could imply very large transfers.

[11] Our use of the term budget balancing is consistent with Holmström [6], who defines budget balancing to occur when total payments to a team involved in the production of a common output equal to the value of the output. Here, the team is the polluters, the value of the common output would be damages, and payments (which are made in reverse due to the negative social value of the common output) are the taxes paid. Others [5, 19] define budget balancing to occur when the social gains from reducing pollution relative to a benchmark level equal to the subsidies paid to polluters.

Consider the lump sum tax,

$$b_i = E\{D(a^*, \eta)\}\rho_i - E\{T(a^*)\}, \qquad \forall i \le n^*, \tag{17}$$

where ρ_i is a distributional parameter ($\rho_i > 0, \Sigma_{i=1}^{n^*} \rho_i = 1$). The following proposition addresses the existence of a budget-balancing solution for this and other cases.

PROPOSITION. *There exists a specification for ρ_i such that, when combined with* (14) *or* (15) *and appropriate lump sum taxes–subsidies for extramarginal firms, the lump sum tax–subsidy defined by either* (17), *$b_i = E\{D(a^*, \eta)\}\rho_i - T(a)$, or $b_i = D(a^*, \eta)\rho_i - T(a)$, $\forall i \le n^*$, ensures* (i) *long-run efficiency, and* (ii) *that an ex ante budget-balancing solution exists. If $V^* > 0$ (where V^* is defined by* (1), *evaluated at the ex ante efficient solution), then the number of specifications for ρ_i is infinite. If $V^* = 0$, then the specification for ρ_i is unique.*

Proof. Define $\rho_i = \pi_i^* / \Sigma_{i=1}^{n^*} \pi_i^*$ and $\tau_i = T(a) + b_i$. Firms $i = 1, \ldots, n^*$ will choose to operate if they expect to produce profitably, i.e., if

$$\pi_i^* - E\{\tau_i\} = \pi_i^* - \beta\pi_i^* = \pi_i^*[1 - \beta] \ge 0, \qquad \forall i \le n^*, \tag{18}$$

where $\beta = E\{D(a^*, \eta)\}/\Sigma_{i=1}^{n^*}\pi_i^* < 1$, with the inequality holding for an interior ex ante efficient solution. Thus, long-run efficiency is guaranteed for $\rho_i = \pi_i^* / \Sigma_{i=1}^{n^*}\pi_i^*$. The aggregate tax under this specification is $\Sigma_{i=1}^{n^*}\tau_i = \Sigma_{i=1}^{n^*}E\{D(a^*, \eta)\}\rho_i = E\{D(a^*, \eta)\}\Sigma_{i=1}^{n^*} \rho_i = E\{D(a^*, \eta)\}$, and the ex ante budget-balancing condition is satisfied.[12]

Finally, if $\Sigma_{i=1}^{n^*}\pi_i^* = E\{D(a^*, \eta)\}$ (i.e., $V^* = 0$), then $\beta = 1$ and each firm expects to earn zero profits after taxes.[13] In this case it is not possible to further redistribute damage payments by changing the specification of ρ_i. If $V^* > 0$, then because $\pi_i^* > \pi_n^*$ for some i, there exists an infinite number of combinations of ϵ_i, $\forall i = 1, \ldots, n^*$ for which $\rho_i = (\pi_i^* / \Sigma_{i=1}^{n^*}\pi_i^*) + \epsilon_i$ also satisfies (18) $\forall i = 1, \ldots, n^*$. ∎

The intuition behind the proposition is straightforward. Tax payments must be made from pretax profits. Therefore, aggregate profits must be at least as great as expected damages in order to pay a tax of the same magnitude.

The discussion to this point has addressed ex ante budget balancing. Ex ante budget-balancing schemes will on average be ex post budget balancing. However, because the realized tax depends on stochastic processes, deviations between expected and realized tax payments will occur from period to period. To illustrate, consider the realized aggregate tax under (14) or (15) and (17),

$$\sum_{i=1}^{n^*} \tau_i = E\{D(a^*, \eta)\} + n^*[T(a^*) - E\{T(a^*)\}]. \tag{19}$$

[12] Miceli and Segerson [8], for the special case of two firms, treat each firm as a marginal firm and find that a budget-balancing solution is only attainable under restricted circumstances. Our budget-balancing solution is made possible by utilizing the differential profit structure of the marginal and inframarginal firms.

[13] If expected net benefits, V, include consumer surplus, then V may be positive even if aggregate profits are less than expected damages in the optimum. In this situation, a budget-balancing solution may not exist even though an interior solution may exist.

With sufficiently large variation, the aggregate tax may be a net subsidy in some years and may well exceed total damages in others.

Ironically, some variability can be reduced by making the lump sum taxes state-dependent. For example, one such tax scheme is defined by (14) or (15) along with

$$b_i = E\{D(a^*, \eta)\}\rho_i - T(a^*), \qquad \forall i \leq n^*. \tag{20}$$

The taxes–subsidies in (20), while lump sum, are state-dependent for both ambient schemes. Ex post, the aggregate tax payment is

$$\sum_{i=1}^{n^*} \tau_i = \sum_{i=1}^{n^*} [E\{D(a^*, \eta)\}\rho_i + T(a^*) - T(a^*)]$$

$$= E\{D(a^*, \eta)\} \sum_{i=1}^{n^*} \rho_i = E\{D(a^*, \eta)\}. \tag{21}$$

This tax system is both ex ante budget balancing and without variation in the final tax payment due to random events. The variation of the lump sum tax–subsidy offsets the stochastic variation of the variable portion of the tax. Consequently, firms are not affected by annual stochastic events.

Finally, both ambient tax schemes will be ex post budget balancing if the lump sum part of the taxes are redefined according to

$$b_i = D(a^*, \eta)\rho_i - T(a^*), \qquad \forall i \leq n^*. \tag{22}$$

However, as with (20), the tax–subsidy defined by (22) is state-dependent. However, (20) and (22) differ by the first term on the RHS. The aggregate tax is

$$\sum_{i=1}^{n^*} \tau_i = \sum_{i=1}^{n^*} [D(a^*, \eta)\rho_i - T(a^*) + T(a^*)]$$

$$= D(a^*, \eta)\} \sum_{i=1}^{n^*} \rho_i = D(a^*, \eta). \tag{23}$$

Thus, the taxes are ex post budget balancing when the lump sum taxes–subsidies are state-dependent and realized damage costs are distributed among firms.[14]

CONCLUSION

Ambient taxes have drawn considerable interest from economists as a mechanism for nonpoint pollution control. In this article, we have examined the optimal design of ambient tax–subsidy schemes under more realistic assumptions about firms' choice sets than prior research. In particular, this analysis recognizes that

[14] Herriges, Govindasamy, and Shogren [5] found that a sufficient degree of risk-aversion is necessary for the budget-balancing ambient tax scheme proposed by Xepapadeas [19] to be feasible. However, it is not clear that their tax scheme satisfies the appropriate entry–exit conditions. Risk-aversion is not a requirement in the present model. The only requirement is that efficient aggregate pretax profits are greater than expected total damages, or total damages, depending on the type of budget-balancing scheme in question.

nonpoint sources have a range of options for changing the probability distribution of their emissions.

The performance of an ambient tax scheme when firms have multidimensional choice sets depends on the instrument's ability to induce firms to optimally consider the impacts of each choice on expected damages. We show that a linear ambient tax in which the tax rate on the ambient base is state-independent can be efficient only when either (i) each firm has less than three choices influencing the distribution of ambient conditions, or (ii) the covariance between marginal damages and the marginal effects of firms' choices on ambient pollution is zero for all firms and inputs. When these conditions are not satisfied, a state-independent linear ambient tax rate cannot provide optimal incentives for firms to manage the environmental risk effects of their production and pollution control choices. Instead, we show that a linear ambient tax with a state-dependent rate equal to marginal damages, and a nonlinear ambient tax where each firm pays an amount equal to total damages can induce efficient input choices.

For all of these cases, it may be necessary to impose lump sum taxes on the extramarginal firms to induce optimal entry and exit. In addition, it may also be necessary to provide the marginal and all inframarginal firms with lump sum subsidies to prevent suboptimal exit.

Finally, because optimally designed ambient taxes may require large transfers either in the form of taxes or subsidies, the budget-balancing implications of the taxes deserve attention. Ambient taxes can only be budget balancing if lump sum taxes–subsidies are imposed on the marginal and all inframarginal firms. If the lump sum portions are state-independent, then the ambient tax schemes are only budget balancing in the long run. In the short run, the stochastic fluctuations in ambient conditions will cause ex post deviation between taxes and damages. The total tax may be a net subsidy or may well exceed total damages in some years, depending on the variability of damages. However, a state-dependent form of the lump sum tax will satisfy the ex ante budget-balancing condition with no stochastic variation and will also eliminate variation in the final tax paid by firms. It is also possible to design ambient taxes to be ex post budget balancing. However, to do so requires adding a random component to the inframarginal lump sum taxes–subsidies. However, as a result, firms would face a degree of randomness in the total tax they actually pay.

APPENDIX[15]

Define the set of all possible firms by S and arbitrarily index firms by $z = 1, \ldots, Z$. Define Ω to be the set of all possible subsets of S (i.e., the power set of S). For each $\Omega_y \in \Omega$ $(y = 1, \ldots, 2^Z)$, expected net benefits are maximized by choosing input levels for each firm to maximize

$$V_y = \sum_{z \in \Omega_y} \pi_z - E\{D(a, \eta)\}, \tag{A1}$$

[15]This appendix builds on Miceli and Segerson's [8] model, which considers the case of only two firms.

where $r_z = 0$, $\forall z \notin \Omega_y$. The first-order necessary conditions for an interior solution are

$$\frac{\partial V_y}{\partial x_{zj}} = \frac{\partial \pi_z}{\partial x_{zj}} - E\left\{\frac{\partial D(a,\eta)}{\partial a}\frac{\partial a}{\partial r_z}\frac{\partial r_z}{\partial x_{zj}}\right\} = 0, \qquad \forall z, j. \tag{A2}$$

Define firm z's input vector that solves (A2) by $\tilde{x}_z$, $\forall z \in \Omega_y$. Plugging $\tilde{x}_z$, $\forall z \in \Omega_y$ into (A1) yields $\tilde{V}_y = \Sigma_{z \in \Omega_y} \tilde{\pi}_z - E\{D(\tilde{a}_y, \eta)\}$, $\forall y$, where $\tilde{a}_y = a(\tilde{r}_1, \ldots, \tilde{r}_z, b, w, \lambda)$, $\tilde{r}_z = r(\tilde{x}_z, \nu_z, \gamma_z)$, $\forall z \in \Omega_y$, and $\tilde{r}_z = 0$, $\forall z \notin \Omega_y$.

The optimal set of firms, Ω_y, is the set of firms that results in maximum net benefits over all possible sets of firms, i.e., it is the set of firms corresponding to $\tilde{V}_{y^*}$, where $\tilde{V}_{y^*}$ satisfies

$$\tilde{V}_{y^*} - \tilde{V}_y > 0, \qquad \forall y \neq y^*. \tag{A3}$$

For simplicity, define $V_{y^*}^* = \Sigma_{z \in \Omega_y^*} \pi_z^* - E\{D(a_{y^*}^*, \eta)\} = \tilde{V}_{y^*}$. Next, the n^* firms in the set Ω_{y^*} are re-indexed. There are a number of ways to index the firms. For simplicity, we choose an index based on deviations from the efficient solution. Define firm $i = n^*$ as the firm that corresponds to

$$\alpha_{n^*} = \min_{n^*}\left\{V_{y^*}^* - \tilde{V}_{y^*(-n^*)}\right\}, \tag{A4}$$

where

$$\tilde{V}_{y^*(-n^*)} = \underset{\{x_{zj},\, \forall z \in \Omega_{y'(-n^*)},\, \forall j\}}{\text{Max}} \left\{\sum_{z \in \Omega_{y^*(-n^*)}} \pi_z - E\{D(a_y, \eta)\}\right\},$$

subject to $r_z = 0$, $\forall z \notin \Omega_{y^*(-n^*)}$, and $\Omega_{y^*} = \{\Omega_{y^*(-n^*)}, n^*\}$. $\tilde{V}_{y^*(-n^*)}$ represents maximum expected net benefits when the nth firm from the optimal set is not included. Thus, α_{n^*} represents the marginal net benefits from the addition of the n^*th firm, and $\alpha_{n^*} > 0$ by (A3). Continue this process iteratively, denoting firm i as the firm that corresponds to

$$\alpha_i = \min_{i < i+1}\left\{V_{y^*}^* - \tilde{V}_{y^*(-i)}\right\}, \qquad \forall i \in \Omega_{y^*}, \tag{A5}$$

where $\tilde{V}_{y^*(-i)} = \text{Max}_{\{x_{zj},\, \forall z \in \Omega_{y^*(-i)},\, \forall j\}}\{\Sigma_{z \in \Omega_{y^*(-i)}} \pi_z - E\{D(a_y, \eta)\}\}$, subject to $r_z = 0$, $\forall z \notin \Omega_{y^*(-i)}$, and $\Omega_{y^*} = \{\Omega_{y^*(-i)}, i\}$. $\tilde{V}_{y^*(-i)}$ represents maximum expected net benefits when firm i from the optimal set is not included. Thus, α_i represents the marginal net benefits at the optimum from the addition of the ith firm, and $\alpha_i > 0$, $\forall i \in \Omega_{y^*}$ by (A3). The largest value of α_i in (A5) is firm $i = 1$, the ith firm corresponding to the next largest value of α_i in (A5) as firm $i = 2$, and so on. Firms denoted by $i = 1, \ldots, n^* - 1$ are defined to be inframarginal firms, although firm n^* is denoted as the marginal firm.

The $Z - n^*$ firms outside the set Ω_{y^*} (i.e., the extramarginal firms) can be indexed iteratively in a similar manner. Define α_i $(i = n^* + 1, \ldots, Z)$ in general by the relationship,

$$\alpha_i = \min_{i > i-1}\left\{V_{y^*}^* - \tilde{V}_{y^*(+i)}\right\}, \qquad \forall i \notin \Omega_{y^*}, \tag{A6}$$

where

$$\tilde{V}_{y^*(+i)} = \underset{\{x_{zj},\, \forall z \in \Omega_{y^*(+i)},\, \forall j\}}{\text{Max}} \left\{ \sum_{z \in \Omega_{y^*(+i)}} \pi_z - E\{D(a_y, \eta)\} \right\},$$

subject to $r_z = 0, \forall z \notin \Omega_{y^*(-i)}$, and $\Omega_{y^*(+i)} = \{\Omega_{y^*}, i\}, \forall i \notin \Omega_{y^*}$. $\tilde{V}_{y^*(+i)}$ represents maximum expected net benefits when the ith firm from outside of the optimal set is included. Thus, α_i represents the marginal net benefits at the optimum from the subtraction of the ith firm, and $\alpha_i > 0, \forall i \notin \Omega_{y^*}$ by (A3). Denote the ith firm that corresponds to the smallest largest value of α_i in (A5) as firm $i = n^* + 1$. Denote the ith firm corresponding to the next smallest value of α_i in (A5) as firm $i = n^* + 2$, and so on.

Several important relationships can be developed from the preceding analysis. First, (A4)–(A6) imply that firms with lower values of i are either more productive and hence earn higher profits, or have a smaller incremental expected damage contribution, or a combination of both may occur. This aspect could be made explicit with the use of dual indices as in Shortle, Horan, and Abler [16].

Finally, it is optimal for the following condition to hold for the marginal firm (by (A4)),

$$\begin{aligned} \alpha_n^* &= V_{y^*}^* - \tilde{V}_{y^*(-n^*)} \\ &= \sum_{i=1}^{n^*} \pi_i^* - \sum_{i=1}^{n^*-1} \tilde{\pi}_i - \left[E\{D(a_{y^*}^*, \eta)\} - E\{D(\tilde{a}_{y^*(-n^*)}, \eta)\} \right] > 0. \quad \text{(A7)} \end{aligned}$$

If n^* is sufficiently large, then $\pi_i^* \approx \hat{\pi}_i, \forall i = 1, \ldots, n^* - 1$ and (A7) is approximately,

$$\alpha_n^* = \pi_{n^*}^* - \Delta_{n^*} E\{D(a_{y^*}^*, \nu)\} > 0, \qquad \text{(A8)}$$

where $\Delta_n^* E\{D(a_{y^*}^*, \eta)\} = [E\{D(a_{y^*}^*, \eta)\} - E\{D(\tilde{a}_{y^*(-n^*)}, \eta)\}]$. This expression is important for characterizing the scale of production for the marginal firm.

For simplicity, the subscripts y^*, $y^*(-i)$, and $y^*(+i)$ are suppressed in the main text.

REFERENCES

1. J. B. Braden and K. Segerson, Information problems in the design of nonpoint source pollution policy, *in* "Theory, Modeling and Experience in the Management of Nonpoint-Source Pollution" (C. S. Russell and J. F. Shogren, Eds.), Kluwer Academic, Boston (1993).
2. R. Cabe and J. Herriges, The regulation of nonpoint-source pollution under imperfect and asymmetric information, *J. Environ. Econom. Management*, **22**, 34–146 (1992).
3. A. M. Freeman, III, "The Measurement of Environmental and Resource Values: Theory and Method," Resources for the Future, Washington, D.C. (1993).
4. L. G. Hansen, "A damage based tax mechanism for regulation of non-point emissions," *Environ. Resource Econom.*, forthcoming.
5. J. R. Herriges, R. Govindasamy, and J. Shogren, Budget-balancing incentive mechanisms, *J. Environ. Econom. Management*, **27**, 275–285 (1994).
6. B. Holmström, Moral hazard in teams, *Bell J. Econom.*, **13**, 324–340 (1982).
7. R. E. Just, D. L. Heuth, and A. Schmitz, "Applied Welfare Economics and Public Policy," Prentice-Hall, Englewood Cliffs, NJ (1982).

8. T. J. Miceli and K. Segerson, Joint liability in torts: Marginal and infra-marginal efficiency, *Internat. Rev. Law Econom.*, **11**, 235–249 (1991).
9. National Research Council, "Soil and Water Quality," Natl. Acad. Press, Washington D.C., (1993).
10. Organization for Economic Cooperation and Development, "Environmental Taxes in OECD Countries," Organization for Economic Development and Cooperation, Paris, (1994).
11. C. R. Plott, Externalities and corrective taxes, *Economica*, **33**, 84–87 (1966).
12. K. Segerson, Uncertainty and incentives for nonpoint pollution control, *J. Environ. Econom. Management*, **15**, 88–98 (1988).
13. J. S. Shortle, The allocative efficiency implications of water pollution abatement cost comparisons, *Water Resources Res.*, **26**, 793–797 (1990).
14. J. S. Shortle and D. G. Abler, Nonpoint Pollution, *in* "International Yearbook of Environmental and Natural Resource Economics" (H. Folmer and T. Teitenberg, Eds.), Edward Elgar Cheltenham, U.K., (1997).
15. J. S. Shortle and J. W. Dunn, The relative efficiency of agricultural source water pollution control policies, *Amer. J. Agricultural Econom.*, **68**, 668–677 (1986).
16. J. Shortle, R. Horan, and D. Abler, "Economic Incentives for Nonpoint Pollution Control: Input vs. Ambient-Based Taxes," selected paper, Annual meeting of the European Association of Environmental and Resource Economists, Tilburg, The Netherlands (June 1997).
17. T. Tomasi, K. Segerson, and J. Braden, Issues in the design of incentive schemes for nonpoint source pollution control, *in* "Nonpoint Source Pollution Regulation: Issues and Policy Analysis" (T. Tomasi and C. Dosi, Eds.), Kluwer Academic, Dordrecht, (1994).
18. U.S. EPA, "Managing Nonpoint Pollution: Final Report to Congress on Section 319 of the Clean Water Act," EPA-506/9-90 (January 1992).
19. A. Xepapadeas, Environmental policy under imperfect information: Incentives and moral hazard, *J. Environ. Econom. Management*, **20**, 113–126 (1991).
20. A. Xepapadeas, Environmental policy design and dynamic nonpoint source pollution, *J. Environ. Econom. Management*, **23**, 22–39 (1992).
21. A. Xepapadeas, Controlling environmental externalities: Observability and optimal policy rules, *in* "Nonpoint Source Pollution Regulation: Issues and Policy Analysis" (T. Tomasi and C. Dosi, Eds.), Kluwer Academic, Dordrecht, (1994).

JOURNAL OF ENVIRONMENTAL ECONOMICS AND MANAGEMENT 22, 134–146 (1992)

The Regulation of Non-Point-Source Pollution Under Imperfect and Asymmetric Information*

RICHARD CABE

New Mexico State University, Las Cruces, New Mexico 88003

AND

JOSEPH A. HERRIGES

Iowa State University, 266 Heady Hall, Ames, Iowa 50011

Received April 8, 1991; revised July 16, 1991

This paper develops a Bayesian framework for discussing the role of information in the design of non-point-source pollution control mechanisms. An ambient concentration tax is examined, allowing for spatial transport among multiple zones. Imposition of the tax requires costly measurement of concentrations in selected zones, and the selection of zones for measurement must be undertaken without perfect information regarding several parameters of the problem. Potentially crucial information issues discussed include: (a) the impact of asymmetric priors regarding fate and transport, (b) the cost of measuring ambient concentration, and (c) the optimal acquisition of information regarding fate and transport. © 1992 Academic Press, Inc.

I. INTRODUCTION

As progress is made in controlling point sources of pollution, non-point-source problems command greater attention. The rising concern with off-site consequences of agricultural chemical application provides a prominent example. But the design of regulatory mechanisms to control non-point-source pollutants brings to prominence a different set of issues than those which arise in controlling point-source pollutants. Answers to the question "What information is available at what cost" should be expected to play a crucial role in determining the best structure and parameters of a regulatory mechanism. Indeed, the useful distinction between point- and non-point-source pollution problems lies in the differing cost structures for the acquisition of information regarding important parameters of the problem.[1] While considerations of the cost of information and informational asymmetries between polluters and regulators have been raised in the discussion of

*Partial support for this research was provided by the Resources and Technology Division of the Economic Research Service, USDA. Views expressed in this paper do not necessarily reflect the view of ERS. The authors thank Kathleen Segerson, Jason Shogren, and two anonymous referees for helpful comments on an earlier draft of this paper. All remaining errors are those of the authors.

[1]In the case of point-source pollution, emissions created by individual polluters are regarded as "measurable." Such magnitudes are regarded as "not measurable" in cases of non-point-source pollution, or measurable only at a cost which is automatically prohibitive. In fact, rather than being "measurable" or "unmeasurable," these magnitudes are almost always subject to measurement, but at a cost and with varying reliability.

0095-0696/92 $3.00

proposed regulatory mechanisms, such considerations have rarely played an explicit role in formal models designing and comparing mechanisms for pollution control.[2]

The purpose of this paper is to examine the role of information structure (i.e., cost, reliability, and distribution among agents) in the design of the ambient concentration tax mechanism recently proposed by Segerson [9]. Therein, the author develops a novel control mechanism for non-point-source pollutants in which firms pay a tax based upon the ambient concentration of a pollutant. This paper develops a Bayesian framework in which to consider information issues which may be crucial to design of such a mechanism. Many of these information issues rest on biogeophysical processes which transform human activity in one place (e.g., the application of an agricultural chemical, which is not a problem so long as the chemical remains near the surface of the field to which it is applied) into chemical concentrations in another place, where they are regarded as a pollution problem. Inclusion of this linkage between production and ambient concentrations (the fate and transport mechanism) requires that the model provide for multiple zones in which the pollutant can cause harm. The paper provides for the possibility that regulators and producers may have different beliefs about parameters of the fate and transport mechanism, and considers the regulator's problem of acquiring information regarding producers' beliefs and perhaps engaging in education intended to alter those beliefs. Finally, implementation of the ambient concentration tax requires the regulator to incur the cost of acquiring information regarding production and control practices of firms, refining priors on the fate and transport mechanism, or estimating ambient concentrations at various potential damage sites. The cost of acquiring this information and the reliability of information acquired is explicitly incorporated into the design of the optimal ambient concentration tax.

II. A SPATIAL MODEL OF NON-POINT SOURCE POLLUTION

A. Notation

Consider a geographical region consisting of N zones. The region is determined by the jurisdiction of the regulator. The zones are established to divide this jurisdiction according to the nature of damage from the pollutant under consideration and production and pollutant transport attributes of the region.[3] Specifically, from the damage perspective, zones are defined so that the aggregate level of the pollutant within a zone determines the damage to society. From the pollution creation perspective, areas must be small enough so that the fate of a pollutant entering the environment from within the zone can be treated as the same, without regard to the precise location of release of the pollutant. This division is based upon the mechanics of transport, including hydrologic characteristics, prevailing winds, etc.[4] Finally, the initial zone divisions are specified so that opportunity cost of pollution abatement is uniform within the zone. For agricultural non-point-

[2] See Rausser and Howitt [8] for a notable early exception.

[3] This notation of zones is similar to the one employed by Tietenberg [12].

[4] Delineation of the zones within a given region is itself a difficult task. Recent work by Gold *et al.* [3], Young *et al.* [15], and Anderson, Opaluch, and Sullivan [1] provide potential tools for this process.

source pollution, for example, this division will depend upon the productivity attributes of the soil.

The non-point-source pollutant of interest originates within the region as a by-product of a single production that can be undertaken in any of the zones, with y_i denoting the level of production in zone i and $y_. \equiv (y_1, \ldots, y_N)'$. Individual firms can reduce the level of pollutant entering the environment from within their zone through abatement effort, denoted by a_i, and $a_. \equiv (a_1, \ldots, a_N)'$.[5] In general, there need not be a one-to-one correspondence between the abatement effort, a_i, and abatement level. For example, in an agricultural context, a_i may represent alternative tillage or rotation practices; a_i and y_i can then be viewed as joint outputs of the farm. The cost of producing y_i with abatement effort a_i is denoted by $C_i(y_i, a_i)$, where C_i is assumed to be a strictly convex function of y_i and a_i.

The combination of $y_.$ and $a_.$ determines the ambient level of pollution in each zone through a transport mechanism. Specifically, let X_i denote the ambient level of pollution in zone i, with $X_. \equiv (X_1, \ldots, X_N)'$. These pollution concentrations are determined by the transformation $X_. = T(y_., a_., \varepsilon)$, where ε is an $M \times 1$ vector of stochastic transport effects.[6] The random component reflects the uncertainty on the part of both regulators and firms regarding the exact nature of the transport mechanism, due to such factors as weather and imperfect knowledge of relevant physical processes. The regulator is assumed to have a prior on the distribution of these unknown factors, denoted by $f_r(\varepsilon)$, while producers are assumed to share a potentially different prior, denoted by $f_p(\varepsilon)$. The priors on ε in turn generate priors on the ambient concentrations that result from a given level of production activity and abatement level within the region. Finally, the damage to society from the pollutant in question is assumed to be a nondecreasing function of the vector of ambient pollution levels and is represented by the function $D(X_.)$, where $D_i \equiv \partial D/\partial X_i \geq 0$, $i = 1, \ldots, N$, with $\partial D/\partial X_i > 0$ for some i.

B. *Taxes on Ambient Pollution Levels under Uncertainty*

Segerson [9] proposed a pollution control mechanism in which firms pay a tax based upon an uncertain ambient level of pollution with a known distribution. Firms and regulators were assumed to share a common prior on the transport mechanism. This section considers a multizone version of such a tax. Specifically, a tax on the producers in zone i is considered where the tax is calculated as the product of the tax rate (t_{ij}) and the deviation of the ambient level of pollution in zone j (X_j) from the regulator's target concentration (X_j^*). The distinction between the regulator and producer priors regarding the transport mechanism is maintained. Initially, the administrative costs associated with the tax are ignored.

[5] For the sake of simplicity, the zones are assumed to be small enough to contain a single firm. It is straightforward, though somewhat cumbersome, to allow for multiple firms within a zone, along the lines of Segerson's [9] multiple pollutors problem in Section 3. Incorporating this change would allow for a long-run analysis, optimizing the number of firms per zone, along the lines of Segerson [9]. However, such an analysis is beyond the scope of the current paper. See Graham-Tomasi and Wiese [13] for a discussion of non-point pollution control and the location issue.

[6] In general, one would expect $\partial X_j/\partial y_i \geq 0$, $\partial X_j/\partial a_i \leq 0$, $\partial^2 X_j/\partial y_i^2 \geq 0$, $\partial^2 X_j/\partial a_i^2 \leq 0$, and $\partial^2 X_j/\partial a_i \partial y_i \leq 0$. That is, X_j is convex in y_i and $(-a_i)$. The random component vector, ε, need not be $N \times 1$ (i.e., with $M = N$), as there may be multiple sources of uncertainty within or across zones affecting the transport mechanism differently.

The regulator's problem is to maximize the expected sum of producer and consumer surplus, less the damages resulting from the ambient pollution levels generated by producers; i.e., $X_{.}$. If given direct control over the levels of production and abatement effort, the regulator would choose $y_{.}$ and $a_{.}$ to solve

$$W^* = \underset{y.,a.}{\text{Max}}\, E_{\text{r}}\left\{\int_0^Y p(s)\,ds - \sum_{i=1}^{N} C_i(y_i, a_i) - D(X.)\right\}$$
$$\equiv \underset{y.,a.}{\text{Max}}\, W(y.,a.), \qquad (1)$$

where $E_{\text{R}}(\ )$ denotes the expectation operator given the regulator's prior distribution on ε (i.e., $f_{\text{r}}(\varepsilon)$), $p(y)$ denotes the demand for the output produced, and $Y \equiv \Sigma_{i=1}^{N} y_i$ denotes system-wide production of y. The corresponding first order necessary conditions are given by

$$0 = p(Y) - [\partial C_i/\partial y_i] - \bar{D}^{\text{r}}_{yi} \qquad i = 1,\ldots,N \qquad (2a)$$

and

$$0 = -[\partial C_i/\partial a_i] - \bar{D}^{\text{r}}_{ai} \qquad i = 1,\ldots,N, \qquad (2b)$$

where

$$\bar{D}^{\text{r}}_{yi} \equiv \sum_{k=1}^{N} E_{\text{r}}\{[\partial D/\partial X_k][\partial X_k/\partial y_i]$$

and

$$\bar{D}^{\text{r}}_{ai} \equiv \sum_{k=1}^{N} E_{\text{r}}\{[\partial D/\partial X_k][\partial X_k/\partial a_i]$$

denote the regulator's expectations regarding the system-wide damage resulting from a marginal change in output and abatement levels in zone i, respectively. Let $y^*_{.}$ and $a^*_{.}$ denote the solutions to (2). Then $\bar{X}^*_{.} \equiv E_{\text{r}}[T(y^*_{.}, a^*_{.})]$ establishes the regulator's mean ambient concentration cutoff level for each zone and $W^* = W(y^*_{.}, a^*_{.})$ denotes the social optimum.

Without direct control over production and abatement levels, the regulator must rely upon the tax mechanism to direct producers toward the socially optimal levels. For the producer in zone i, the ambient tax takes the form of $t_{i.}(X_{.} - X^*_{.})$, where $t_{i.} \equiv (t_{i1},\ldots,t_{iN})$. Each firm is assumed to be a price taker and risk neutral, maximizing its expected profits and taking the level of production and abatement in other zones as given. Thus, the firm in zone i solves

$$\pi_i(t_{i.}) \equiv \underset{y_i,a_i}{\text{Max}}\, E_{\text{p}}\{py_i - C_i(y_i, a_i) - t_{i.}(X_{.} - X^*_{.})\}$$
$$= \underset{y_i,a_i}{\text{Max}}\left\{py_i - C_i(y_i, a_i) - E_p[t_{i.}X_{.}] - t_{i.}X^*_{.}\right\}, \qquad (3)$$

where $E_{\text{p}}(\ )$ denotes the expectation operator given the producer's prior distribution on ε, $f_{\text{p}}(\varepsilon)$. Assuming that an interior solution exists satisfying the usual second order conditions, firm i's optimal output and abatement levels $\hat{y}_i(t_{i\cdot})$ and $\hat{a}_i(t_{i\cdot})$ will solve the first order conditions[7]

$$0 = p - \partial C_i/\partial y_i - \sum_{j=1}^{N} t_{ij} E_{\mathrm{p}}\{T_{yij}\}$$

$$= p - \partial C_i/\partial y_i - \sum_{j=1}^{N} t_{ij} \bar{T}^{\mathrm{p}}_{yij} \qquad (4a)$$

$$0 = -\partial C_i/\partial a_i - \sum_{j=1}^{N} t_{ij} E_{\mathrm{p}}\{T_{aij}\}$$

$$= -\partial C_i/\partial a_i - \sum_{j=1}^{N} t_{ij} \bar{T}^{\mathrm{p}}_{aij}, \qquad (4b)$$

where $T_{yij} \equiv \partial X_j/\partial y_i$ and $T_{aij} \equiv \partial X_j/\partial a_i$ denote, respectively, the marginal impact of production and abatement effort in zone i on the ambient pollution level in zone j and

$$\bar{T}^{\mathrm{p}}_{yij} \equiv E_{\mathrm{p}}\{T_{yij}\}$$

$$\bar{T}^{\mathrm{p}}_{aij} \equiv E_{\mathrm{p}}\{T_{aij}\}$$

denote the prior means producer's form regarding these marginal effects. In general, $T(y_{\cdot}, a_{\cdot})$ is a nonlinear function of its arguments. However, if T is linear, or approximately so near the optimal levels for $y_{\cdot}$and $a_{\cdot}$, then the concentration taxes (i.e., the t_{ij}'s) influence production and abatement cffort only through the tax indices

$$\bar{T}^{\mathrm{p}}_{yi} \equiv \sum_{j=1}^{N} t_{ij} \bar{T}^{\mathrm{p}}_{yij}$$

$$\bar{T}^{\mathrm{p}}_{ai} \equiv \sum_{j=1}^{N} t_{ij} \bar{T}^{\mathrm{p}}_{aij}.$$

This suggests that the optimal concentration tax matrix, developed below, need not be unique in a multizone setting.[8]

[7]In general, the solution to the maximization process in (3) need not be an interior one. First, the necessary second order conditions may not hold when (4a) and (4b) are satisfied because T is not a convex function of y_j and a_j. Second, with the imposition of the ambient tax, the farm may no longer be profitable, leading to exit from the market and a discontinuity in the objective function at points (y_j, a_j) such that profits are zero for a given $t_{j\cdot}$. This problem is discussed further in Section III.

[8]This should not be surprising given that there are N^2 policy tools (i.e., that t_{ij}'s) and only 2N variables to control (i.e., the y_i's and a_i's). A similar situation arises in Segerson [9], where the tax mechanism incorporates both a marginal tax rate (t) and a lump sum penalty (k) for exceeding the regulator's cutoff level ($\bar{x}$). A unique tax is developed only when an additional objective, optimal industry size, is added. In the multizone setting, optimizing industry size within each zone would add N additional equilibrium conditions, reducing but not eliminating the non-uniqueness of $t^*_{\cdot\cdot}$.

Comparing Eqs. (2) and (4), it is clear that the regulator can induce the socially optimal levels of production and abatement effort if there exists a $t_{..} \equiv (t'_{1.}, \ldots, t'_{N.})'$, say $t^*_{..}$, such that

$$\sum_{j=1}^{N} t^*_{ij} \overline{T}^{\mathrm{p}}_{yij} = \overline{D}^{\mathrm{r}}_{yi} \tag{5a}$$

and

$$\sum_{j=1}^{N} t^*_{ij} \overline{T}^{\mathrm{p}}_{aij} = \overline{D}^{\mathrm{r}}_{ai}. \tag{5b}$$

The left hand-side of each equation indicates, for production and abatement effort respectively, the change in marginal tax burden producers expect to be generated by a change in the tax rate t_{ij}, while the right hand-side measures the marginal benefit of the tax rate in terms of reducing pollution damages.

A number of special cases of this problem are of interest. In particular, suppose that a_i measures effluent emission controlled. Then $(-T_{aij})$ can be interpreted as a generalized transfer coefficient, depending upon both the level of production and abatement in zone i. If, in addition, it is assumed that the level of output does not directly influence the ambient concentrations (i.e., $T_{yij} \equiv 0$), then the optimal tax is determined solely by Eq. (5b), a system of only N equations with N^2 unknowns (i.e., the t_{ij}'s). With no direct relationship between the level of production and ambient pollution levels, the level of production $(\hat{y}_i)$ becomes an implicit function of abatement effort $(\hat{a}_i)$ through Eq. (4a). The tax now induces change in the concentration of pollution in each zone only through its impact on the level of abatement. Maintaining these restrictions yields the following special cases.

Case 1: *The Single Polluter/Single Damage Site.* In the case of a single polluter and a single damage site (i.e., $N = 1$), the optimal tax, t^*, becomes

$$\begin{aligned} t^* &= \overline{D}^{\mathrm{r}}_a / \overline{T}^{\mathrm{p}}_a \\ &= E_{\mathrm{r}}[(\partial D/\partial X)(\partial X/\partial a)]/E_{\mathrm{p}}[\partial X/\partial a]. \end{aligned} \tag{6}$$

When the regulator and producers have the same prior on the transport mechanism, this is equivalent to Segerson's [9] Eq. (5a). However, if the producers do not perceive that they have a significant influence on ambient concentration at the damage site (i.e., $E_{\mathrm{p}}[\partial X/\partial a]$ is small relative to $E_{\mathrm{r}}\{\partial X/\partial a\}$), then t will have to be large in order to efficiently reduce the level of pollution damage.[9]

[9]This assumes that the tax does not become so large as to drive the producer out of the market (see Section IIIA). It is also possible that producers may have higher transport expectations than the regulators (i.e., $E_{\mathrm{p}}[\partial X/\partial a]$ is large relative to $E_{\mathrm{r}}\{\partial X/\partial a\}$). In this case, the tax required to achieve W^* is reduced. If the regulator fails to recognize this asymmetry in transport priors, the concentration objective will be surpassed.

Case 2: *Linear Damage Function, Multiple Zones.* Suppose $D(\ )$ is linear in the X_i's, with $\partial D/\partial X_i \equiv \alpha_i$. Then the optimal tax rates for the multiple zone model are defined by:

$$\sum_{k=1}^{N} t_{ik}^{*} \bar{T}_{aij}^{\mathrm{p}} = \bar{D}_{ai}^{\mathrm{r}}$$

$$= \sum_{k=1}^{N} \alpha_k \bar{T}_{aij}^{\mathrm{r}} \qquad i = 1, \ldots, N, \tag{7}$$

where $\bar{T}_{aij}^{\mathrm{r}} \equiv E_{\mathrm{r}}\{T_{aij}\}$ denotes the regulator's expectation regarding the generalized transfer coefficient. Again, the matrix of t_{ij}'s solving (7) need not be unique. However, one solution can be found by noting that Eq. (7) identifies the diagonal elements in the matrix relationship

$$t_{\cdot\cdot}^{*}(\bar{T}_{a\cdot\cdot}^{\mathrm{p}})' = A(\bar{T}_{a\cdot\cdot}^{\mathrm{r}})',$$

where $\bar{T}_{a\cdot\cdot}^{\mathrm{p}} \equiv \{\bar{T}_{aij}^{\mathrm{p}}\}$, $\bar{T}_{a\cdot\cdot}^{\mathrm{r}} \equiv \{\bar{T}_{aij}^{\mathrm{r}}\}$ and $A \equiv \alpha_{\cdot} \otimes \iota_N$, with ι_N being an $N \times 1$ vector of ones and $\alpha_{\cdot} \equiv (\alpha_1, \ldots, \alpha_N)$. If $\bar{T}_{a\cdot\cdot}^{\mathrm{p}}$ is invertible, then[10]

$$t_{\cdot\cdot}^{*} = A\left[(\bar{T}_{a\cdot\cdot}^{\mathrm{r}})^{-1} \bar{T}_{a\cdot\cdot}^{\mathrm{p}}\right]'. \tag{8}$$

If producers and regulators have the same prior means with regards to the transfer coefficients (i.e., $\bar{T}_{a\cdot\cdot}^{\mathrm{r}} = \bar{T}_{a\cdot\cdot}^{\mathrm{p}}$), then Eq. (8) reduces to $t_{ij}^{*} = \alpha_j$. That is, the marginal tax rate for all zones with respect to concentration impacts on zone j is simply the marginal damage cost from the increased concentration (α_j).

III. INFORMATION

In general, the burden of information required to implement the tax derived in Section II is significant. The regulator must know the nature of the firm's costs, the nature of demand, the ambient level of pollution in each zone, and the nature of the damages in area j from the ambient level of pollution. In addition, the regulator must be able to evaluate expectations defined by the producer's prior distribution on the transport mechanism, as well as evaluate expectations defined by its own prior distribution on the transport mechanism. In this section, consideration is given to impact of information structure on the ambient tax mechanism developed above.

[10]The matrix $\bar{T}_{a\cdot\cdot}^{\mathrm{p}}$ will be singular if the number of zones, defined in accordance with the principles set out above, exceeds the rank of $\bar{T}_{a\cdot\cdot}^{\mathrm{p}}$, which reflects producer perceptions. This suggests an aggregation of zones for the purpose of defining taxes, but not necessarily for defining damages or production costs.

A. Education Costs

As illustrated in the case of a single polluter and single damage site (Case 1), the discrepancy between the prior beliefs of the regulator and the producer can have a significant impact on the optimal tax policy. In the extreme, if producers in a given zone, say i, believe they have no influence on concentration levels, the tax becomes a discrete policy tool. With $T_{yi\cdot} \equiv 0$ and $T_{ai\cdot} \equiv 0$, the first order conditions in Eqs. (4a) and (4b) are independent of the ambient taxes. As a result, with $\partial C_i/\partial a_i > 0$, the firm will set $a_i = 0$, and the level of production will be determined independently of t_{ij} by the first order condition $P = \partial C_i(y_i 0)/\partial y_i$. The policymaker must then choose between enduring the damage caused by pollution emanating from zone i or driving the producers out of the market entirely by setting taxes at a level $t_{i\cdot}$ such that $\pi(t_{i\cdot}) < 0$.

This problem is illustrated graphically in Fig. 1 for the single zone case. For ease of exposition, abatement effort is assumed to be zero (i.e., $a = 0$), so that the firm influences concentration levels within the zone only through changes in the level of

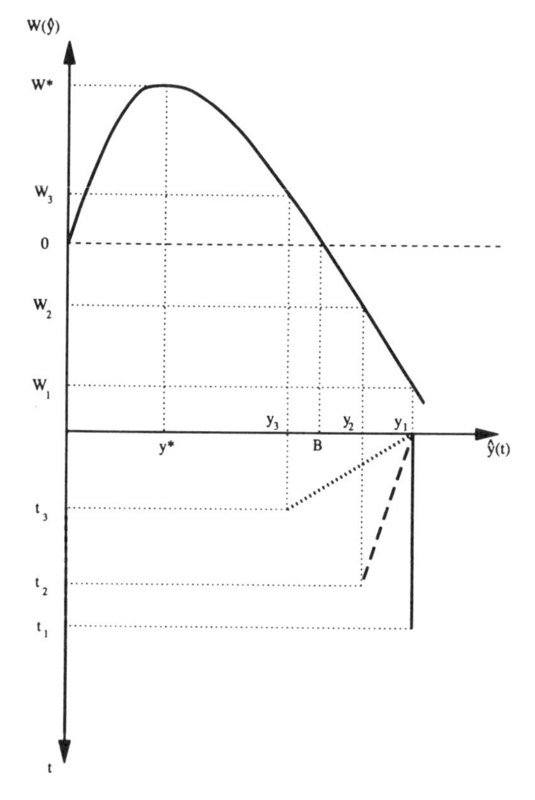

FIG. 1. The impact of producer priors and yield discontinuities on tax program flexibility.

production. Total societal net benefits, W, can then be written as a function of output, y, with

$$W(y) \equiv E_r\left\{\int_0^y p(s)\,ds - C(y,0) - D[T(y,0,\varepsilon)]\right\}$$

and $W(0) = 0$. This relationship is illustrated in the upper quadrant of Fig. 1.

The lower quadrant of Fig. 1 depicts $\hat{y}(t)$ as a function of the ambient tax level, t. If $\bar{T}_y^p \equiv 0$, then the firm perceives its tax burdens to be independent of its production level and $\hat{y}(t)$ is determined by the solid line in the lower half of Fig. 1. That is, $\hat{y}(t)$ remains at a constant level (i.e., $\hat{y}(t) = y_1$) for $t \leq t_1$, where t_1 solves $py_1 - C(y_1, 0) = t(X - X^*)$. Once t exceeds t_1, the tax burden becomes sufficient to drive the firm out of the market (i.e., $\hat{y}(t) = 0$, $t > t_1$). Under these circumstances, the ambient tax mechanism becomes a crude policy tool, only allowing the regulator to choose between (a) continued production and pollution (with $W = W(y_1)$) and (b) the termination of production (with $W = W(0) = 0$). The former will be chosen as long as $y_1 < B$ in Fig. 1, while the latter will be preferred if $y_1 > B$.[11]

A similar problem emerges when $\bar{T}_{yij}^p$ and $\bar{T}_{aij}^p$ are small relative to their true values or those perceived by the regulating agency. Again, Fig. 1 illustrates the situation for the single zone scenario. If $\bar{T}_y^p$ is small, $\hat{y}(t)$ will change little as the tax level is increased, as with the dashed line in Fig. 1. Eventually, however, the taxes will reach a level t_2 that will drive the firm from the market. In this case, the optimal policy will again be to raise taxes beyond t_2, forcing the firm out of operation. If $\bar{T}_y^p$ is larger, as in the case of the dotted line, then continued operation may be optimal, with output reduced from y_1 to y_3 using an ambient tax level of t_3. The range of policy alternatives, however, remains narrow, and the optimal tax policy achieves a social welfare level, W_3, substantially below the global maximum $W^* = W(y^*)$.[12]

The above arguments suggest that the regulator's ability to ascertain and alter the firm's prior beliefs about the transport mechanism is likely to be crucial to success in designing and implementing the ambient tax mechanism. If producers initially believe their activities have little effect on ambient pollution and these reduction costs are high, emission standards or restrictions on technology, typically viewed as less efficient policy tools, may prove to be the more cost-effective policy mechanisms. Furthermore, in the multizone context we have seen that t_i^* need not be unique, providing regulators with flexibility in their efforts to achieve W^*. If an ambient mechanism is used, positive ambient taxes should set for those transport

[11]Given the usual curvature assumptions for D and T, y_1 will lie to the right of y^* in Fig. 1.

[12]Segerson [9] also considers a nonstochastic penalty applied when the level of concentration exceeds the targeted concentration level. Problems similar to those illustrated in Fig. 1 arise for this policy tool. For producers with small marginal transport expectations, the optimal fixed penalty required to induce socially optimal production and abatement may also yield negative profits. It is true, however, that the extent of the problem may differ in the case of a fixed penalty versus a tax. This is because, while the tax and the penalty need to have the same *marginal* impact on the firm at the optimum, they need not have the same *total* tax burden. It is the latter that causes the "exit" problem. As suggested by an anonymous referee, the addition of a random lump sum penalty, along the lines proposed by Xepapadeus [14] and Rasmusen [8], could be used to mitigate or eliminate this problem if the producers are risk averse.

linkages (i.e., "*ij*" pairs) for which regulators and pollutors share a common or similar prior, or at least reliance should be placed on those linkages which producers believe they can influence (i.e., $\overline{T}^{p}_{yij}$ and $\overline{T}^{p}_{aij}$ large).

B. Monitoring Costs

The analysis presented in Section III.A assumes that information on the true level of $X.$ is known without cost to the regulator. In fact, "...determining groundwater pollution and monitoring groundwater quality are extremely difficult and expensive." (Ng [5], p. 777). The decision to impose an ambient tax must be considered jointly with the cost of obtaining the necessary measures of pollution concentration within each zone. Thus, the regulator's problem needs to be extended to include the cost of measuring ambient pollution in every area *j*, on which a tax will be based.

Let $\delta_j = 1$ if the regulator chooses to measure the pollutant's concentration in zone *j*, with $\delta_j = 0$ otherwise. Thus, with $\delta_j = 0$, $t_{ij} = 0\ \forall i$. Furthermore, let χ_j denote the cost of measuring ambient concentration in area *j* and $W(t..) \equiv W[\hat{y}(t..), \hat{a}(t..)]$. The regulator's problem is then to solve

$$
\begin{aligned}
W^{**} &\equiv \underset{t..,\,\delta.}{\mathrm{Max}}\left[W(t..\Delta) - \delta.\chi.\right] \\
&\quad \text{s.t. } t_{ij} \geq 0 \qquad i, j = 1, \ldots, N \\
&= \underset{t..,\,\delta.}{\mathrm{Max}}\left[W(t.., \delta)\right] \\
&\quad \text{s.t. } t_{ij} \geq 0 \qquad i, j = 1, \ldots, N, \qquad (9)
\end{aligned}
$$

where $\chi. \equiv (\chi_1, \ldots, \chi_N)'$, $\delta. \equiv (\delta_1, \ldots, \delta_N)$, $\Delta \equiv \mathrm{diag}\{\delta_i\}$, and $W(t.., \delta.) \equiv W(t..\Delta) - \delta.\chi..$ Once monitoring costs are included, the regulator must decide which receptor sites should be monitored. This decision will be based, in part, on the regulator's priors on ambient concentration levels, $X..$ These priors could be formed on the basis of two sorts of information. First, if the regulator has knowledge of $y.$ and $a.$, then priors on the transport mechanism induce a prior distribution on concentration levels. Second, whether $y.$ and $a.$ are known or not, the regulator may believe that ambient concentration levels are correlated in a way related to their spatial relationships. If so, then measurement at a given site will cause the regulator to revise priors on nearby concentrations. The possibility of spatial correlation casts the regulator's problem as one of optimal search.[13] The initial choice of sites to measure is based on priors about what will be found in the measurements, and subsequent measurement decisions are based on priors informed by the results of earlier measurement.

The choice of $\delta.$ in (9) will also depend upon the priors held by producers. The measurement of a given site has two benefits to the regulator.[14] First, it increases

[13]Whether the search will be sequential (i.e., measuring ambient concentrations one site at a time), fixed-sample-size (i.e., a one-time choice of $\delta.$), or variable-sample-size (i.e., sequentially choosing the number of sites to measure) will depend upon the degree of perceived correlation between the X_j's and the discount factor with respect to time. The higher the perceived correlation, the greater the attraction will be for sequential search. See Morgan and Manning [4], Cressie and Morgan [2], and Olson [6].

[14]The choice of $\delta.$ impacts the producer's profit by determining which damage zones are taxed.

the flexibility of the tax policy by allowing $t_{\cdot j} > 0$ once site j is measured. Second, it provides information with which priors on the transport mechanism can be updated. Depending upon the structure of producer priors, however, the former benefit may quickly become zero. For example, if T is linear in $y_{\cdot}$ and $a_{\cdot\cdot}$, then, as indicated in Section II.B.1, the taxes influence firm behavior only through the tax indices $\bar{T}^{\mathrm{P}}_{yi}$ and $\bar{T}^{\mathrm{P}}_{aj}$. As long as $\bar{T}^{\mathrm{P}}_{y \cdot j}$ and $\bar{T}^{\mathrm{P}}_{a \cdot j}$ are nonzero for two j's, two sites will exhaust all flexibility benefits from monitoring additional sites. The corresponding $t_{\cdot j}$'s can be set so as to achieve the levels of $\bar{T}^{\mathrm{P}}_{yi}$ and $\bar{T}^{\mathrm{P}}_{ai}$ that will induce the optimal $y_{\cdot}$ and $a_{\cdot\cdot}$.

C. Knowledge of Pollutant Fate and Transport

For a given source area, i, imposition of the tax for transfer to receptor area j will involve monitoring and education costs discussed above, as well as a reduction in the sum of producer and consumer surplus generated by production in area i. Since T is not known with certainty, imposition of the tax may result in too much or too little pollution reduction at the receptor site, even when the producer's abatement response to the tax is perfectly anticipated by the regulator. Thus, the regulator's prior distribution on ε, which determines the transport mechanism, induces a prior distribution on the net social value of extending the tax to account for transport from area i to area j.

Let

$$V(t_{\cdot\cdot}, \delta_{\cdot}, \varepsilon) = \int_0^{Y(t_{\cdot\cdot}\Delta)} p(y)\, dy - \sum_{i=1}^{N} C_i[\hat{y}_i(t_i.\Delta), \hat{a}_i(t_i, \Delta)] - D(X_{\cdot}) - \delta_{\cdot}\chi_{\cdot\cdot} \tag{10}$$

The $W(t_{\cdot\cdot}, \delta_{\cdot})$ of Eq. (9) is then given by

$$W(t_{\cdot\cdot}, \delta_{\cdot}) = \int_{\Omega} V(t_{\cdot\cdot}, \delta_{\cdot}, \varepsilon) f_{\mathrm{r}}(\varepsilon)\, d\varepsilon,$$

where Ω denotes the state space of ε. Equation (9) describes the "no data" problem of choosing an action, $(t_{\cdot\cdot}, \delta_{\cdot})$, to maximize the expected value of $V(\ )$. Regardless of this initial choice of $(t_{\cdot\cdot}, \delta_{\cdot})$ based on current beliefs regarding ε, it may be desirable to acquire additional information about the transport mechanism to better inform subsequent regulatory decisions.[15] Suppose the regulator has the option to undertake a research project at a cost of ψ with outcome $z \in Z$, related to ε by the conditional distribution $h(z|\varepsilon)$. Using the outcome of the research project, the posterior expectation of V, conditional on z, is given by

$$W_{\mathrm{p}}(t_{\cdot\cdot}, \delta_{\cdot}, z) = \int_{\Omega} V(t_{\cdot\cdot}, \delta_{\cdot}, \varepsilon) f_{\mathrm{r}}(\varepsilon|z)\, d\varepsilon.$$

[15] In the case of agricultural non-point-source pollution of groundwater, substantial resources are now being devoted to such an acquisition of information. Olson [6] analyzes the similar problem of information acquisition on carcinogenicity to inform regulatory decisions.

The regulator should fund the research project if

$$\psi < \int_Z \operatorname*{Max}_{t..,\delta.} W(t..,\delta.,z)g(z)\,dz - \operatorname*{Max}_{t..,\delta.} W(t..,\delta.),$$

where

$$g(z) \equiv \int_\Omega h(z|\varepsilon)f_r(\varepsilon)\,d\varepsilon.$$

This simple formulation of the regulator's problem of information acquisition in support of the Segerson tax neglects the multiperiod duration of the benefit of new information and considers a single research project of fixed size and scope. In fact, acquisition of information on the physical processes influencing fate and transport of pollutants is best regarded as a long-term investment, with benefits enduring over several periods. This suggests that the regulator's discount rate could be crucial to decisions regarding the desirability of research projects. Furthermore, the scope of research is clearly endogenous. Not only is there flexibility in the total budget, ψ, to be devoted to research on fate and transport, but there is a trade-off between the quality of information generated by the project and its geographic coverage. Thus, one project's information, characterize by $h(z|\varepsilon)$, could offer low variance of z given ε for a restricted set of components of the vector z. An alternative project with the same budget could offer an $h(z|\varepsilon)$ with higher variance for a less restricted set of components of z.

IV. SUMMARY AND CONCLUSIONS

This paper has examined the role of information structure in the design of a particular mechanism for controlling non-point-source pollution. The tax mechanism considered is based on ambient concentration of pollutants, and therefore must rely on the acquisition of information regarding concentrations at appropriately designated sites. The mechanism avoids routine acquisition of information concerning production and abatement practices of individual firms, but requires at least some information on fate and transport of the pollutant, as would any likely regulatory scheme. The paper develops a Bayesian framework which allows discussion of several potentially important information acquisition issues. It is shown that producers' beliefs regarding the fate and transport mechanism can be crucial to the path of influence of the mechanism (abatement effort, production amount, or exit) as well as for choice of the mechanism's design parameters. The desirability of the mechanism considered for any specified non-point-source pollution problem may hinge on the regulator's ability to ascertain producers' beliefs about pollutant fate and transport, or to influence these beliefs through education or extension. The paper analyzes considerations which must enter into the regulator's choice of sites for monitoring of ambient concentrations. Again, the value of a monitoring program depends on the control mechanism in which the resulting information will be used. When the cost of monitoring is not trivial, the framework provides an explicit statement of the tradeoff between the cost of monitoring and the value of information generated by monitoring, and raises the issues which must be considered in designing an optimal monitoring program. As was shown to be the case for education and monitoring, the regulator's decision regarding the optimal level of funding for research into the fate and transport mechanism will necessarily depend on how the results of such research will be used. In other words, the value of information on the fate and transport mechanism depends on the nature of the control mechanism through which the information will be applied, as well as the precision and scope of the information generated and the nature of the pollution problem at hand.

This paper has expanded on issues of information structure in the design of a particular mechanism. The larger issue, and a clear next step in investigating the role of information in the design of regulatory mechanisms, is consideration of the choice among alternative regulatory mechanisms. Since regulatory mechanisms differ in their information requirements, and costs of acquiring and processing information differ among the different contexts in which regulation may be considered, it should be expected that the balancing of information costs against allocative merits of mechanisms will not lead to the superiority of any single structure of mechanism for all contexts. The economics of regulation under imperfect information should provide a framework within which to consider the suitability of various mechanisms for regulatory contexts with different information structures. Especially in the information intensive business of regulating non-point-source pollutants, comparisons of alternative mechanisms must be undertaken within this framework.

REFERENCES

1. G. Anderson, J. Opaluch, and W. Sullivan, Nonpoint agricultural pollution: Pesticide contamination of groundwater supplies, *Amer. J. Agr. Econ.* **67**, 1238–1243 (1985).
2. N. Cressie, and R. B. Morgan, "The VPRT, a Sequential Testing Procedure Dominating the SPRT," revision of preprint 86-17, Iowa State Statistical Laboratory (1989).
3. A. J. Gold, S. Tso, P. V. August, and W. R. Wright, Using soil surveys to delineate stratified drift deposits for groundwater protection. *J. Soil Water Conserv.* **44**, 232–234 (1989).
4. P. Morgan, and R. Manning, Optimal search, *Econometrica* **53**, 923–944 (1985).
5. L. Ng, A DRASTIC approach to controlling groundwater pollution, *Yale Law J.* **98**, 773–791 (1989).
6. L. Olson, The search for environmental quality: The economics of screening and regulating environmental hazards, *J. Environ. Econom. Management* **19**, 1–18 (1990).
7. E. Rasmusen, Moral hazard in risk-averse teams, *RAND J. Econom.* **18**, 428–435 (1987).
8. G. C. Rausser and R. Howitt, Stochastic Control of Environmental Externalities, *Ann. Econom. Social Measurement* **4**, 271–292 (1975).
9. K. Segerson, Uncertainty and incentives for nonpoint pollution control, *J. Environ. Econom. Management* **15**, 87–98 (1988).
10. J. S. Shortle and J. W. Dunn, The relative efficiency of agricultural source water pollution control policies, *Amer. J. Agr. Econom.* **68**, 668–677 (1986).
11. W. O. Spofford, A. J. Krupnick, and E. F. Wood, Sources of uncertainty in economic analysis of management strategies for controlling groundwater contamination, *Amer. J. Agr. Econom.* **68**, 1234–1239 (1986).
12. T. H. Tietenberg, Derived decision rules for pollution control in a general equilibrium space economy, *J. Environ. Econom. Management* **1**, 3–16 (1974).
13. T. Graham-Tomsai and A. Wiese, "Land Use and Incentive Schemes for Nonpoint Pollution Control in a Spatial Equilibrium Setting, Staff paper, Department of Agricultural and Applied Economics, University of Minnesota (1990)
14. A. P. Xepapadeas, Environmental policy under imperfect information: Incentives and moral hazard, *J. Environ. Management* **20**, 113–126 (1990).
15. R. A. Young, C. A. Onstad, D. D. Bosch, and W. P. Anderson, AGNPS: A nonpoint-source pollution model for evaluating agricultural watersheds, *J. Soil Water Conserv.* **44**, 168–173 (1989).

[6]

The On-Farm Costs of Reducing Groundwater Pollution

Scott L. Johnson, Richard M. Adams, and Gregory M. Perry

Agricultural chemicals are a source of groundwater pollution in some areas. Regulatory options to reduce such nonpoint pollution imply costs to producers. By integrating plant simulation, hydrologic, and economic models of farm-level processes, this study evaluates on-farm costs of strategies to reduce nitrate groundwater pollution. The empirical focus is on intensively managed, irrigated farms in the Columbia Basin of Oregon. Results suggest that changes in timing and application rates of nitrogen and water reduce nitrate pollution with little loss in profits. Once such practices are adopted, further reductions in nitrates can be achieved only at increasing costs to producers.

Key words: bioeconomic modeling, groundwater pollution, nitrates, on-farm costs.

The adoption of pesticides and inorganic fertilizers by U.S. agriculture since World War II has kept food cost relatively low. However, environmental costs are associated with some of these inputs, including pollution of groundwater by agricultural chemicals. Among the sources of groundwater pollution in rural areas, agriculture is potentially the most serious long-term problem because (*a*) groundwater pollution cannot be traced readily to particular individuals or locations, and (*b*) the area vulnerable to pollution is extensive (CAST). About 50 million people rely on groundwater in areas identified as vulnerable to agricultural groundwater pollution (Lee and Nielsen).

The most common agricultural chemical pollutant is nitrogen in the form of (water-soluble) nitrates. Elevated nitrate levels in groundwater are attributed to the low relative cost of nitrogen and other chemical fertilizers and the ease with which nitrates move in soil. While few cases of death or severe illness are linked directly to agricultural contamination, the human health consequences of nitrate exposure include methemoglobinemia (blue-baby disease) in infants and gastric cancer in adults (Bower). In addition, the potential for surface water pollution from groundwater is also an important environmental concern; approximately 30% of surface water streamflow is from groundwater sources (Saliba).

Public concern over the potential health consequences of groundwater pollution is motivating the U.S. Environmental Protection Agency (EPA) and state environmental agencies to intensify regulatory activities in the area of nonpoint pollution. Regulatory options to reduce nonpoint pollution imply a range of costs for agricultural producers. For example, reduction of nitrate leaching requires modification of farmers' management practices that, in turn, may lead to reduced profits. Regulations that require major reductions in nitrate groundwater pollution may result in significant shifts in crop production between regions of the United States, possibly increasing food costs for consumers. On the other hand, elimination of nitrate pollution may be possible in some regions with relatively little effect on farm profits and consumer costs if producers have a range of alternative management practices and crops from which to choose. Quantification of these potential economic and environmental tradeoffs is important information in the societal debate on this issue.

The reseach reported here builds upon the conceptual and empirical approaches found in some existing studies to develop a bioeconomic

Scott L. Johnson is an economist, South Coast Air Quality Management District, and former graduate student, Oregon State University; Richard M. Adams and Gregory M. Perry are, respectively, a professor and an associate professor, Department of Agricultural and Resource Economics, Oregon State University.

Technical Paper No. 9435 of the Oregon Agricultural Experiment Station.

The research described in this paper was partially supported by the U.S. Geological Survey under USGS award number 14-0300001-G1486 to Oregon State University. The views and conclusions contained in this document are those of the authors and should not be interpreted as necessarily representing the official policies, either expressed or implied, of the U.S. government.

The authors gratefully acknowledge the assistance of Bob Berrans in data collection and the helpful comments of Bob Taylor on an earlier draft of this manuscript.

analysis of agricultural groundwater pollution. The overall objective is to assess the on-farm economic effects of strategies to reduce nitrate groundwater pollution. The assessment reflects the linkages between producer behavior (management practices), crop yields, and groundwater pollution by integrating plant simulation, hydrologic, and economic models of farm-level processes. Using this framework, the analysis (*a*) identifies possible changes in farm management and farm income to meet reductions in nitrate leachate and (*b*) evaluates the effectiveness of input restrictions and taxes commonly proposed as solutions to groundwater pollution.

Related Literature

The economic and policy issues in regulating agricultural nonpoint pollution are well documented. For example, the conceptual dimensions of nonpoint pollution are developed in Langham, Sharp and Bromley, and Griffin and Bromley. Within the last five years, additional theoretical evaluations of nonpoint pollution regulatory issues are found in Saliba, Shortle and Dunn, Milon, and Segerson.

In practice, regulations often focus on definition and implementation of "best management practices" (BMPs). Assessing the benefits and costs of changes aimed at controlling nonpoint pollution from agriculture is difficult because of the complex linkages between the physical and economic environment. However, an understanding of these biophysical processes is critical to the performance of empirical economic analyses.

A number of economic studies have included the hydrological/biophysical aspects of nonpoint pollution. Early examples are found in Jacobs and Casler and Park and Shabman. Other examples include Anderson, Opaluch, and Sullivan and Gardner and Young. Anderson, Opaluch, and Sullivan specifically note the substantial informational requirements of linking nonpoint agricultural pollution of groundwater to actual management policies and economic impacts. Additional analyses concerning economic "targeting" of nonpoint pollution abatement and optimal spatial management are found in Braden, Hericks, and Larson and in Bouzaher, Braden, and Johnson.

Most empirical studies address surface water pollution. Exceptions include Gardner and Young, Stevens, and Knapp et al. For groundwater pollution, an accurate economic assessment of the biophysical processes concerning leachate requires input from several disciplines (e.g., engineering, geohydrology, plant science). The site-specific nature of nonpoint pollution events also suggests that such information be available at the farm or regional level (Stevens; Anderson, Opaluch, and Sullivan). Few empirical studies have successfully used biophysical simulation models in assessing the economic implications of groundwater quality regulations.

Study Area

The empirical focus is on nitrate groundwater pollution in a two county area of the Columbia Basin in Oregon, an important agricultural area in the Pacific Northwest. The region's climate is semi-arid, with hot, dry summers and cold winters. Mean annual precipitation is approximately nine inches, requiring irrigation for most crop production. The two counties (Umatilla and Morrow) contain 244,000 acres of irrigated land, of which 137,000 acres are irrigated using center pivot systems (Miles). The principal crops are alfalfa, potatoes, winter wheat, and field corn. Most of this production occurs on light-textured soils (208,000 acres classified as sands) (Johnson and Makinson, Hosler). Irrigated production on such soils has significant potential for nitrate groundwater pollution as attempts to keep soil moisture at near-optimum levels for physical output can result in excessive water applications and the leaching of water (and nitrates) below the root zone. The Oregon Department of Environmental Quality reports that nitrate levels in eleven of twenty-five test wells in the two-county area exceed current U.S. EPA standards (10 mg N/l), with some nitrate concentrations as high as 80 milligrams N/l (Pettit).

Farms in the study area are highly capitalized and intensively managed. For example, the effectiveness of water and fertilizer application rates on potato growth is monitored weekly via infrared aerial photography, petiole sampling, and other advanced techniques. Fertilizer input levels and yields are high; for potatoes, total fertilizer applications typically range from 675 to 900 kilograms per hectare, of which 60% or more is nitrogen. Application of fertilizers and water are controlled primarily through center pivot sprinkler irrigation systems ("fertigation").

While the empirical results generated here will not exactly represent adjustment costs in other agricultural areas, they can suggest (*a*) the importance and feasibility of modeling accurately the biophysical aspects of agricultural nonpoint pollution and (*b*) the magnitude and costs of leachate reductions achievable under intensely managed systems.

Methodology

The integrated approach used in this analysis includes the linking of (*a*) a plant simulation model to predict crop yields under different input levels, (*b*) a dynamic optimization model to optimize irrigation and fertilization decisions for each crop, and (*c*) a linear programming model to assess rotational (crop mix) implications for a representative farm in the study area. Each of these component models is discussed in this section.

The crop simulator is from the CERES family of plant simulation models (Ritchie, Godwin, and Otter-Nacke; Hodges et al., Jones and Kiniry). CERES models account for water and fertilizer use by a crop, amounts of water and fertilizer leached throughout the season, and resulting yield. CERES models for potatoes, corn, and wheat are adapted for use in the study area.[1] These crops account for 80% of crop acreage in the study area.

The CERES models estimate daily potential photosynthesis based on weather, accumulated biomass, leaf area, and genetics; CERES then uses water and (or) nitrogen stress estimates to calculate actual photosynthesis. Treatment of water and nitrogen balances in CERES are somewhat asymmetric, i.e., insufficient water and (or) nitrogen will inhibit plant growth, thereby reducing final yield, but water and nitrogen applications in excess of optimal levels generally do not inhibit yields. Most of CERES's stress calculations are based on a form of the law of the minimum, expressed as

$$(1) \qquad B = \text{Min}\,(f(Sa), f(N), M),$$

where B is biomass from a single days growth, $f(Sa)$ is the maximum biomass as imposed by soil moisture levels, $f(N)$ is the maximum biomass under a given soil nitrogen level, and M is the maximum biomass imposed by other factors such as weather and genetics (Waggoner and Norvell; Lanzer, Paris, and Williams).

In addition to the plant growth simulator, the CERES models include routines to estimate soil water and nitrogen balances. These routines simulate the vertical movement of water and nitrates between layers of the root zone as well as other aspects of the nitrogen cycle. Together, the routines provide daily predictions of water and nitrate movement out of the root zone into the vadose zone. The topsoil and vadose layers in the study area are predominantly sands or other light textured soils, allowing relatively free movement of water into the aquifer.

The second component of the assessment framework is a two-state variable dynamic optimization model for scheduling irrigation and fertilization decisions for each crop. Although a forward recursive dynamic algorithm is used in the solution process, the model formulation is conceptually similar to that of the open-loop stochastic control model used in Zavaleta, Lacewell, and Taylor. The dynamic optimization model determines daily water and nitrogen applications to maximize per acre returns above variable costs, subject to restrictions imposed on the system. The dynamic optimization model utilizes the CERES crop simulator to predict yields and returns from different irrigation and fertilization strategies. The model identifies an optimal irrigation and fertilization strategy by considering returns less water and nitrogen costs.

The third component in the assessment is a farm-level linear programming (LP) model. This component is used to examine shifts in crop mix on a representative farm that may result from restrictions on overall nitrate leachate levels. The farm represents the economic, cultural, soil, and other conditions typically found in the area, including the high level of technology currently used in potato production. The technical coefficients linking nitrogen, water use, and yields in the LP model are taken from the optimal solutions obtained by the dynamic optimization model.

Linking the Models

The output from the CERES plant simulation models is used as input in the dynamic optimization models (in the form of the transition relations). The outputs from the dynamic optimization model, such as yields, nitrogen, and water use and associated leachates for various regulatory options are then placed in a whole-farm context via the LP model. The optimization model is discussed below to clarify these linkages.

The general dynamic optimization problem can be represented by the following equations:

[1] A CERES model for alfalfa, the other major crop in the study area, was not available.

1066 November 1991 Amer. J. Agr. Econ.

$$(2)\quad R(f(t), w(t), V(t), P, t) = \max_{w \leftarrow W,\, f \leftarrow F} \{MR_t(w(t), f(t), Sa(t), Ntr(t), P, V(t), t) + R(f(t-1), w(t-1), V(t-1), P, t-1)\},$$

$$(3)\quad MR_t(\cdot) = MY(Sa(t), Ntr(t), V(t), t)(Py - Ch) - f(t)Pf - df(t)Lf - w(t)Pw - dw(t)Lw,$$

$$(4)\quad bf(t) = 0 \text{ for } f(t) = 0;\ = 1 \text{ otherwise},$$

$$(5)\quad bw(t) = 0 \text{ for } w(t) = 0;\ = 1 \text{ otherwise},$$

$$(6)\quad Ntr(t+1) = K(f(t), Ntr(t), V(t), t),$$

$$(7)\quad Sa(t+1) = I(w(t), Sa(t), V(t), t),$$

$$(8)\quad Sa(0) = Sa_0,$$

$$(9)\quad Ntr(0) = Ntr_0,$$

$$(10)\quad LL \leq Sa(t) \leq Sat\ t = 0,1, \ldots, T,$$

$$(11)\quad Nmin \leq Ntr(t) \leq Nmax\ t = 0,1, \ldots, T,$$

$$(12)\quad Wmin * bw(t) \leq w(t) \leq Pc(t)\ t = 0,1, \ldots, T,$$

$$(13)\quad 0 \leq f(0) \leq Mxf, \text{ and}$$

$$(14)\quad Fmin * bf(t) \leq f(t) \leq c * w(t)\ t = 1,2, \ldots, t.$$

Equation (2) provides the basic forward recursion relationship for the dynamic optimization algorithm. Using this relationship, the producer is maximizing a before-tax-net-return function for a given hectare [R] with respect to the two decision variables, irrigation [$w(t)$] and fertilizer [$f(t)$] quantities, two state variables, soil moisture [$Sa(t)$] and nitrogen [$Ntr(t)$], exogenous random factors [$V(t)$], and input and output prices [P]. The net return function is the cumulative sum of the daily marginal return functions [$MR_t(\cdot)$] up to and including stage t. The marginal return function (3) is defined by the marginal yield function (the incremental change in yields) when moving between stages [$MY(\cdot)$], the output price [Py], variable harvest cost [Ch], water [Pw] and fertilizer [Pf] costs, and fixed irrigation [Lw] and fertilizer [Lf] costs. Equations (4) and (5) define binary variables used to indicate if any irrigation [$bw(t)$] or fertilization [$bf(t)$] has occurred on a given day. All costs incurred up to the first decision period (such as land, capital, and planting costs) are fixed. Only variable costs associated with irrigation and fertilization activities affect the decision set. The model uses 1988 price and cost data.[2]

Equations (6) and (7) define the transition functions from one stage and state to the next stage for both soil moisture [$Sa(t)$] and nitrogen [$Ntr(t)$], as predicted by CERES. Equations (8) and (9) provide the initial conditions for the state variables. Equations (10) through (14) define the constraints and boundary conditions for the decision and state variables. Specifically, equations (10) and (11) define the boundary conditions for the state space at time t, with LL and Sat the lower and upper limits on soil moisture and $Nmin$ and $Nmax$ the lower and upper limits on soil nitrogen, respectively. Equations (12) through (14) provide the possible decision space at time t. These equations restrict irrigation and fertilization decisions to be either zero or above some minimum amount ([$Wmin$] and [$Fmin$]) and restrict the irrigation and fertilization amounts to be less than or equal to specified levels.[3] Equation (12) ensures that irrigation quantities do not exceed pumping capacity ($Pc(t)$), (13) restricts preplant fertilizer to some maximum rate based on toxicity, and (14) requires that fertilizer applied after planting not exceed a fixed proportion (concentration) of water applied in the same period.

[2] The output prices (less per unit harvesting and marketing variable costs) are \$0.074, \$0.16, and \$0.11 per kilogram of potatoes, winter wheat, and field corn. The costs of irrigation water and nitrogen fertilizer were \$0.16 per hectare-millimeter and \$0.34 per kilogram, respectively.

[3] This integer constraint (of requiring a minimum amount of irrigation if irrigation quantity is greater than zero) is based on physical limits on the equipment (e.g., maximum speed of the circles) and the fact that farmers will not make applications of water or nitrogen less than some minimum level.

This formulation of a producer's seasonal decision problem is based on a dynamic programming [DP] solution algorithm; however, it does not have all of the desirable properties of DP. First, DP assumes that the marginal product derived from moving from one state and stage to another state in the next stage is invariant to the path taken to reach that state. In the formulation used here, that assumption does not hold. Second, CERES provides only the final yields from a given irrigation and fertilization pattern. Therefore, marginal (incremental) yields must be inferred through comparison of final yields of feasible states. As a result, the dynamic optimization model does not yield solutions that are (necessarily) globally optimal.

Producer Behavioral Considerations

An important requirement in this assessment is to portray accurately the irrigation and fertilization decisions faced by farmers. On the surface, it appears that farmers using irrigation could eliminate the majority of nitrate leachate by more careful management of nitrogen and water applications. However, farmers in the study area confront a complex problem which includes uncertainty in upcoming weather events, periods of high evapotranspiration, limited pumping capacity, and heterogeneity in the physical environment. These factors contribute to nitrate pollution under irrigated conditions and thus need to be considered in a modeling approach.

Leaching will not occur unless soil moisture levels exceed field capacity. Imperfect knowledge about soil moisture levels in the crop root zone at the time of the irrigation decision may result in excessive application of water. Similarly, imperfect knowledge about soil fertility levels may result in excess nitrogen applications. In formulating the optimization model, however, it is assumed that farmers know with certainty the current fertility and moisture states of their fields. While a strong assumption for grain crops, most potato farmers in the area monitor fertility and moisture levels with the intensity assumed here.

Even with knowledge about current nitrogen and moisture levels, uncertainty concerning future events may result in leaching. For example, a heavy rain immediately after irrigation can cause leaching. Changes in temperature, wind, and other climate variables can also change irrigation efficiency, resulting in too much or too little water entering the soil profile. Soil heterogeneity can cause nitrate leaching; i.e., a field that is relatively homogenous in soil type may still have substantial variability in water holding capacity. Thus, a farmer who irrigates to ensure that the most drought-prone part of a field is never stressed will overwater the rest of that field.

Several features are included in the optimization model to address these issues. Irrigation and fertilizer decisions are made each day based on expected, rather than actual, weather throughout the remainder of the growing season, but final outcomes of the decisions reflect actual weather. The use of expected weather is necessary because the CERES models require weather data for the entire growing season to compute yields from a given irrigation pattern. Irrigation efficiency is treated as a normally distributed random variable with the optimization model using expected efficiency when making decisions. To simulate potential soil heterogeneity, routines were added to CERES to allow for subfields (with distinct soils and yields) receiving the same management strategies. Decisions are based on the weighted average of these subfields.

Optimal fertilization and irrigation rates are compared to actual producer fertilizer and irrigation data to suggest the extent to which producers are overapplying nitrogen or water. The subsequent empirical analyses are based on actual producers' practices within the study area. Specifically, actual field data on fertilizer rates and frequencies, irrigation schedules, yields, and soil nitrate levels were used to establish base model conditions for each crop.

Modeling Crop Rotation

The LP model used in this study is a standard crop mix model similar to that formulated by El-Nazer and McCarl. As noted by El-Nazer and McCarl, the only factors restricting cropping choices in this area are land and rotational considerations. The high profitability of potatoes encourages producers to structure farm capital so as not to restrict production choices. In addition, the long growing season and low rainfall tend to minimize timing restrictions.

The LP formulation includes basic federal farm program provisions including acreage set-asides and deficiency payments. The model's objective is to find the optimal crop mix within the context of various constraints on farm-level nitrate pollution rates. It is assumed that base acreage for government programs does not limit acreage planted to any program crop.

The LP is restricted to seven rotation strate-

1068 November 1991 Amer. J. Agr. Econ.

Table 1. Simulated Nitrogen, Water, Yield and Profit Levels, by Crop

Crop/Analysis	Pred. Yield	Profit	Application Water	Application Nitrogen	Leachate Water	Leachate NO_3	Change in Profits
	(kg/ha)	($/ha)	(mm/ha)	(kg/ha)	(mm/ha)	(kg/ha)	($/ha)
Wheat							
Current practices	8,779 [9,076]	1,207	605	298	66	1.53	
Optimal solution:							
w/Shano Silt	9,176	1,280	394	303	11	0.92	73
w/Quincy Sand	9,102	1,253	531	280	45	3.17	46
w/25% N reduction	8,156	1,128	531	210	50	3.44	−79
w/N tax	8,603	1,107	468	244	10	1.00	−173
w/pollution tax	8,870	1,208	485	260	10	1.00	−72
Field corn							
Current practices	11,424 [11,460]	1,010	720	320	42	2.37	
Optimal solution:							
w/Shano Silt	12,019	1,100	456	383	0	0.00	90
w/Quincy Sand	11,992	1,067	605	391	24	2.07	57
w/25% N reduction	10,342	858	605	293	25	1.98	−152
w/N tax	12,004	929	613	382	45	3.71	−81
w/pollution tax	11,987	1,049	636	392	9	0.81	−39
Potatoes							
Current practices	58,900 [58,986]	4,081	711	434	18	5.11	
Optimal solution							
w/Quincy Sand	61,007	4,224	799	400	14	2.30	143
w/25% N reduction	47,504	3,262	799	300	14	2.25	−819
w/N tax	60,241	4,038	728	390	15	3.09	−186
w/pollution tax	59,798	4,120	724	407	6	1.12	−104

gies.[4] All rotations are currently practiced in the area and restrict potatoes on a given field to no more than one out of four years to minimize disease and pest problems. The dynamic optimization model provides the coefficients for water and fertilizer costs and quantities, yield, and leachate quantities for each crop in each rotation. For three of the crops, coefficients are optimal for given crop and leachate constraint levels, a characteristic that is different from most LP crop-mix models. The exception is alfalfa, for which no CERES model was available. The alfalfa yield coefficients are based on average yield data from representative farms in the region. It is assumed that no nitrates leach while a field is planted in alfalfa.

Base Case and Regulatory Options

The above methodology is used to investigate the changes in yields, profits and nitrate leachate for potatoes, winter wheat, and field corn under six fertilizer-water application situations. The six analyses reported here include (*a*) a base case that replicates the effect of current irrigation and fertilization practices on profits and leachate, (*b*) an optimal solution based on the irrigation and fertilization patterns determined by the dynamic optimization models, (*c*) an across-the-board 25% reduction in applied nitrogen for the optimal case, (*d*) a nitrate leachate constraint placed on the optimization model, (*e*) a nitrogen input tax, and (*f*) a pollution (Pigovian) tax. All solutions were generated under 1988 weather conditions for the area and expected weather as generated by a weather generator (Richardson and Wright).

Results

The results from applying the CERES-optimization models for these crops and conditions to the above input and leachate alternatives are presented in table 1. The table contains nitrogen and water applications, yields, nitrate leachate, and profits for the base case, optimal case, and

[4] The seven rotations are [wheat (w), corn (c), wheat (w), potatoes (p), alfalfa (a)]: w-c-w-p; c-a-a-a-p; w-a-a-a-p; c-a-a-a-a-p; a-a-a-p; a-a-a-a-p; and w-a-a-a-a-p.

under each regulatory scenario.[5] The base case, representing actual producer behavior, provides a benchmark against which the effects of alternative nitrogen and water application strategies can be evaluated. Because some corn and wheat in the area are grown on silty soil, the optimization models for these crops were run with both a sandy and a silty soil. The use of two soils provides information on the effect of water-holding capacity on leaching rates. The actual yield (reported by the famers sampled in the study area) for each base case is shown in brackets under predicted yields. The results of the base models for corn, wheat, and potatoes indicate current leachate levels of approximately 2.4, 1.5, and 5.1 kilograms per hectare of nitrates, respectively.

While no data on actual leachate levels exist for the study area, the leachate levels modeled under current practices were unexpectedly low for all three crops when compared with levels reported in other regions. For example, Hergert found that actual leaching rates for irrigated corn in Nebraska on sandy soil ranged from 12 to 146 kilograms of nitrogen per hectare depending on the weather year and irrigation strategy. In part, the low levels predicted in this study can be attributed to (*a*) relatively homogenous soils, (*b*) use of center pivot irrigation systems, (*c*) very few summer rainfall events, and (*d*) a high level of current irrigation and fertility management. Nonetheless, there is evidence that CERES understates absolute nitrate leachate levels under these conditions.[6] However, the relative ranking between different model solutions are assumed correct for each crop.

The first point of comparison is between the base case and the solution of the optimization model for each crop. As is evident from the table, movement to optimal timing of nitrogen and water applications resulted in less total water and slightly less total nitrogen applied, substantially less total nitrogen leachate, a slight increase in yields, and greater profits for all three crops. Thus, if the application amounts and timing predicted by the optimization model were followed under the 1988 crop year conditions, producers could, in some cases, reduce water and/or nitrogen applications, increase profits, and reduce leachate. However, the results for the optimization model are "best case" estimates for a given weather year. Alternative weather years may not yield the same outcome.

Additional optimization models were run for corn and wheat using silty soils, which are less prone to leaching. Yields and profits increase relative to production on sandy soil, while water usage and nitrate leachate decrease because it is more costly to minimize water stress and percolation on soils with low versus moderate water holding capacity.

Because of the *ex post* nature of the dynamic optimization, the optimal solutions use information about weather events that may not be available to farmers. However, movement from current management strategies closer to those identified as optimal is possible even without perfect information about future weather events. The primary change in management observed in the optimal solution is the number of fertilizer applications. Specifically, the number of nitrogen applications is typically two to three times higher under the optimal solutions than for current practices.[7] Quantities of nitrogen per application were smaller under the optimal solution. In effect, higher profits and lower nitrate leachate levels can be achieved if farmers time applications to coincide more closely with plant needs.

Restrictions on Inputs

One objective of this study is to investigate the on-farm economic effects of possible regulatory restrictions on nitrogen, water, and other inputs. To explore this type of restriction, nitrogen applications were reduced by 25% in both the current practice and optimal solutions. This case corresponds to a situation where producers reduce application levels by a fixed percentage (here 25%), but follow the same schedule recommended by the optimization model. For corn,

[5] Because of the significant computational time required to obtain an optimal solution to the two state variable dynamic optimization problem, models were solved on three different computers. The computers used for this study were a VAX 8700, a Definicon 785, and a Floating Point System 164/264. The basic dynamic optimization analyses would require approximately 30, 90, and 98 hours of computer time for wheat, potatoes, and corn if all were run on a Definicon 785. Finally, to minimize any problems with varying numeric precision, all solutions for a given crop in a given rotation were solved on the same operating system.

[6] Recent experiments with CERES for similar crops and conditions in Washington resulted in higher leachate levels (N. K. Whittlesey, Washington State University, personal communication). Given the difficulty of validating leachate routines in biophysical simulators for all possible soil-water-crop situations, the utility of such models is greatest in ranking potential leachate changes, not in predicting absolute leachate levels.

[7] The number of nitrogen applications for the current practices compared with the base optimization model was 5 vs. 10 for wheat, 6 vs. 24 for corn, 12 vs. 24 for potatoes following alfalfa, and 12 vs. 32 for potatoes following grain.

a 25% reduction decreased nitrogen leachate by approximately 4% from the optimization model. Yields were reduced about 14%, resulting in a profit loss of approximately $200 per hectare. For wheat, this nitrogen reduction increased leachate by about 9%.[8] Yields were reduced about 10%, resulting in a profit loss of approximately $125. For potatoes, the nitrogen reduction reduced leachate by about 2%. Yields were reduced about 22%, resulting in a profit reduction of over $900 per hectare. In short, imposing such a constraint on nitrogen applications under efficient timing of nitrogen and water applications imposed large costs on farmers, with little change in nitrate leachate.

To explore further the effect of nitrogen reductions, the optimization models were solved to determine the minimum level of nitrate leachate possible without terminating crop growth. While not reported in table 1, these analyses resulted in substantial reductions (above 70%) but not total elimination of leachate for all crops. These reductions were accomplished through significant reductions in water application and more moderate decreases in nitrogen applications. The analysis suggests that some level of nitrate leachate will result from the production of these crops, even under optimal management. The results also emphasize the importance of water management in an overall strategy to minimize nitrate leachate rates.

Constraints on nitrate leachate rates, by definition, assure decreases in pollution rates independent of the value of the crop. However, the optimization models indicate clear upper bounds on leachate restrictions that can be imposed on any crop. These upper bounds are determined by the number and size of rainstorms and the number of days with relatively high evapotranspiration within the growing season. For example, in the base wheat model, the minimum amount of nitrate leachate was 0.98 kilograms of nitrogen per hectare. Lower leachate rates could not be reached without killing the crop. Further, limiting leachate has little practical value because it is costly to monitor nonpoint pollution rates. Nevertheless, such limitations provide a means for defining (in general terms) the "best management practices" to achieve given target reductions in pollution rates. These preferred practices can be used to formulate regulatory guidelines for reducing nitrate leachate from high-risk soils.

Taxes

Various forms of taxes are widely suggested as a means to induce changes in farmers' behavior (Griffin and Bromley, Shortle and Dunn, Segerson, Gardner and Young, and Stevens). A simple input tax tested here involved a 100% tax on nitrogen. The tax reduced profits for wheat, corn, and potatoes by 14%, 7%, and 4%, respectively. Taxes had negligible impacts on water use in the corn model and only minor impacts on nitrogen use for the corn and potato models. Conversely, a nitrogen tax in the wheat model moderately reduced both water and fertilizer use so that a 68% reduction in nitrate levels was achieved.

The short run (within a growing season) derived-demand price elasticities for nitrogen fertilizer implied here range from −0.02 to −0.13 for the 100% price increase. Such low elasticities are expected because the price elasticity of derived demand varies directly with the share of total production cost associated with a given factor and the elasticity of substitution between a given factor and other inputs. In this case, water and nitrogen are the only variable inputs. Furthermore, the factor shares for nitrogen in the base models are small, varying from 3% to 11% of total variable costs. The low demand elasticities calculated from the modeled nitrogen use imply that high input taxes would be needed for most crops to meaningfully reduce pollution.

Taxes on leachate (Pigovian taxes) are expected to be more efficient economically than input taxes, because leachate taxes allow greater latitude in producer adjustment. To test this, a Pigovian tax based on mitigation costs was applied to each optimization model. Specifically, a tax of $26.42 per kilogram of nitrate leached into the vadose zone was used, based on estimated costs to remove nitrates from drinking water (Walker and Hoehn). This leachate tax results in slight to moderate reductions in profits ranging from 3% to 6% compared to the base optimization model. However, this leachate tax reduces leachate levels more than the 100% input tax. The exception is wheat, where the leachate levels under each tax are similar (because the leachate level in the input tax analysis was already within a few % of the minimum leachate level). As expected, the Pigovian tax analyses had higher profits than the direct nitrogen tax analysis.

[8] The simulated increase in leachate rates occurred because the early season nitrogen stress limited the ability of the plant to use nitrogen later in the season, increasing the opportunities for mineralized nitrogen to leach.

Table 2. Effects of N Leachate Restrictions on Crop Mix

Analysis	Net Returns ($)	Nitrate Leachate	Shadow price N Leachate	Crop Mix					
				Potato (Grain)	Potato	Wheat	Corn	Alfalfa	Idlement
		(Kg/N)	($/Kg)	(hectares)					
Base	523,000	1,526			190	267	152		141
750 kg N leachate	485,000	750	67		190	276	152[a]		141
500 kg N leachate	447,000	500	220		190	276[a]	152[a]		141
250 kg N leachate	372,000	250	449	43[a]	147[a]	214[a]	118[a]	129	105

[a] Acreage receives reduced N and/or water to achieve N restriction.

Like restrictions on leachate, Pigovian taxes have limited practical value because of the costs required to monitor actual pollution rates for each farm (Segerson). Leachate taxes would also impose added farm management costs related to basing input decisions on complex pollution forecasting models (Shortle and Dunn). However, analysis of Pigovian taxes is useful for forming socially optimal pollution goals and in generating irrigation and fertilization strategies, assuming the tax levels roughly approximate the true marginal social cost of nitrate pollution.

Whole-Farm Effects

The results of the farm-level LP model reported in table 2 suggest that the wheat-corn-wheat-potato rotation is the profit-maximizing long-run equilibrium solution when no leachate restrictions are imposed. Such a rotation is similar to that currently practiced in the study area. For the 750-hectare representative farm, the rotation results in an annual average nitrate leachate rate of 1,525 kilograms of nitrogen. A leachate restriction of approximately 50% (750 kg) is achieved without crop mix changes through reductions in nitrogen and water applications to corn. As more restrictive total leachate constraints are imposed, the model shifts toward production activities that restrict leachate in both corn and wheat. Further restrictions cause the model to select activities that restrict leachate in potatoes. Finally, the model shifts to a potato-alfalfa-alfalfa-alfalfa rotation. The costs of these restrictions (reduced profits) increase at an increasing rate, as reflected in the shadow price of leachate. For example, to achieve the last 250-kilogram reduction imposes a profit penalty of $75,000, whereas the profit penalty for the first 750-kilogram reduction was only $38,000.

This analysis indicates that the lowest cost method of reducing pollution across a whole farm in the study area is to restrict nitrate pollution on grain fields; the profits gained in potato production by allowing an additional kilogram of nitrogen leachate are much larger than when an additional kilogram of nitrogen leaches under corn or wheat production. Moreover, as the nitrate constraint becomes more restrictive, land rents fall significantly because of reduced profits arising from the restrictions on pollution and shifts in crop rotations. These whole-farm results are obtained under optimal timing of nitrogen and fertilizer applications; the cost of potential leachate reductions could be higher if measured against different management intensities.

Conclusions

These analyses, for intensively managed, irrigated conditions on fields with low to moderate water-holding capacities, suggest some nitrogen leachate reductions can be accomplished with little loss in profits. Specifically, by changing the timing and application rates of nitrogen and water, profits for some crops increase while reducing total nitrogen application levels and resultant leachate. However, once these efficiencies are obtained, further reductions in nitrate leachate bring increasing costs to producers. Crop mixes as modeled here are relatively insensitive to most taxes and regulations, given the need to maintain at least one high-value crop (potatoes) in the rotation.

These results are based on specific fields for well-managed, highly capitalized farms. They do not necessarily reflect management decisions or field conditions of other producers of irrigated crops. The leachate and profit benefits of improved irrigation and fertilizer scheduling may be greater for farms with alternative irrigation systems. Conversely, if producers are already managing intensely their water and fertilizer applications, leachate reductions may not be achievable without profit penalties. This analysis is also representative of only three crops and does not address the affect of different irrigation technologies or climates on optimal fertilization strategies. However, the integrated methodology does reflect the relationships between producer behavior and the physical and biological dimensions of groundwater pollution. Such an integrated approach would appear critical to understanding the economics of nonpoint pollution control in other agricultural settings.

[Received April 1990; final revision received November 1990.]

References

Anderson, G. D., J. J. Opaluch, and W. M. Sullivan. "Nonpoint Agricultural Pollution: Pesticide Contamination of Groundwater Supplies." *Amer. J. Agr. Econ.* 67(1985)1238–43.

Bouzaher, A., J. B. Braden, and G. V. Johnson. "A Dynamic Programming Approach to a Class of Nonpoint Source Pollution Control Problems." *Manage. Sci.* 36(1990):1–15.

Bower, H. *Groundwater Hydrology*. New York: McGraw-Hill Book Co., 1978.

Braden, J. B., E. E. Herricks, and R. S. Larson. "Economic Targeting of Nonpoint Pollution Abatement for Fish Habitat Protection." *Water Resour. Res.* 25(1989):2399–2405.

Braden, J. B., G. V. Johnson, A. Bouzaher, and D. Miltz. "Optimal Spatial Management of Agricultural Pollution." *Amer. J. Agr. Econ.* 71(1989):404–13.

Council for Agricultural Science and Technology (CAST). *Agricultural and Groundwater Quality*. CAST Rep. No. 103. Ames IA, 1985.

Edwards, S. F. "Option Prices for Groundwater Protection." *J. Environ. Econ. and Manage.* 15(1988):475–87.

El-Nazer, T., and B. A. McCarl. "The Choice of Crop Rotation: A Modeling Approach and Case Study." *Amer. J. Agr. Econ.* 68(1986):127–36.

Gardner, R. L., and R. A. Young. "Assessing Strategies for Control of Irrigation-Induced Salinity in the Upper Colorado River Basin." *Amer. J. Agr. Econ.* 70(1988):37–49.

Griffin, R. C., and D. W. Bromley. "Agricultural Runoff as a Nonpoint Externality: A Theoretical Development." *Amer. J. Agr. Econ.* 64(1982):547–52.

Hergert, G. W. "Nitrate Leaching through Sandy Soil as Affected by Sprinkler Irrigation Management." *J. Environ. Quality* 15(1986):272–78.

Hodges, T., T. Mogusson, S. L. Johnson, and B. Johnson. *Substor Potato Model*. FORTRAN program, 1989.

Hosler, R. E. *Soil Survey of Morrow County Area, Oregon*. Washington DC: U.S. Department of Agriculture, Soil Conservation Service, 1983.

Jacobs, J. J., and G. L. Casler. "Internalizing Externalities of Phosphorous Discharges from Crop Production to Surface Water: Effluent Taxes Versus Uniform Restrictions." *Amer. J. Agr. Econ.* 61(1979):309–72.

Johnson, D. R., and A. J. Makinson. *Soil Survey of Umatilla County Area, Oregon*. Washington DC: U.S. Department of Agriculture, Soil Conservation Service, 1988.

Jones, C. A., and J. R. Kiniry. *CERES-Maize: A Simulation Model of Growth and Development*. College Station TX: Texas A&M University Press, 1986.

Knapp, K. C., B. K. Stevens, J. Letey, and J. D. Oster. "A Dynamic Optimization Model for Irrigation Investment and Management Under Limited Drainage Conditions." *Water Resour. Res.* 26(1990):1335–43.

Langham, M. R. "Theory of the Firm and the Management of Residuals." *Amer. J. Agr. Econ.* 54(1972):315–25.

Lanzer, E. A., Q. Paris, and W. A. Williams. *A Nonsubstitution Dynamic Model for Optimal Fertilizer Recommendations*. Davis CA: Giannini Foundation, 1987.

Lee, L. K., and G. Nielsen. "Groundwater: Is It Safe to Drink?" *CHOICES*, no. 3 (1988), pp. 4–7.

Miles, S. D. "Oregon Irrigated Acres, 1987." Dep. Agr. and Resour. Econ., Oregon State University, 1987.

Milon, J. W. "Interdependent Risk and Institutional Coordination for Nonpoint Externalities in Groundwater." *Amer. J. Agr. Econ.* 68(1986):1229–33.

Nielsen, E. G., and L. K. Lee. *The Magnitude and Costs of Groundwater Contamination from Agricultural Chemicals*. Washington DC: U.S. Department of Agriculture, Economic Reseach Service, 1987.

Park, W. M., and L. A. Shabman. "Distributional Constraints on Acceptance of Nonpoint Pollution Controls." *Amer. J. Agr. Econ.* 64(1982):455–62.

Pettit, G. "Assessment of Oregon's Groundwater for Agricultural Chemicals." Dep. Environ. Quality, State of Oregon, 1988.

Richardson, C. W., and D. A. Wright. *WGEN: A Model for Generating Daily Weather Variables*. Washington DC: U.S. Department of Agriculture, Agricultural Research Service, 1984.

Ritchie, J. T., D. C. Godwin, and S. Otter-Nacke. *CERES-Wheat: A Simulation Model of Wheat Growth and Development*, 1985.

Saliba, B. C. "Irrigated Agricultural and Groundwater Quality—A Framework for Policy Development." *Amer. J. Agr. Econ.* 67(1985):1231–37.

Segerson, K. "Uncertainty and Incentives for Nonpoint Pollution Control." *J. Environ. Econ. and Manage.* 15(1988):88–98.

Sharp, B. M. H., and D. W. Bromley. "Agricultural Pollution: The Economics of Coordination." *Amer. J. Agr. Econ* 61(1979):591–600.

Shortle, J. S., and J. W. Dunn. "The Relative Efficiency of Agricultural Source Water Pollution Control Policies." *Amer. J. Agr. Econ.* 68(1986):668–77.

Stevens, B. K. "Fiscal Implications of Effluent Charges and Input Taxes." *J. Environ. Econ. and Manage.* 15(1988):285–96.

Waggoner, P. E., and W. A. Norvell. "Fitting the Law of the Minimum Applications and Crop Yields." *Agronomy J.* 71(1979):352–54.

Walker, D. R., and J. P. Hoehn. "Rural Water Supply and the Economic Cost of Groundwater Contamination: the Case of Nitrates." Dep. Agr. Econ. Staff Pap. No. 88-67, Michigan State University, July 1988.

Zavaleta, L. R., R. D. Lacewell, and C. R. Taylor. "Open-loop Stochastic Control of Grain Sorghum, Irrigation Levels, and Timing." *Amer. J. Agr. Econ.* 62(1980):785–92.

Q53 Q58 69-77 (95)

[7]

Regulating Nonpoint Source Pollution Under Heterogeneous Conditions

Gloria E. Helfand and Brett W. House

Because of difficulties in measuring effluent from nonpoint pollution, proposals for regulating agricultural runoff often suggest instruments applied to inputs or management practices. When pollution functions vary across sources, uniform input instruments cannot achieve a least-cost pollution reduction, but efficient instruments may be difficult to administer. In this paper we analyze lettuce production on two soils in California's Salinas Valley to consider empirical costs associated with uniform input taxes and regulations. The results suggest that uniform instruments may not be costly relative to an efficient baseline. Though taxes are more efficient, farmers have higher profits with regulations.

Key words: input instruments, lettuce production, nitrate pollution, nonpoint source pollution, pollution restrictions.

Nonpoint source pollution, such as agricultural runoff, has not received much focus in federal water pollution control laws. The characteristics of nonpoint source pollution have made it particularly difficult to control. Because, by definition, it does not enter water bodies at a defined point, it is difficult to observe the effluent, making effluent-based regulation almost impossible. Additionally, even if the effluent were observable, underground water flows can travel sufficient distances that it may not be possible to relate the effluent to its source(s). For these reasons, proposed regulatory approaches to nonpoint source pollution have focused on indirect instruments, especially controls on input use and management practices.

The theoretical literature on controlling nonpoint source pollution focuses on achieving optimal solutions. Papers by Griffin and Bromley and by Shortle and Dunn compare the effects of effluent instruments (either standards or incentive methods) and input instruments (again, either standards or incentive methods). Under certainty, Griffin and Bromley find that any of these instruments can achieve a specified goal at least cost, though the number of regulatory instruments necessary varies; if there are J pollution sources, each using K inputs, the necessary number of instruments is one for an effluent tax, J for effluent regulations (if sources are not identical), and $J \times K$ in the case of both input taxes and input regulations. Shortle and Dunn focus on one source of pollution (or, by extension, a number of identical sources) when uncertainty about the level of effluent, weather, and a regulator's knowledge of farmer behavior affects policy decision making. In this case, incentive instruments applied to management practices (which could be inputs) are more efficient than incentive instruments applied to (expected) effluent or to regulations applied to either effluent or management practices. While they do not extend their results to heterogeneous effluent sources, they argue that incentives applied to management practices are likely to be more efficient in this case as well.

While optimal instruments will achieve a specified pollution target at least cost, they may not always be easy to implement. For instance, it is likely to be very difficult, if not impossible, to charge each user of an input a different price for that input, as optimality would require if the users are not identical. The possible ease of implementing second-best instruments must be weighed against the increased social costs that they impose. The magnitude of these social costs is, of course, empirical and must be con-

---

The authors are, respectively, associate professor and member of the Giannini Foundation, and graduate research assistant, Department of Agricultural Economics, University of California, Davis.

This work originated in joint research with Ken Tanji in the Department of Land, Air, and Water Resources at the University of California, Davis. The authors are grateful for the contributions of Douglas Larson, Bruce Warden, and Louise Jackson to this project, and to two anonymous reviewers and the associate editor for their helpful comments.

Amer. J. Agr. Econ. 77 (November 1995): 1024–1032

sidered in the context of a case study.

In this paper we analyze empirically the effects of different regulatory instruments when effluent sources vary, both in their crop production and in their pollution production functions. A uniform set of input incentives will not achieve a social optimum. As noted by Griffin and Bromley, it is necessary to have as many sets of input incentives for a social optimum as there are effluent sources. If uniform regulatory instruments are applied—for instance, if the same set of input taxes are used for all sources—least-cost pollution cleanup is possible only under highly restrictive conditions. Of course, the magnitude of losses due to imperfect regulatory instruments is an empirical issue, and is likely to be application-specific. The losses due to the use of second-best regulatory instruments when pollution sources vary in characteristics are estimated in this paper using data related to crop production and nitrate effluent in the Salinas Valley of California.

Input Instruments Versus Pollution Instruments

The well-known advantage of applying incentive mechanisms to effluent from heterogeneous sources is that marginal costs of abatement are equalized across polluters, thus yielding a least-cost method of achieving a specified effluent level. Holterman demonstrated that taxes on inputs can be used to achieve the same social optimum as an output tax. As Griffin and Bromley showed, though, separate taxes need to be identified not only for each input that affects the externality, but also for each source of pollution.

From a social perspective, with fixed prices, the objective is to maximize profits while taking into account the damages due to production. The question then becomes whether various regulatory instruments can replicate those conditions. A uniform tax on each firm's effluent can come close to this social optimum. The effluent tax achieves the optimum if marginal damages are constant regardless of the source of the additional pollution (such as with an additive damage function). Marginal damages from pollution sources might vary, as in the case of effluent entering a river at different places with different assimilative capacities. In the case of varying marginal damages, a uniform effluent tax will not achieve the social optimum (see, for instance, Howe). In other situations, such as when pollution sources are highly localized, constant marginal damages across sources is an acceptable assumption, yielding the usual result on the efficiency of pollution taxes.

Taxes on inputs require more stringent conditions to achieve the social optimum. For such a solution, even if marginal damages are constant across sources, different sets of input taxes are required for each source as long as each source's pollution function is different.

For nonpoint source pollution, because effluent flows are difficult to observe, incentives related to inputs are more feasible.[1] Still, uniform input incentives achieve the social optimum only under quite unlikely circumstances (unless conditions in an area are highly homogeneous). Nonuniform input incentives, though theoretically capable of achieving the social optimum, could be difficult to implement: the inputs could not be subject to resale, or those facing lower input taxes could buy large quantities and resell to those who would otherwise face higher taxes; and the amounts used by individuals would have to be monitored and priced individually.[2] For an input such as irrigation water, which is often produced by a water district and provided directly to a user by the district, individual prices might be possible. For an input such as fertilizer, in contrast, it could be virtually impossible to maintain separate prices for different users, since the good can easily be resold. If input prices cannot be individualized, then the social optimum is almost certainly not attainable. The relative efficiency of other instruments is then an empirical exercise. Our goal in this paper is to quantify the potential cost effects of these other suboptimal instruments. These suboptimal instruments are presented in the next section.

Alternative, Second-Best Control Approaches

Instruments focused on input levels and practices have been proposed not only in theory but in practice as well. For instance, so-called Best Management Practices (BMPs), farming practices which reduce erosion or other undesirable runoff, are mandatory in some states and localities. Pepin County, Wisconsin, has experi-

[1] As they were in the Shortle and Dunn model, management practices can be considered a variant on an input in this model.

[2] For this reason, a system of marketable permits for inputs would not achieve a social optimum.

mented with reducing property taxes on farmers who adopt soil-conserving BMPs. Olmsted County, Minnesota, has developed a mandatory program for farmers to adopt erosion-reducing management practices. The Central Platte Natural Resource District, Nebraska, is beginning to incorporate design standards on fertilizer use into its agricultural pollution control programs (Thompson, chaps. 3 and 7).

Because of the relative infeasibility of practices that could achieve a social optimum, it is interesting to consider alternative, less efficient instruments that could be used to achieve a pollution reduction. These include the following:

1. *Identical input taxes for all pollution sources.* To avoid the problem of resale of an input from low-tax purchasers to high-tax purchasers, regulators could develop a uniform set of input taxes. This system could be made identical to a set of marketable input permits, which would limit production of a polluting input but permit sales of the input within that constraint.

2. *Identical reductions in inputs contributing to pollution on a percentage basis for all sources.* In this scenario, all input users would be required to "roll back" their use of polluting inputs by a specified percentage. This approach is analogous to requiring all polluters to reduce emissions by a specified amount, except that the rollback is measured in terms of inputs. It should be noted that a reduction in pollution of x percent can be achieved by rollbacks in inputs of x percent only if all relevant pollution functions are homogeneous of degree one; otherwise, the percent reduction in inputs required will be greater (less) than the percent reduction in the pollution target if the pollution functions exhibit decreasing (increasing) returns to scale.

3. *Identical taxation of single inputs.* Often, agencies with regulatory authority over one input may not have authority over others. For instance, pesticide use may be regulated by state or federal departments of agriculture, while irrigation water prices are controlled by state or local water districts or the federal Bureau of Reclamation. It may be more difficult to get all agencies involved with inputs to coordinate their policies. Instead, consideration of administrative feasibility suggests focusing efforts on taxing only one input to production.

4. *Identical restriction on single inputs.* As noted above, different regulatory authorities control different aspects of the production process. Incentive approaches are not always used; for instance, pesticide use is frequently strictly regulated, and water providers may limit the amount that each grower in the district may receive. These restrictions may be identical across heterogeneous producers.

Although it is well known that these instruments are inefficient, an important question is whether there are substantial gains from an approach designed to achieve the social optimum. Because these are all imperfect methods for regulating nonpoint source pollution, the relative efficiency of these measures is purely empirical. This question will be examined using crop and pollution production data for heterogeneous soil conditions in the Salinas Valley of California.

Each of the regulatory approaches described above can be modeled when information can be developed on crop and pollution production functions for producers. If output price is held fixed, social welfare is maximized by maximizing aggregate profits subject to the restrictions imposed by the different instruments. Nonlinear programming can be used to solve both for the optimal level(s) of the regulatory instrument(s) as well as the levels of inputs that will be used by each producer; once these levels are derived, all other relevant variables can be found as well. A more detailed description of the solution methods can be obtained from the authors.

An Empirical Example: Lettuce in California's Salinas Valley

Data rarely exist on the relationship between inputs and pollution for nonpoint source pollution, making empirical studies involving pollution impossible to conduct. However, agronomic models, which simulate plant responses to management inputs and practices, can be adapted to generate data on the pollution expected from a management regime, as well as data on crop yield. With this adaptation, such models can be used to examine the effects of policies to reduce pollution generation. The Erosion/Productivity Impact Calculator (EPIC) (Sharpley and Williams) has been adapted to estimate nitrate effluent from crop production.

One drawback to the use of an agronomic model is that it is too cumbersome to be used directly in an optimization framework. EPIC simulates plant growth on a daily basis over the course of a specified period of time (which can be as short as one rotation, or as long as decades). For any set of inputs and management practices, it then generates crop yield and pollution. Translating this information into data appropriate for an objective function must be done outside the model, making searches for optimal responses to policies difficult.

EPIC has been employed for various projects of agronomic and economic importance. Taylor, Adams, and Miller studied economic incentives to offset nonpoint source pollution from agriculture using EPIC linked to linear programming models. EPIC has been used in dynamic programming models to derive optimal decision rules on agricultural practices (Bryant, Mjelde, and Lacewell; Segarra; Young and van Kooten). Putman, Williams, and Sawyer used EPIC to estimate the impact of soil erosion on resource productivity and fertilizer requirements for the 1985 Resource Conservation Act appraisal.

Many studies using EPIC view a specific combination of inputs and management practices in fixed proportions with output, and then use different management practices in a linear programming model to examine choice of management practices and cropping patterns. Alternatively, if management practices are not expected to vary but input use might, EPIC can be used to generate data on input use, yield, and pollution by varying input levels and calculating the resulting yield and pollution. These data can then be used in the same way as data from agronomic experiments: regression analysis can be used to find functional forms and parameters that fit the data, and those functional forms can be used in the optimization scenarios described above. This approach allows for variations in input levels, while the linear programming approaches provide for fixed input relationships but permit variations in management practices and cropping patterns. In this case study (Salinas Valley lettuce production), changes in management practices and cropping patterns are less likely in the short run than changes in input use levels (Louise Jackson, personal communication); thus, the production function approach is more appropriate.

In this case study, nitrate contamination due to lettuce production on two soil types in California's Salinas Valley is used to demonstrate the issues. Buildup of nitrates in groundwater is raising concerns in the area about drinking water quality (since groundwater is used for drinking water). One of the main sources of nitrates in the soil is from the fertilizers applied, and its mode of transport is water. Nitrate leached below the crop's roots is unavailable for plant uptake, resulting in potential movement to water aquifers below. While other inputs are used in the production of lettuce, water and nitrogen appear to be the key variables both for nitrate production and for head lettuce production. Both inputs tend to be applied in excess of the plants' physiological needs because of the risk of underapplying inputs in parts of the field. The management practices by which water and fertilizer are applied can have significant effects on runoff as well; they are omitted from the current analysis because their discrete-choice nature makes them difficult to analyze in this framework, and because, as noted above, these changes are unlikely in the short run.

Heterogeneity enters this study through the use of different soil types. Head lettuce is commonly grown on two of the Salinas Valley's major soil types: Mocho, a silty loam, and Pacheco, a silty clay loam. The clay loam, being less porous, requires less irrigation water and nitrogen fertilizer. As a consequence, less nitrate is leached beyond the crop root zone. Farmers commonly know on what type of soil their lettuce is grown and apply water and fertilizer nutrients accordingly. As a result, nitrate leaching will differ between fields, not only because of the physical attributes of the soil, but also because of different applications of the inputs contributing to nitrate leaching. Thus, the effluent function on soil i can be modeled as $E_i = g(W, N; S_i)$, where W and N refer to water and nitrogen, and S_i to soil type. The production function is modeled as $Y_i = f(W, N; S_i)$.

EPIC does its analysis on a per hectare basis. The analysis assumes an equal distribution of head lettuce fields grown in the valley for each of these two major soil types; thus, results below are based on per hectare analysis for each type. Field data collection and calibration of EPIC have been performed on the two soils for actual spring and summer double-cropped lettuce farming conditions using 1990 weather data (Jackson et al.).

Approximately 750 observations of nitrate and crop production were generated for each of these soil types by changing the relative amounts of irrigation and fertilization applications, and running these inputs through EPIC. Observations were based on input levels above and below the "baseline" conditions (as observed currently by lettuce farmers) of irrigation and fertilization applications. These input, yield, and pollution data are then used econometrically to estimate crop yield and nitrate leaching production functions using ordinary least squares in SHAZAM version 7.0. For crop production, the square-root functional form sta-

Table 1. Estimated Crop Production and Pollution Functions

	Production Function			
Variable	*Y* (Mocho)	*Y* (Pacheco)	NO_3 (Mocho)	NO_3 (Pacheco)
Intercept	0.9117	2.296	−26.06	43.82
	(14.5)	(79.7)	(−6.8)	(2.64)
Nitrogen (*N*)	3.63E−3	−1.95E−3	−0.152	−0.17
	(−10.4)	(−9.67)	(−6.09)	(−1.8)
Water (*W*)	−2.9E−3	−1.58E−3	0.158	0.313
	(−46.9)	(−47.4)	(40.8)	(19.3)
$N \times W$	1.93E−6	1.53E−6	3.63E−4	4.66E−4
	(15.9)	(23.2)	(14.4)	(16.6)
$N^{1/2}$	5.06E−2	2.49E−2		−1.11
	(7.17)	(6.19)		(2.02)
$W^{1/2}$	0.156	7.32E−2		−7.28
	(44.9)	(43.9)		(0.90)
Adj. R^2	0.77	0.77	0.96	0.96
F	503	497	5,634	4,011
χ^2 ($\beta = 0$)	1,430	1,302	2,386	2,464

Note: t-statistics are in parentheses.

tistically outperformed other forms used for both soils, and it is used in this paper. With *N* for nitrogen application and *W* for water application, this function is $Y = \beta_0 + \beta_1 N + \beta_2 W + \beta_3 NW + \beta_4 N^{1/2} + \beta_5 W^{1/2}$. The Breusch-Pagan/Godfrey test identified heteroskedasticity in the crop production functions.[3] This was taken into account in the estimation procedure by using Antle's flexible moment approach for variances of errors estimated as second-order polynomials (Antle).

The nitrate pollution function differs for the two soil types. For the Mocho soil, a restricted version of the quadratic and square-root function (α_4 and α_5 equal to zero in the following equation) is used to describe nitrate (NO_3) leaching. For the Pacheco soil, a square-root specification fit the data best; i.e., $NO_3 = \alpha_0 + \alpha_1 N + \alpha_2 W + \alpha_3 NW + \alpha_4 N^{1/2} + \alpha_5 W^{1/2}$. The Breusch-Pagan/Godfrey test did not indicate the presence of heteroskedasticity for nitrate leaching in either soil type.[4] Results of the crop and nitrate pollution production function estimations are given in table 1. The functions are highly significant, with coefficient estimates generally significant at the 1% confidence level, and goodness of fit indicating R-squared values of 0.77 for the lettuce production functions and 0.96 for the nitrate leaching functions. Chi-squared tests for model significance against the alternative of no relationship are highly significant.

With input prices of $0.23/mm per hectare for irrigation water and $0.70/kg for nitrogen fertilizer, and output price net of harvesting costs of $1,350/ton, the quasi-rent before policy imposition is $4,224.43 and $4,250.68 per hectare for lettuce on Mocho and Pacheco soils, respectively.[5] Optimal input levels prior to the policies are 723 and 478 mm per hectare of water, and 85 and 52 kg per hectare of nitrogen for Mocho and Pacheco soils, respectively. Yields, defined as the dry weight lettuce biomass available for harvesting, without constraints, are 3.296 and 3.257 tons of lettuce on Mocho and Pacheco soils, respectively.

Total pollution is calculated as the sum of runoff from each hectare.[6] The policy instruments described above will be compared based on total nitrate runoff being constrained to 20% less than the "baseline." This reduction in pollution was suggested by researchers at the California State Water Resources Control Board.

[3] The chi-squared test statistic for Pacheco soil is 133, well above the critical value of 15.1 at the 99% significance level. For Mocho soil, the test statistic is 82.

[4] The chi-squared test statistic for the Pacheco soil is 3, well below the critical value of 11.1 at the 95% significance level. For Mocho soil, the test statistic is 8.

[5] Quasi-rent is total revenue less variable costs due to water and nitrogen. Other inputs contributing to lettuce production are not included in this analysis; as a result, quasi-rents here do not represent profits.

[6] Though, as noted above, total pollution need not be a linear function of pollution from individual sources, it seems reasonable in this case, where the runoff ends up in a common aquifer.

Table 2. Combined Results of Different Policies, Averaged Over the Two Soil Types on a Per Hectare Basis

	Unconstrained (Baseline)	Individual Taxes	Uniform Taxes	Uniform Rollback	Single Tax		Single Standard	
					Nitrogen	Water	Nitrogen	Water
Welfare cost ($)		6.67	6.78	6.86	90.16	6.83	90.39	12.94
Cost as % of unconstrained quasi-rent ($)		0.157	0.159	0.162	2.130	0.161	2.130	0.305
Yield (tons/ha)	3.277	3.255	3.254	3.255	3.170	3.255	3.170	3.252
Nitrate (kg/ha)	67	53.1	53.1	53.1	53.1	53.1	53.1	53.1
Transfers ($)		101.76	100.74		47.44	100.89		

Results

The policy instruments analyzed are different methods of achieving the 20% pollution reduction by reducing water and nitrogen inputs in crop production, as described above. In brief, they are (*a*) separate input taxes for each soil type and each input (a solution which achieves the social optimum); (*b*) tax both inputs to production, with taxes uniform across soil types; (*c*) require a uniform percentage rollback in levels of input use; (*d*) tax either water or nitrogen uniformly across soil types; and (*e*) restrict either water or nitrogen use uniformly across soil types. The socially optimal solution to (*a*) is presented only as a benchmark of comparison; as noted previously, this solution is likely to be infeasible in practice. Table 2 presents aggregate welfare costs of the policies as reductions in quasi-rents on a per hectare basis (with tax payments as transfers rather than losses); tables 3 and 4 present the results of these analyses for typical hectares of Mocho and Pacheco soils, respectively.

Notably, three of the second-best policies—the water tax, the policy of identical input taxes for both soil types, and the uniform reduction in all input use—are nearly as efficient as the use of individual input taxes. When the complexity and cost of getting separate instruments "right" for both inputs and soil types are considered, one of these alternative, uniform measures may be a preferable method of achieving reduced nitrate runoff.

The policies with the largest welfare losses are the nitrogen tax and a control on nitrogen application; for all other policies, the welfare loss is modest in relation to the overall revenues from lettuce production. In addition, the nitrogen tax creates the greatest loss in revenues for both soil types. Even those losses are modest, at $85/ha (2%) for a Pacheco soil farmer, and $190/ha (4.6%) for a Mocho soil farmer.

The lettuce grown on Mocho soil has slightly higher yields than the lettuce grown on Pacheco soils, but the quasi-rents are greater on Pacheco soil because water and nitrogen use is much lower. Because Mocho soil is more porous and sandy, yield is more sensitive to lower applications of water and nitrogen. Since pollution is being reduced by input reductions, yield, and thus quasi-rents, are reduced more on Mocho soil.

For the socially optimal policy, the tax bill for those farmers growing their crop on Mocho soil is much greater than those on Pacheco soil for reasons discussed earlier. The farmers with Mocho soil would have to pay 1.2 times the Pacheco water tax and 3.8 times the Pacheco nitrogen tax. If both inputs are taxed uniformly for each soil type, the taxes for both water and nitrogen are lower for Mocho and higher for Pacheco relative to the socially optimal policy. Farmers with Mocho soil benefit, and farmers with Pacheco soil suffer, from this policy.

The uniform rollback produces the highest quasi-rents for either soil type under the pollution constraint (except for the water standard on Pacheco soil, which is not binding). Even though this policy is less efficient than all but one of the taxing policies, the tax transfers have a significant effect on the money the farmers actually can keep. Unless tax revenues can be transferred back to farmers in lump-sum compensation, the costs in efficiency loss due to the rollback (less than 40¢ per hectare) are small compared to their distributional advantage.

If only one input is taxed, the necessary taxes are generally much greater than if both inputs are taxed. Indeed, to achieve the reduction in

Table 3. Effects of Different Policies on Mocho Soil, per Hectare

	Unconstrained (Baseline)	Individual Taxes	Uniform Taxes	Uniform Rollback	Single Tax		Single Standard	
					Nitrogen	Water	Nitrogen	Water
Quasi-rent ($)	4,224.43	4,082.17	4,093.21	4,214.61	4,034.70	4,093.05	4,084.64	4,198.55
Water (mm/ha)	723	645	651	636	657	650	654	585
Nitrogen (kg/ha)	85	73	74	75	12	77	9	70
Yield (tons/ha)	3.296	3.271	3.274	3.269	3.16	3.275	3.142	3.246
Nitrate (kg/ha)	97	82	83	80	82	83	78	70
Rollback (%)				12%				
N tax/standard		$0.088/kg	$0.061/kg		$6.157/kg		9 kg/ha	
W tax/standard		$0.198/mm-ha	$0.184/mm-ha			$0.192/mm-ha		478 mm/ha

Table 4. Effects of Different Policies on Pacheco Soil, per Hectare

	Unconstrained (Baseline)	Individual Taxes	Uniform Taxes	Uniform Rollback	Single Tax		Single Standard	
					Nitrogen	Water	Nitrogen	Water
Quasi-rent ($)	4,250.68	4,176.09	4,166.91	4,246.68	4,165.20	4,166.66	4,209.70	4,250.68
Water (mm/ha)	478	411	403	421	439	402	443	478
Nitrogen (kg/ha)	52	45	43	45	4	45	9	52
Yield (tons/ha)	3.257	3.238	3.235	3.241	3.18	3.236	3.198	3.257
Nitrate (kg/ha)	37	26	24	27	29	24	29	37
Rollback (%)				12%				
N tax/standard		$0.023/kg	$0.061/kg		$6.157/kg		9 kg/ha	
W tax/standard		$0.166/mm-ha	$0.184/mm-ha			$0.192/mm-ha		478 mm/ha

pollution by taxing only fertilizer, an extremely high tax must be used to induce the required low levels of nitrogen use. That low level of nitrogen use has a major effect on yields. Although the actual tax bill for taxing only water is twice the level of the tax bill for nitrogen use, yields drop less under this policy than under the nitrogen tax, because water does not need to be reduced as drastically and thus affects yield less. Thus, if only a single input is to be taxed, water appears to be the preferable input.

Analogously, if a single input is to be constrained, limiting water use causes much less reduction in both efficiency and in quasi-rents. If water use is constrained, the limitation is binding only for lettuce grown on Mocho soil, imposing on farmers with that soil all the costs of meeting the pollution target. For nitrogen use restrictions, both farm types are constrained to use low (perhaps infeasible) levels of nitrogen fertilizer. Even though the efficiency losses with the water standard are relatively high, the effects on quasi-rents are zero for Pacheco soils, and only a uniform rollback has higher quasi-rents for Mocho soils.

From an efficiency perspective, policies that focus on multiple inputs are generally superior to single-input policies, and tax policies (except the nitrogen tax) are superior to regulatory strategies. From the perspective of individual farmers, though, the tax transfers are large enough that either the uniform rollback or a water standard is more profitable than any tax approach.

Overall, the results indicate that, in this setting, uniform instruments do not lead to large losses relative to a socially optimal option. Even the most inefficient policies (taxing or restricting fertilizer use) result in only a 2% welfare loss relative to quasi-rents; the cost of other policies is below one-half of 1%. This result is consistent with theory which argues, based on the envelope theorem, that small deviations from optimality result in second-order effects (Akerlof and Yellen). The input restrictions (with the exception of the nitrogen standard) cause far less reduction in quasi-rents than the tax instruments. Higher costs are imposed on lettuce producers on Mocho soil because the soil is more porous, permitting nutrients and water to leach more readily below the crop root zone. Because this soil type is more pollution intensive, it is reasonable that the policies analyzed here generally lead to greater discouragement of lettuce production on this soil.

These results are based on only two soil types, with identical farmers on each type. As more variation of conditions is added, these results can be expected to change.

Conclusions

In this paper we provide an empirical evaluation of policies to reduce nonpoint source pollution regulation under heterogeneous conditions. Because optimal instruments may be difficult to implement or enforce, a number of imperfect instruments are presented as alternatives. Because production functions typically exhibit diminishing marginal product with respect to input levels, and pollution functions typically exhibit increasing marginal product, combinations of instruments are likely to achieve a specified pollution level at lower cost than single-input instruments. Additionally, incentive approaches such as taxes are generally more desirable than input restrictions, because they give producers a choice over how to respond to the instrument. The actual magnitudes of these welfare effects, though, is empirical.

As nonpoint source pollution receives greater policy attention, a number of alternative instruments will be considered for reducing runoff. It is possible that these instruments will be applied uniformly in some areas even when underlying conditions, such as soil type, differ. The possible costs of using uniform instruments in nonuniform conditions could be quite high; on the other hand, the costs of using nonuniform instruments, in terms of monitoring and enforcement costs, could also be quite high. Which set of costs is higher is, of course, an empirical question. It is reasonable to expect that the sum of these costs is reduced if pollution restrictions are developed for watersheds or relatively uniform regions, rather than nationally.

This study has shown that several uniform regulatory instruments can achieve a pollution target at relatively low social cost in the Salinas Valley of California. While this result will of course vary with the particular case study and the degree of heterogeneity, it nevertheless indicates that perfect instruments are not necessary to achieve a pollution target, and that imperfect instruments may not always impose high social cost. Without an explicit model of transactions costs, it is impossible to rank the efficiency of these instruments in actual practice; consideration of ease of enforcement of these instruments could lead to alteration of some of the results presented. As areas under study become more heterogeneous, though, it is likely that the social cost of uniform instruments will increase.

[Received September 1994;
final revision received August 1995.]

References

Akerlof, G.A., J.L. Yellen. "Can Small Deviations from Rationality Make Significant Differences to Economic Equilibria?" *Amer. Econ. Rev.* (September 1985):708–20.

Antle, J. "Testing the Stochastic Structure of Production: A Flexible Moment-Based Approach." *J. Bus. and Econ. Statist.* 1(April 1988):192–201.

Bryant, K.J., J.W. Mjelde, and R.D. Lacewell. "An Intraseasonal Dynamic Optimization Model to Allocate Irrigation Water Between Crops." *Amer. J. Agr. Econ.* 75(November 1993):1021–29.

Griffin, R.C., and D.W. Bromley. "Agricultural Runoff as a Nonpoint Externality: A Theoretical Development." *Amer. J. Agr. Econ.* 64(August 1982):547–52.

Holterman, S. "Alternative Tax Systems to Correct for Externalities, and the Efficiency of Paying Compensation." *Economica* 43(February 1976):1–16.

Howe, C.W. "Taxes Versus Tradable Discharge Permits: A Review in the Light of the U.S. and European Experience." *Environ. and Resour. Econ.* 4(April 1994):151–69.

Jackson, L., L. Stivers, B. Warden, and K. Tanji. "Crop Nitrogen Utilization and Soil Nitrate Loss in a Lettuce Field." *Fertilizer Res.* 37(1994):93–105.

Putman, J., J. Williams, and D. Sawyer. "Using the Erosion-Productivity Impact Calculator (EPIC) Model to Estimate the Impact of Soil Erosion for the 1985 RCA Appraisal." *J. Soil and Water Conserv.* 43(July-August 1988):321–25.

Segarra, E. "Optimizing Nitrogen Use in Cotton Production," in Proceedings, 1989 Summer Computer Simulation Conference, Society for Computer Simulation, Austin TX, 24-27 July 1989, pp. 727–31.

Sharpley, A., and J. Williams. *EPIC—Erosion/Productivity Impact Calculator: 1. Model Docu-*

mentation. Washington DC: U.S. Department of Agriculture Technical Bulletin No. 1768, 1990.

SHAZAM User's Reference Manual Version 7.0. Vancouver: McGraw-Hill, 1993.

Shortle, J.S., and J.W. Dunn. "The Relative Efficiency of Agricultural Source Water Pollution Control Policies." *Amer. J. Agr. Econ.* 68(August 1986):668–77.

Taylor, M.L., R.M. Adams, and S.F. Miller. "Farm-Level Response to Agricultural Effluent Control Strategies: The Case of the Willamette Valley." *J. Agr. and Resour. Econ.* 17(July 1992):173–85.

Thompson, P. *Poison Runoff: A Guide to State and Local Control of Nonpoint Source Water Pollution*. New York: Natural Resources Defense Council, 1989.

Young, D., and G. van Kooten. "Economics of Flexible Spring Cropping in a Summer Fallow Region." *J. Production Agr.* 2(April-June 1989):173–78.

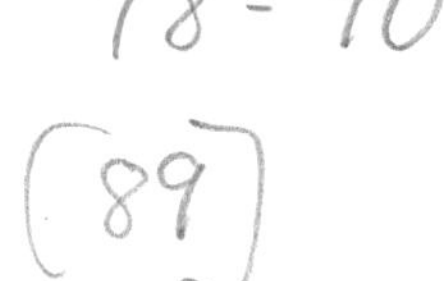

[8]

JOURNAL OF ENVIRONMENTAL ECONOMICS AND MANAGEMENT **17**, 22–34 (1989)

Regulating Environmental Health Risks under Uncertainty: Groundwater Contamination in California[1]

ERIK LICHTENBERG

Department of Agricultural and Resource Economics, University of Maryland, College Park, Maryland, and Western Consortium for Public Health, Berkeley, California

DAVID ZILBERMAN

Department of Agricultural and Resource Economics, University of California, Berkeley, California

AND

KENNETH T. BOGEN

University of California, Lawrence Livermore National Laboratory, Livermore, California

Received December 1, 1986; revised April 22, 1988

Our method for incorporating uncertainty into environmental health risk policy determination corresponds closely to the legal and political strictures governing these policies. An application to a case of groundwater contamination demonstrates that (1) the desirability of developing new water supplies versus cleaning up existing ones varies across locations; (2) the stringency of any given policy varies inversely with the margin of safety required; (3) the cost premiums imposed by greater aversion to uncertainty may be quite large; and (4) the marginal cost of risk reduction decreases significantly as aversion to uncertainty grows, implying that assessments of policies based on average risk will tend to overestimate allowable risk. Comparing an efficient program to the current inefficient approach to risk management indicates that the cost of inefficiency is not great. © 1989 Academic Press, Inc.

Concerns about potential risks to human health and safety have become increasingly prominent in public discussions of environmental problems. Issues such as toxic wastes, groundwater contamination, chemical spills, and pesticide residues have emerged as important public policy concerns.

These health hazards generally are expressed as risks, that is, as probabilities that the adverse outcome may occur. One of the most prominent—and troublesome—features of environmental health problems is that there tends to be considerable uncertainty about the magnitude of these health risks. The most widely cited sources of uncertainty include: (1) the chronic or subchronic nature of many of these hazards (cancers, birth defects, genetic damage); (2) the lack of epidemiological evidence and the consequent use of animal bioassay models to assess dose–response relations; and (3) the complexity of environmental contamination and

[1]Although the information described in this article has been funded wholly or in part by the U.S. Environmental Protection Agency under assistance agreement CR811200-03 to the Western Consortium for the Health Professions, it has not been subjected to the Agency's required peer and administrative review and therefore does not necessarily reflect the views of the Agency and no official endorsement should be inferred. The authors would like to thank Richard Jackson, Robert C. Spear, and two anonymous referees.

0095-0696/89 $3.00

transport processes and the limitations imposed on data collection and environmental modeling by the need to make regulatory decisions in a timely manner.

It will be useful to clarify some terminology since the terms "risk" and "uncertainty" are used somewhat differently in a health context than is traditional in economics. By health risk we mean the probability that an individual selected randomly from a population contracts an adverse health effect (i.e., the probability of mortality or morbidity). The relationship between a health risk and the variables that generate it is not known with certainty so that the health risk estimates used for policy evaluation are subject to error. The term uncertainty will be used here as a measure of the magnitude of this error.

Policy analyses relating to environmental health problems, then, need to deal not just with risk but with risk compounded with uncertainty. Such uncertainty is generally nonnegligible. This paper develops a framework for incorporating uncertainty about health risks into regulatory decisions by combining a detailed, probabilistic model of the health risk generation process with a practical mechanism for incorporating policymakers' preferences regarding uncertainty.

EVALUATING RISK AND UNCERTAINTY

In empirical applications, economists have tended to evaluate proposed environmental health regulations using benefit–cost analyses employing estimates based on average risk. A key drawback of this approach is that it completely ignores uncertainty. As an alternative, two of us have proposed a procedure for deriving uncertainty-compensated trade-offs between risk and social cost and implicit value-of-lifesaving measures reflecting both uncertainty about risk and decision-makers' preferences regarding that uncertainty (Lichtenberg and Zilberman [6]).

Our approach essentially involves combining a probabilistic health risk assessment—a type of risk assessment that has become increasingly widespread in recent years—with a safety-rule decision mechanism. The resulting estimates of uncertainty-compensated trade-offs between risk and social cost then can be used for policy determinations that rely on formal decision criteria (benefit–cost, risk–benefit) or for subjective evaluation of regulatory alternatives.

This procedure has a number of appealing characteristics. First, it makes use of the full range of information available in a practical manner. Second, this approach is more amenable to interdisciplinary cooperation. It utilizes the kinds of information produced by risk analysis and is equivalent to using confidence intervals for statistical decision-making which is the method preferred in the natural sciences. Third, because safety rules have been used in a variety of economic applications, they are well understood. Moreover, they have been shown to give good approximations of expected utility decisions in several empirical contexts (Thomson and Hazell [8]).

Finally, the safety rule approach corresponds quite closely to the terms of much of the relevant legislation as well as the strictures of practical politics. Both require regulators to balance social cost against protection of public health with a sufficient margin of safety. It also corresponds to a "disaster avoidance" approach to decision-making that is widespread among the public and the regulatory community. Thus, this approach describes the preference structures of decision-makers in many contexts—e.g., both legislation and compliance regulations for a wide range of environmental problems (Beavis and Walker [1]).

The general approach can be expressed formally as follows. Consider a set of M policies. Let $X = (X_1, \ldots, X_M)$ denote the extent to which each policy is used. The costs imposed on society by choosing a set of policies X, denoted $\tilde{C}(X)$, is assumed to be a monotonically increasing function of X: The greater the extent to which any policy is employed, the greater the social cost is assumed to be. Social cost in these cases consists of items like the cost of cleanup, government monitoring and enforcement, protective measures taken by the public, and producer and consumer market welfare costs.

The safety rule aspect can be expressed as a condition specifying that risk R be constrained to remain below a given maximum allowable level (or risk standard) R_0 within a given margin of safety P, that is, that R exceeds R_0 no more than a fraction $1 - P$ of the time. This can be written formally as:

$$\Pr\{R \leqslant R_0\} \geqslant P \quad \text{or} \quad \Pr\{R \geqslant R_0\} \leqslant 1 - P. \tag{1}$$

The regulatory decision problem then can be expressed as the choice of a set of optimal policies $X_1^*, \ldots, X_M^*$ to minimize the total social cost of meeting the safety rule:

$$\min \tilde{C}(X) \qquad \text{subject to (1)}. \tag{2}$$

The solution to this problem is a set of policies, $X_1^*, \ldots, X_M^*$, that can be characterized in terms of the total social cost $\tilde{C}(X^*)$ of achieving a risk standard R_0 with a margin of safety P.

It is easy to see that the margin of safety P essentially represents society's (or the decision-maker's) aversion to uncertainty concerning actual risk. The higher the margin of safety, the more averse to uncertainty the decision-maker can be said to be. A useful benchmark is the standard generally used for scientific reliability, for example, 0.95, which might be said to typify "normal" social aversion to uncertainty for environmental contamination cases. Greater than normal aversion to uncertainty would then be represented by a higher margin of safety, e.g., 0.99.

Exploration of the characteristics of policies derived in this manner requires the use of more specific models. Elsewhere [6], we examine the case of the risk-assessment model proposed for regulatory purposes by Crouch and Wilson [3]. This model assumes that risk can be represented by a multiplicative combination of parameters, e.g., contamination rates, transport rates, exposure rates, and dose-response rates. It was assumed, in addition, that all parameters were distributed lognormally, an assumption that entails little or no loss of generality since one can derive similar formulations using Chebyshev's inequality or other methods.

In this case it becomes convenient to work with the logarithm of the health risk, denoted r, which is distributed normally with mean $\mu_r(X)$ and standard deviation $\sigma_r(X)$. Under this specification, the safety rule constraint (1) becomes

$$\mu_r(X) + F(P)\sigma_r(X) \leqslant r_0, \tag{3}$$

where r_0 is the log risk standard and $F(P)$ is the critical value of the standard normal distribution exceeded only with probability $1 - P$.

The necessary and sufficient conditions for cost minimization are the constraint (3) and the M equations

$$\frac{\partial C}{\partial X_k} + v\left[\frac{\partial \mu_r}{\partial X_k} + F(P)\frac{\partial \sigma_r}{\partial X_k}\right] \geqslant 0, \qquad k = 1,\ldots, M, \tag{4}$$

where v, the Lagrange multiplier, represents the marginal cost of risk reduction with margin of safety P. It is evident from Eq. (4) that the least-cost mix of regulatory policies will be a portfolio of activities, some of which emphasize reductions in mean risk and some, reductions in uncertainty about risk.

So far, the risk standard has been taken as given. In general, though, policy analyses are aimed at choosing standards as well as instruments. For this reason, it is preferable to examine both instrument choices and social costs associated with the entire range of feasible risk standards, that is, a trade-off curve between social cost and risk under any given margin of safety.

Such trade-off curves can be derived by solving the cost minimization problem (2) for every relevant risk standard under a fixed margin of safety. The solution for each risk standard represents the cost-efficient portfolio of policies for that risk standard and margin of safety, $X^*(R_0, P)$. Substitution into the social cost function $\tilde{C}$ yields an uncertainty-adjusted cost curve for risk reduction $C(R_0, P)$. This uncertainty-adjusted cost curve $C(R_0, P)$ bears the same relationship to the cost function $\tilde{C}(X)$ that the indirect utility (expenditure) function bears to the direct utility function. Points above the curve represent policies that entail greater costs of achieving a given risk standard with a given margin of safety. Thus, the trade-off curves derived in this manner can be used to identify inefficient policies and to estimate potential losses from imposing inefficient policies.

Elsewhere [6], we examined the characteristics of these trade-off curves for the case of the multiplicative, lognormal risk model with two policy choices. We showed, among other things, that (1) $\partial C/\partial r_0 < 0$, they are downward sloping in the risk standard and (2) $\partial C/\partial P > 0$, a higher margin of safety (greater aversion to uncertainty) shifts these cost curves for risk reduction up and to the right.

These cost curves can be used as a basis for informal comparisons of regulatory strategies. Alternatively, they can be used in formal decision methodologies like benefit–cost analysis or risk-benefit analysis as defined by Starr [7]. Either way, this approach differs from current standard economic practice in that it defines cost in terms of risk with a given margin of safety, that is, it employs uncertainty-adjusted marginal and average costs of risk reduction.

The slope of this cost curve, the Lagrange multiplier v, represents the marginal cost of risk reduction with a margin of safety P. If one imputes a benefit–cost rationale to the choice of a risk standard, this marginal cost measure implicitly gives an estimate of social willingness to pay for risk reduction with a given margin of safety and can thus be used to construct uncertainty-adjusted estimates of the value of saving a life implicit in any risk standard. Alternatively, it can be used to enforce the consistency of policies concerned with all risks to life, since minimizing the cost of reducing risks to human health and safety from all sources implies equal marginal costs of risk reduction across sources of risk. For example, one could specify that policies relating to different contaminants be chosen to equalize the imputed value of life with a given margin of safety. Because a more lax risk standard (higher acceptable risk) implies a lower cost, the marginal cost of risk reduction is negative. We have shown for the case of a multiplicative lognormal risk model with two policy choices that marginal cost of risk reduction is decreasing (increasing in absolute value) in the risk standard ($\partial v/\partial r_0 < 0$) and in aversion to uncertainty ($\partial v/\partial P < 0$).

The fact that marginal cost associated with policy choices may depend on the margin of safety as well as average risk may help explain some seeming inconsistencies in environmental health and safety regulation. For example, there appear to be great disparities between the value of life saving implicit in nuclear power plant regulation and regulation of coal production. However, these comparisons are made on the basis of average risk. When aversion to uncertainty is taken into account, these implicit values of life may well be comparable simply because the uncertainties involved in nuclear power are so much greater.

THE CASE OF DIBROMOCHLOROPROPANE CONTAMINATION OF GROUNDWATER IN FRESNO COUNTY

The practical usefulness of this approach can be illustrated using a case study involving regulation of groundwater contaminated by the pesticide 1,2-dibromo-3-chloropropane (DBCP) in Fresno County, California. This groundwater serves as a drinking water supply.

DBCP was used as a soil fumigant to control nematodes on a variety of important crops in the United States. Having been implicated in the generation of adverse reproductive effects (abnormally low or absent sperm counts) in chemical plant operatives in 1977 and having been found to be carcinogenic in mice and rats in several studies conducted between 1972 and 1977, it was banned for all agricultural uses by the Environmental Protection Agency (EPA) in 1979. That same year, it was detected in well water used for drinking water in the San Joaquin Valley, California, and in Southern California. After an epidemiological study found positive associations between DBCP contamination levels and cancer incidence in Fresno County, California, the California Department of Health Services and local water agencies took measures to keep DBCP contamination of drinking water below an "action level" of one part per billion (ppb).

The excess cancer risk faced by an individual drawn at random from the population of Fresno County was assumed to result from a multiplicative process like the one discussed earlier. The effective dose of DBCP received by a randomly selected individual was assumed to be the product of four parameters: G, the lifetime time-weighted average concentration of DBCP in drinking water; U, the sampling error involved in estimating G from available environmental monitoring data; A, the lifetime time-weighted average consumption of water; and F, a factor that transforms animal doses into their human equivalents. Risk was then assumed to be proportional to dose so that the excess cancer risk was modeled as

$$R = GUAFQ, \qquad (5)$$

where Q is a dose–response potency parameter.

A probability distribution was specified and estimated for each of the five parameters G, U, A, F, and Q. Briefly, the values of these parameters were estimated as follows; further discussion can be found in Bogen [2].

Contamination, G, was measured by a weighted average of DBCP in the drinking water at service connections in each of the 108 census tracts in Fresno County on the basis of the 1979 survey mentioned above. The probability that an individual drawn at random from the population of the county as a whole drank water of any given contamination level G was assumed to equal the proportion of the total population of the county living in census tracts with an average concentration of DBCP in drinking water wells equal to G. In 36 tracts (accounting for about 32% of the population), the concentration of DBCP was below the analytical detection limit of the devices used, 0.005 ppb; for these tracts, G was set equal to 0.0025, the midpoint of the nondetectable range.

The error involved in estimating GU was modeled separately for these two types of tracts. For tracts where $G = 0.0025$, U was assumed to be uniformly distributed between 0 and 2 (implying that GU was uniformly distributed between 0 and 0.005). For tracts where $G \geqslant 0.005$, U was assumed to be normally distributed such that $E(GU) = G$ and $\sqrt{V(GU)}$ was proportional to G. The estimate $U = 1 + 0.0827Z$, where Z is a standard normal random variable, was derived using data provided by Dr. Richard Jackson of the California Department of Health Services.

Lifetime consumption of water, A, was represented by a symmetric, triangular density varying between 80 and 120% of mean intake, as estimated using data presented by the International Commission on Radiological Protection [5]. This assumption yields an intake range of 1,560–2,340 ml/70 kg/day for males and 1,300–1,960 ml/60 kg/day for females. It was assumed that 49% of the population of Fresno County was male.

There are currently two main competing hypotheses about the proper way to estimate interspecies dose equivalence, F. EPA regards dose per unit of surface area per day as the appropriate frame of reference (EPA [9]). This suggests that F should equal the ratio of human to test animal weight raised to the $\frac{1}{3}$ power; for a 70 kg human and 0.5 kg rat (the relevant test animal), this procedure gives a value $F = 5.192$. Alternatively, it has been argued that dose per unit of body weight per day should be regarded as the appropriate frame of reference (see, for instance, Hogan and Hoel [4]); in this case, one would select $F = 1$. Since there is no hard evidence clearly supporting one approach over the other, it was assumed that both approaches are equally likely to be correct.

The dose–response parameter Q was estimated by applying a multistage dose–response model to data from a two-year feeding study of rats used by EPA in its risk assessment of DBCP (EPA [10]). Maximum likelihood estimates of the dose–response parameters were estimated for each of the three significant tumor types appearing in male rats. Confidence limits were derived for these estimates using Monte Carlo simulations. They were then used to obtain the corresponding probability distribution of Q which represents composite tumorigenic potency.

A Monte Carlo approach then was used to estimate the probability distribution of risk R under the assumption that all of the parameters were independent except G and U which were found to be positively correlated with an empirical covariance of 0.0008.

Since the pesticide is no longer used, the only regulatory options remaining involve reducing contamination levels in drinking water prior to delivery. Only two technologies turn out to be cost efficient, each applicable to a different type of situation. In urban areas, the cheapest solution has been to drill new wells that produce clean water. The cost of the test and production wells required, spread over the 100 households each well typically serves, averages about $300 per person, and the wells typically have a lifetime of 10 years. In rural areas lacking community water systems, in contrast, the cheapest solution has been to install a filtration system on each individual well. The cost of such a system, taking into account the (nondiscounted) cost of replacement filters and servicing over a 10-year period, averages about $750 per person.

To take this cost differential into account fully, tracts were distinguished according to type of system (rural vs. urban) as well as level of contamination. Thus, a single census tract that was partially urban and partially rural was treated as two separate tracts with the same level of contamination but different control costs. Similarly, two or more separate census tracts that were identical in terms of contamination and type of system were treated as a single tract in the analysis. This resulted in 66 functionally distinct tracts in Fresno County (which has 108 census tracts).

The risk/cost trade-off set was calculated by taking the tract with the lowest per capita cost of contamination reduction, assuming that contamination was reduced to a trace amount via the relevant technology (new well for urban systems and filtration for rural systems). The probability distribution of excess cancer risk was recalculated by Monte Carlo simulation using the resulting distribution of contamination. The additional cost was calculated by multiplying the population of the tract by the relevant per capita cost. This procedure was repeated, cumulating cost at each step, until all census tracts contained only trace amounts of DBCP in delivered drinking water. (A formal derivation of this procedure is given in the Appendix.)

This procedure yields a set of correspondences between total cost and a probability distribution of risk, that is, a risk/cost trade-off set that incorporates uncertainty. Uncertainty-adjusted cost curves for risk reduction can be derived from this set by plotting total cost against the appropriate critical values of the probability distribution for excess cancer risk.

For comparative purposes, a risk/cost trade-off set was estimated for uniform standards. This latter approach, which corresponds to the "action level" approach actually employed, requires that contamination not exceed a given standard anywhere in the county. The cost-efficient approach, in contrast, imposes a double standard: a more stringent one in urban areas where per capita cost of control is less expensive and a more lax one in rural areas where control is more expensive. The algorithm used to estimate these trade-offs under uniform standards differed from the cost-efficient approach only in that contamination was reduced according to the level of contamination (the most contaminated tract being first) rather than per capita cost of contamination reduction.

Figure 1 depicts these cost/risk trade-offs for a mean risk standard and for risk standards with 95 and 99% margins of safety under the least-cost approach. As expected, the cost of risk reduction is decreasing in the risk standard and increasing in aversion to uncertainty. It is apparent that policy decisions will be quite sensitive to the choice of a margin of safety (aversion to uncertainty), although this sensitivity declines as risk standards become more stringent (risk decreases). For example, the risk standards achieved by lax policies (policies with costs under $10 million) with 95 and 99% margins of safety are about 5 and 18 times the level of the correspond-

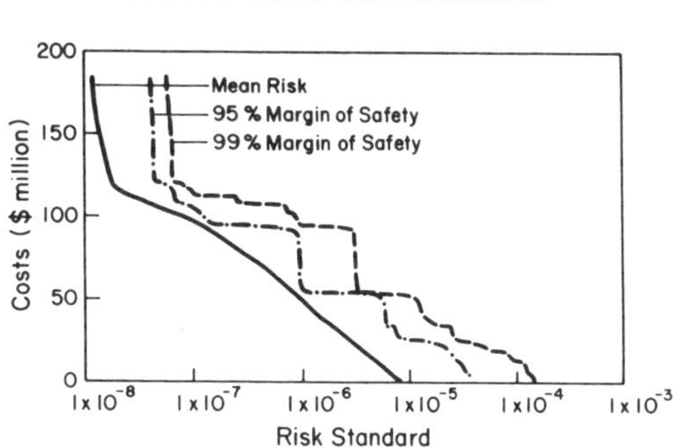

FIG. 1. Risk/cost trade-offs—least cost approach.

ing mean risk standards. For more stringent policies (policies with costs over $110 million), these differentials drop to about three and five times the corresponding mean risk standards—smaller, but still noteworthy.

To facilitate further analysis, the data shown in Fig. 1 were used to estimate an uncertainty-adjusted cost curve for risk reduction $C(R_0, P)$. The functional form was selected on the basis of goodness of fit, as measured by adjusted R^2. The margins of safety corresponding to the mean risk levels were estimated by linear interpolation between the risk levels attained with 50 and 95% margins of safety. All were close to 60%; the range was 57 to 61%.

The best fit regression estimate was:

$$\begin{aligned} C(R_0, P) = &- 250.5920 - 33.73152 \ln R_0 - 0.6197606 \ln^2 R_0 \\ &(22.06670) \quad (3.536180) \quad (0.1361573) \\ &+ 497.1813 \ln P + 513.8630 \ln^2 P + 9.56498 \ln P \ln R_0. \\ &(36.30448) \quad (53.10413) \quad (1.501724) \end{aligned} \tag{6}$$

Standard errors of the coefficient estimates are shown in parentheses; the R^2 was 0.9681. Cost is measured in millions of dollars.

Consider first the impact of aversion to uncertainty on the cost of risk reduction. It is evident from Fig. 1 that meeting a given risk standard with a higher margin of safety entails a greater cost. This increase in cost can be considered a premium for uncertainty aversion akin to the risk premium of the standard literature on risk and uncertainty. It can be verified that this "uncertainty premium," $W = \partial C/\partial P$, is positive for all (R_0, P) in the relevant range.[2] It can be shown that $\partial W/\partial R_0 > 0$, that is, that the uncertainty premium decreases as the risk standard becomes more

[2]Specifically, $W > 0$ for margins of safety ranging from 0.66 under the most lax risk standard (4 in 10,000) to 0.72 under the strictest risk standard (4 in 100,000,000). It is negative in the neighborhood of the margin of safety corresponding to neutrality toward uncertainty (0.57–0.61) because it increases more rapidly as P increases than a quadratic function can account for. Since the relevant range of margins of safety for uncertainty-averse decision-makers is typically 0.90–0.99, this approximation error can be ignored.

TABLE I
Numerical Examples: Uncertainty Premium and Marginal Cost

	Risk standard		
	5/100,000	1/100,000	1/1,000,000
A. Uncertainty premium (million dollars per 1 percentage point increase in P) ($W = [497.1813 + 1027.423 \ln P + 9.56498 \ln R_0]/100P$)			
95% margin of safety	3.681	3.519	3.287
99% margin of safety	3.961	3.805	3.583
B. Per capita marginal cost of risk reduction (dollars per person to reduce risk of cancer by 1 in 10,000) ($V = [33.73152 + 1.2395212 \ln R_0 - 9.56498 \ln P]/(10{,}000) \times (508976)\ R_0$)			
Mean risk	104	477	4222
95% margin of safety	86	393	3360
99% margin of safety	84	385	3281

stringent (allowable risk decreases). As the data in Table I indicate, this decrease becomes more pronounced as the risk standard becomes more stringent and as the aversion to uncertainty declines. It can also be verified that $\partial W/\partial P > 0$—the uncertainty premium increases as the aversion to uncertainty increases. The data in Table I show that this difference increases as the risk standard becomes more stringent and decreases as the aversion to uncertainty increases.

In sum, the cost of aversion to uncertainty increases as allowable risk increases and as aversion to uncertainty increases—results consistent with the standard literature on risk and uncertainty.

The marginal cost of risk reduction, expressed in absolute value terms for analytical convenience, is $V = |\partial C/\partial R_0|$. It can be verified that $\partial V/\partial R_0 < 0$ for all (R_0, P) in the relevant range, that is, that the marginal cost of risk reduction increases as the risk standard becomes more stringent. As the data in Table I show, the marginal impact of a decrease in the risk standard increases as the risk standard becomes more stringent and as the aversion to uncertainty declines. It can also be verified that $\partial V/\partial P < 0$, that is, that the marginal cost of risk reduction decreases as the aversion to uncertainty increases. For example, the data in Table I indicate that the per capita marginal cost of risk reduction is 21 to 26% higher under neutrality toward uncertainty than under a 95% margin of safety and 23 to 29% higher than under a 99% margin of safety. Clearly, the impact of the aversion to uncertainty on the marginal cost of risk reduction increases both absolutely and relatively as the risk standard becomes more stringent and as the aversion to uncertainty increases. In fact, for a risk standard of 1 in 10 million, the marginal cost of risk reduction under neutrality toward uncertainty exceeds that under a 99% safety margin by almost 35%. Differences of such magnitudes could explain a large proportion of the seeming discrepancies in values of life saving implicit in different policies, e.g., regulation of nuclear power vs coal production or groundwater contamination vs smoking.

Revealed preference studies of workers' or consumers' valuation of health and safety risks have estimated the value of reducing risk of death by 1 chance in 10,000 at about \$65 to \$875 (1984 dollars) for risks on the order of 1 in 10,000 to 1 in 1000

(Violette and Chestnutt [11]). The optimal risk standards implied by these estimates range between 8.0 in 100,000 and 5.3 in 1,000,000 under neutrality toward uncertainty and between 6.5 in 100,000 and 4.2 in 1,000,000 (19 and 21% lower, respectively) under 95 and 99% margins of safety.

To allow comparison of the least-cost and "action level" (uniform) approaches, a cost function like Eq. (6) was obtained by regressing the total cost under the uniform approach on the implied risk standard and margin of safety. The estimated cost function was

$$\begin{aligned} T(R_0, P) = &- \underset{(17.33024)}{284.3540} - \underset{(2.780061)}{39.84614} \ln R_0 - \underset{(0.1075869)}{0.882668} \ln^2 R_0 \\ &+ \underset{(28.11291)}{503.6917} \ln P + \underset{(39.83769)}{471.7926} \ln^2 P + \underset{(1.186343)}{11.77267} \ln P \ln R_0. \end{aligned} \tag{7}$$

Standard errors of the coefficient estimates are shown in parentheses; the R^2 was 0.9789. Cost is measured in millions of dollars, as before.

The estimated additional cost due to the use of uniform standards was

$$\begin{aligned} Z(R_0, P) &= T(R_0, P) - C(R_0, P) \\ &= -33.76197 - 6.114620 \ln R_0 - 0.2629074 \ln^2 R_0 \\ &\quad + 6.510375 \ln P - 42.07043 \ln^2 P + 2.207727 \ln P \ln R_0. \end{aligned} \tag{8}$$

Values of $Z(R_0, P)$ in the relevant range are shown in Fig. 2. It is readily apparent that the least-cost and uniform approaches are essentially identical both for very lax and very stringent risk standards (very high and very low acceptable risk). This occurs because the most (least) contaminated tracts are those with the lowest (highest) per capita costs of contamination reduction. The cost of inefficiency $Z(R_0, P)$ is greatest for risk standards in the middle of the relevant range where the disparities between contamination levels and per capita costs of contamination reduction have the most weight. Increases in aversion to uncertainty shift the entire

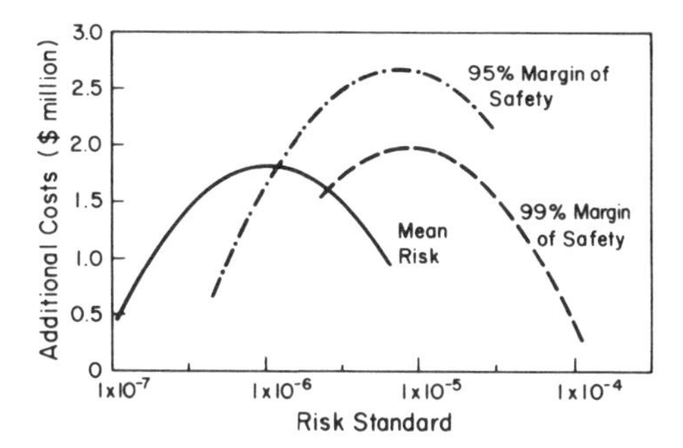

FIG. 2. Cost of inefficiency.

curve to the right, increasing the costs of inefficiency for more lax risk standards and increasing them for more stringent ones.

The maximum costs of inefficiency under neutrality toward uncertainty and under 95 and 99% margins of safety were, respectively, $1.81, $2.68 and $1.98 million, corresponding to 4.1, 6.1, and 3.6% of the cost of achieving the same risk standards using the least-cost approach. It is therefore evident that the social losses from using the uniform approach are not very great.

CONCLUSION

In recent years, the emphasis in environmental health policy has shifted from an ex post, or reactive, mode to an ex ante view which stresses prevention. Definitive proof of health effects is typically unobtainable under such an approach, making it necessary to base policy on systems analyses of risk production processes known as quantitative risk assessments. The procedures followed in making risk assessments are based on the best scientific knowledge available; however, because of the limited time in which they must be completed, they generally cannot attain the levels of predictive reliability considered necessary in scientific work. Therefore, the estimates of risk available for policy determination contain significant uncertainties that render decision-making problematic.

This paper presents a decision methodology that takes these problems into account: It is designed for use with quantitative risk assessments and incorporates uncertainty—and attitudes toward uncertainty—explicitly. The methodology has attractive features conceptually, and the empirical application presented here demonstrates its practical usefulness.

The theoretical and empirical analyses suggest three important directions for further research in the area of environmental health policy:

1. The analyses took the margin of safety (level of aversion to uncertainty) as given; however, the mechanisms by which a consensus value of the margin of safety could or should be established deserves further study.The stakes of such research are clearly significant. For example, a shift from setting mean risk standards to a 99% margin of safety in the DBCP example implies using a value of life in benefit–cost analyses that may be as much as 35% lower.

2. The analysis demonstrates that uncertainty-averse regulators place relatively large cost premiums on uncertainty reductions. This suggests that improving the accuracy of estimates has value in and of itself. The implications of this for resource allocation in research deserve further study.

3. The empirical example illustrated the potential for increased efficiency through the use of state-dependent policies, that is, standards that vary according to local conditions. This suggests the desirability of further research on integrating monitoring strategies into regulation.

APPENDIX

This appendix derives the algorithm used to compute the least-cost strategy for the DBCP case from the regulatory decision problem. Let n_i represent the population of the ith tract; $N = \Sigma n_i$, the total population of Fresno County; K_i, the per capita cost of eliminating contamination in the ith tract ($300 in urban tracts and $750 in rural ones); and G_i, the level of contamination in the ith tract. Let d_i represent the proportion of the ith tract's drinking water supply in which contamination is eliminated. The total cost of risk reduction is $\Sigma d_i n_i K_i$. Average contamination in the county is $\Sigma(1 - d_i) n_i G_i$.

Since the other four parameters of the risk model (U, A, F, and Q) are not susceptible to control, it will be convenient to express the product $UAFQ$ as a single random variable H with a continuous probability distribution of $S(h)$ and density $s(h)$. Since risk necessarily lies in the unit interval and since $G_i > 1$ for many tracts, it must be the case that $0 < H < 1$ as well. It will be convenient to treat G and H as independent, a not unreasonable assumption because of the small size of the correlation between G and U (0.001) and because of the particular nature of that correlation.[3] Under this assumption, the probability that risk R does not exceed a maximum allowable standard R_0 is

$$\begin{aligned}\Pr\{R \leqslant R_0\} &= \Pr\{H\Sigma(1 - d_i)n_i G_i/N \leqslant R_0\}\\ &= \Pr\{H \leqslant R_0 N/\Sigma(1 - d_i)n_i G_i\} \qquad \text{(A.1)}\\ &= S(T),\end{aligned}$$

where $T = R_0 N/\Sigma(1 - d_i)n_i G_i$ expresses the risk per unit of average contamination at the maximum allowable standard R_0. The safety rule constraint, then, is

$$S(T) \geqslant P, \qquad \text{(A.2)}$$

where P is the margin of safety.

The regulatory decision problem thus involves choosing a set of proportions of each tract's drinking water supply to be cleaned up (i.e., d_i) to minimize $\Sigma d_i n_i K_i$ subject to the safety rule constraint (9) and to the constraint that $0 \leqslant d_i \leqslant 1$. Letting λ denote the shadow price of the safety rule constraint and a_i the shadow price of the constraint $0 \leqslant d_i \leqslant 1$, the necessary conditions for a minimum are

$$a_i \geqslant n_i\{G_i[\lambda T s(T)/\Sigma(1 - d_i)n_i G_i] - K_i\} \qquad \text{(A.3)}$$

plus the constraints (A.2) and $0 \leqslant d_i \leqslant 1$. If $d_i \geqslant 0$ then

$$a_i = n_i\{G_i[\lambda T s(T)/\Sigma(1 - d_i)n_i G_i] - K_i\} \geqslant 0 \qquad \text{(A.4)}$$

which implies that

$$K_i/G_i \leqslant \lambda T s(T)/\Sigma(1 - d_i)n_i G_i. \qquad \text{(A.5)}$$

If $d_i < 1$, then $a_i = 0$ and condition (A.5) will hold with equality.

[3] G and U are correlated only because of the assumption that the distribution of U in tracts with contamination below the detection limit differs from that in tracts with contamination above the detection limit. However, U is independently identically distributed for all levels of contamination above the detection limit and similarly below the detection limit. This suggests that the correlation between G and U should affect regulatory decisions only at contamination levels very near the detection limit, if at all, that is, in a very narrow range; because low contamination implies a high per capita cost of contamination reduction K_i/G_i, this suggests further that any errors caused by use of the algorithm should be concentrated among the highest cost, lowest risk elements of the trade-off set.

Note that except for λ condition (A.5) consists entirely of constants (K_i, G_i, n_i, N, $S(T)$, $\Sigma\, d_i n_i G_i$). Moreover, λ is itself determined by the parameters of the system, including the maximum allowable risk standard and margin of safety. This implies that the optimal way to meet a given risk standard with a given margin of safety will be to reduce contamination completely in all tracts where the per capita cost of contamination reduction K_i/G_i is less than a constant determined by the risk standard R_0 and the margin of safety P (specifically, $\lambda TS(T)/\Sigma\, d_i n_i G_i$). It is possible that partial contamination reduction ($d_i > 1$) will be optimal but only in the event that condition (A.5) holds with equality in the marginal (highest admissible cost) tract.

The algorithm for constructing uncertainty-adjusted cost/risk trade-offs follows directly. Let the 66 tracts in the county be arranged in increasing order of per capita cost of contamination reduction K_i/G_i so that K_1/G_1 is the lowest and K_{66}/G_{66} is the highest. If contamination is eliminated (reduced to a minimum) in tracts 1 through J, the total cost can be calculated as

$$\sum_{i=1}^{J} n_i K_i, \tag{A.6}$$

and the probability distribution of risk can be calculated by applying Monte Carlo simulation using $S(h)$ and the distribution of G implied by the contamination reduction in tracts 1 through J. Applying this procedure for $J = 1, \ldots, 66$ yields a set of correspondences between total cost and a probability distribution of risk, that is, a set of risk/cost trade-offs that incorporate uncertainty.

REFERENCES

1. B. Beavis and M. Walker, Achieving environmental standards with stochastic discharges, *J. Environ. Econ. Manag.* **10**, 103–111 (1983).
2. K. T. Bogen, "Uncertainty in Environmental Health Risk Assessment: A Framework for Analysis and Application to a Risk Assessment Involving Chronic Carcinogen Exposure," Doctoral Dissertation, University of California, Berkeley, School of Public Health (1985).
3. E. Crouch and R. Wilson, Regulation of carcinogens, *Risk Anal.* **1**, 47–57 (1981).
4. M. D. Hogan and D. G. Hoel, Extrapolation to man, in "Principles and Methods of Toxicology" (A. W. Hayes, Ed.), Raven Press, New York (1982).
5. International Commission on Radiological Protection, "Report of the Task Group on Reference Man," No. 23, pp. 11, 338–341, 358–360, Pergamon, New York (1975).
6. E. Lichtenberg and D. Zilberman, Efficient regulation of environmental health risks, *Quart. J. Econ.*, in press.
7. C. Starr, Social benefit versus technological risk: What is our society willing to pay for safety, *Science* **165**, 1232–1238 (1969).
8. K. J. Thomson and P. B. R. Hazell, Reliability of using the mean absolute deviation to derive efficient *E*, *V* farm plans, *Amer. J. Agr. Econ.* **54**, 503–506 (1972).
9. U.S. Environmental Protection Agency, Proposed guidelines for carcinogen risk assessment, *Federal Register* **49**, 46294–46301 (1984).
10. U.S. Environmental Protection Agency, Quantitative risk assessment of dibromochloropropane (DBCP), Office of Drinking Water, EPA, Washington, DC (1981).
11. D. M. Violette and L. G. Chestnutt, Valuing reductions in risks: A review of the empirical estimates, U.S. Environmental Protection Agency, Washington, DC (1983).

Part II
Salinity and Water Allocation

[9]

Adoption of Improved Irrigation and Drainage Reduction Technologies under Limiting Environmental Conditions

ARIEL DINAR*
Department of Economics and The Institute for Deser Reserach Ben-Gurion University of the Negev, Beer Sheva, 84105, Israel

MARK B. CAMPBELL
Management Systems Research, Sacrameto, CA, USA

and

DAVID ZILBERMAN
Department of Agricultural and Resource Economics University of California, Berkeley, CA 94720, USA

Abstract. Modern irrigation technologies have been suggested as a means of conserving scarce water and reducing environmental pollution caused by irrigated agriculture. This paper applies an economic model of technology selection that provides a general framework to analyzing adoption of irrigation technologies under various environmental conditions. Data from the San Joaquin Valley of California is used to verify the theoretical relationships. Results suggest key variables to be considered by policy makers concerned with adoption of modern irrigation technologies. Among these variables are crop prices, water technology costs, farm organization characteristics, and the environmental conditions of the farm or the field. Policy implications were discussed and analyzed.

Key words. Adoption, irrigation technology, environment, drainage pollution, policy.

I. Introduction

The competition among agricultural, urban and environmental[1] users of scarce water, and the quality degradation of this water, has become a worldwide problem. The ability of government and private industry to develop new water sources is dwindling due to increasing marginal costs and budget deficits. Water-saving, pollution-reducing irrigation technologies may play an increasingly important role in balancing the equation between the supply and demand for water quality and quantity (National Research Council, 1989).

The economic literature on the adoption of irrigation technologies suggests several key variables which address the behavior of farm operators and

* This research was conducted while the first author was a visiting scholar with the Dept. of Agricultural Economics, University of California, Davis, and USDA-ERS, USA.

Environmental and Resource Economics **2**: 373—398, 1992.

the levels of irrigation technology adoption. Among these variables are input and output prices, the decision unit dimensions (field and farm size), human capital, and locational variables such as land quality and weather (Caswell, 1991).

Several studies have provided empirical relationships among these variables. Caswell and Zilberman (1985) found an increased likelihood among growers to adopt drip and sprinkler versus furrow irrigation systems as the cost of irrigation water went up. Growers using ground water are more likely to adopt modern irrigation technologies than growers using surface water. On the other hand Negri and Brooks (1990) argue that the selection probabilities of irrigation technologies are inelastic with respect to price of water, but labor wages have the greatest impact. Their findings are based on a data set wherein water was a fixed allocatable input but labor was a variable input. Dinar and Yaron (1990) demonstrated that water cost affected the adoption of modern irrigation technologies. Dinar and Yaron (1992) suggest that diffusion of irrigation technologies is affected by water price, crop yield price and subsidy level for irrigation equipment.

Several studies have addressed the role of input quality in the adoption of modern irrigation technologies. A theoreitcal framework was established by Caswell and Zilberman (1986) emphasizing the capacity of modern irrigation technologies to increase irrigation effectiveness. This tendency is likely to be greater on low-quality (low water retention) lands. These hypotheses were generally confirmed by Lichtenberg (1989) for center pivot irrigation in Nebraska, by Caswell and Zilberman (1985) for drip irrigation and sprinklers in central and northern California, by Dinar and Yaron (1992) for Citrus in Israel, and by Negri and Brooks (1990) for a variety of crops around the U.S.

Dinar and Yaron (1990) have also shown that modern irrigation technologies such as solid-set sprinklers, micro-sprinklers and drip are successfully applied to larger portions of citrus groves in regions with high evaporation rates and poor environmental conditions (such as salinity of water and soil) in Israel; owners of orchard crops located on relatively low quality land with more salinity-sensitive rootstock and under more restrictive water allotments are more likely to be early adopters of the modern irrigation technologies.

The human capital variables (such as age, experience, and education) were found to significantly affect the level of adoption in many empirical studies (Feder *et al.*, 1985, and Dinar and Yaron, 1990). Generally, higher levels of human capital are likely to induce adoption, although this finding is strongly influenced by the specific human capital variable measured. In some cases, such as experience, a higher level may diminish the likelihood to adopt. Lichtenberg *et al.* (1991) estimate factors affecting adoption of nonpoint-source pollution runoff control practices in Maryland agriculture. Their results discriminate between small and large farms, and conclude that separate policies should be designed to reach small versus large farmers. Feder *et al.* (1988), have developed a tenure model to explain investment in land conservation technologies. Their conclusion, based on a regional analysis of practices in Thailand, was that tenure farmers are more likely to invest in capital technologies than non-tenure farmers.

None of the above studies have dealt with adoption of irrigation technologies and management options under combinations of poor drainage, salinity, and environmental constraints. This paper includes these aspects in both the conceptual and empirical frameworks in an attempt to explain differences in tendencies to adopt improved irrigation technologies and management practices.

The framework suggested by Dinar and Zilberman (1991) is applied to survey data on irrigation technology adoption from the San Joaquin Valley of California (SJV) (Dinar and Campbell, 1990), The study provides estimates of the likelihood of growers to adopt irrigation and drainage reduction practices under varying conditions of salinity, soil type, cost of water, and the severity of drainage problems prevailing in the west side of the SJV. The empirical problem addressed here, although applied to the SJV, is of a general nature. In so far as the conditions of reduced water availability and quality apply to other regions (Utton and Teclaff, 1978), the results in this paper should be of interest to decision makers in other countries as well.

The paper proceeds as follows: First, the basic features of the conceptual economic model are described, including the general behavioral patterns resulting from different economic and environmental conditions. The empirical specifications of the model, a description of the variables used in the analysis, and the hypotheses made with regard to the relationships among them, follow. The results of our analysis are presented at both farm and field level. Finally the policy implications are discussed extensively followed by conclusions and suggestions for future investigation.

II. Conceptual Framework

Pollution resulting from agriculture can be included in the agricultural production process by regarding the pollution as a joint output of a production relationship which is uniquely determined by the choice of inputs (Just, 1991). This can be demonstrated using the following relationship $y = \mathbf{f}(X)$, $Z = \mathbf{h}(X)$ where X is a vector of inputs and Z is the amount of pollution resulting from the production process, and $\mathbf{h}$ is a function describing how pollution is depending on the input mix used. The modeling framework in this paper consists of several components that basically respond to the above general relationships. It includes a production function; an irrigation effectiveness function; a drainage function; and input and output prices which are discussed in the following sections (see also Dinar and Zilberman, 1991).

CROP-WATER PRODUCTION FUNCTION

Let $y = f(e, i, c)$ where e is effective irrigation, the amount of water not lost due to runoff, deep percolation, or evaporation, which is used by the crop for growth. It is assumed that $f_e > 0$ and $f_{ee} \leqslant 0$. The variable i is a technology index. Each technology i combines managerial effort with a physical equipment setup. The technologies are enumerated according to their costs, and costlier technologies are assumed to have higher irrigation effectiveness. Traditional technology, say furrow irrigation, is denoted by $i = 0$ while $i \geqslant 1$ may denote a highly developed drip system. It is assumed that $f(e, i + 1, c) > f(e, i, c)$. The variable c denotes weather. It can be measured in degree days or pan evaporation levels. Higher c represents higher temperature (and wind). It is assumed that higher values of c increase yield but also increase evaporation. This relationship holds to a given temperature unique to each crop; and temperatures above this level reduce yield. Thus, for the relevant range, $f_c > 0$ and $f_{cc} \leqslant 0$.

EFFECTIVE IRRIGATION

Let $e = h(a, i, q, c)$ where a denotes effective water application, and q denotes quality, which can be either a measure of water or soil salinity, or of soil water retention. Therefore, higher quality results in higher water effectiveness, and it is reasonable to assume that $h_q > 0$ and $h_{qq} \leqslant 0$.

It is also reasonable to assume that more capital intensive technology improves irrigation effectiveness — $h(a, i + 1, q, c) > h(a, i, q, c)$ — thus, from the water use effectiveness perspective, capital intensive technologies serve to augment the soil in its role as a medium for crop growth. Other reasonable assumptions are that $h_a > 0$ and $h_{aa} \leqslant 0$, and that $h_c < 0$, and $h_{cc} \leqslant 0$.

Effective irrigation tends to increase with transition to a more capital intensive technology. The gain in irrigation water effectiveness assocatied with that transition is likely to decline with quality and increase with weather, i.e.,

$$\partial[h(a, i+1, q, c) - h(a, i, q, c)]/\partial q \leqslant 0, \quad \text{and}$$
$$\partial[h(a, i+1, q, c) - h(a, i, q, c)]/\partial c \geqslant 0.$$

DRAINAGE GENERATION

Drainage per unit area is denoted by z, with the drainage generation function being $z = g(a, i, q, c)$. The term drainage can denote runoff or deep percolation. It is assumed to decline with quality, $g_q < 0$ and $g_{qq} \leqslant 0$. Obviously, drainage increases as more water is applied, $g_a > 0$ and $g_{aa} \geqslant 0$. It is assumed (Letey *et al.*, 1990) that more capital intensive technologies tend to generate less drainage, $g(a, i + 1, q, c) < g(a, i, q, c)$.

PRICES OF OUTPUT, WATER, AND DRAINAGE

Prices of output, water, and drainage are given by P, W, and V, respectively. Drainage is an externality and V can be viewed as drainage tax or a subsidy for drainage reduction. The fixed cost per acre associated with each technology is k_i. This fixed cost includes annualized investment and maintenance costs that are independent of water quantity applied. The technologies are ordered according to their capital intensity. Assume therefore $k_{i+1} > k_i$.

THE FARMER CHOICE PROBLEM

The farmer decision problem is to select an irrigation technology, $i \in I$, and water application level. His choice problem can be presented as

$$\max_{a,\, \delta_i} \sum_{i=0}^{i} \delta_i * \{P * f[h(a, i, q, c), i, c] - W * a - V * g(a, i, q, c) - k_i\}$$

where δ_i is a dichotomous variable with a value of 0 if technology i is not selected and 1 when it is selected.

The farmer optimization problem involves a joint discrete-continuous choice. For analytic simplicity, let the analysis be conducted in two stages: First, optimal water use will be determined for each distinct technology; and second, profits will be compared across technologies. Obviously, the technology with the highest profit will be selected and its water use will prevail. If none of the technologies yield positive profit, the farmer will not operate at all.

Optimal Water Use within a Technology

Let π_i denote maximum quasi-rent per unit area obtained under technology i. It is determined by the choice of applied water a_i with technology i, i.e., by

$$\pi_i = \max_{a_i} \{P * f[h(a, i, q, c), i, c] - W * a - W * g(a, i, q, c) - k_i\}. \quad (1)$$

The optimal condition associated with this problem is

$$P * f_e * h_a - W - V * g_a = 0. \quad (2)$$

Water will be applied at a level where the value of the marginal productivity of applied water is equal to the marginal cost of applied water. The value of the marginal productivity of applied water is the product of output price, the marginal productivity of effective water (f_e), and the marginal effectiveness of applied water (h_a). The marginal cost of applied water is denoted by

$$U = W + V * g_a$$

and is the sum of the market price of water and the marginal cost of drainage ($V * g_a$) associated with water application.

The second-order condition for optimality is

$$D = P * f_{ee} * (h_a)^2 + P * f_e * h_{aa} - V * g_{aa} < 0.$$

Assumptions regarding concavity of the production function and of the effective irrigation functions, and convexity of the drainage function, assure that the second-order condition holds. With this condition, one can obtain properties of optimal choice of applied irrigation water for a given technology.

First, note that the differentiation of [1] yields $\mathrm{d}a/\mathrm{d}W = 1/D < 0$. Thus, the elasticity of applied water (demand for water) is denoted by

$$h_{a,W} = -[\mathrm{d}a/\mathrm{d}W] * [W/a] = -[1/D] * [W/a].$$

One can also generalize to obtain elasticity of water use with respect to water cost,

$$h_{a,U} = -[1/D] * [U/a].$$

Using this definition and differentiating [1] with respect to P yields

$$\mathrm{d}a/\mathrm{d}P = -[1/P] * [f_e * h_a] = -[1/D] * [U/P]. \quad (3)$$

For further considerations let a general elasticity be denoted as

$$\eta_{\Psi,x} = |\Psi_x X/\Psi|,$$

and the elasticity of its derivative be denoted as

$$\eta_{\Psi,x_x} = |\Psi_x X/\Psi|_x,$$

where Ψ is some function and X is an argument of this function.

Using this notation, total differentiation of [1] with respect to P yields

$$\eta_{a,P} = \eta_{a,U} * [a/P].$$

Water use elasticity with respect to quality, is obtained by differentiating [1] with respect to q

$$\mathrm{d}a/\mathrm{d}q = -[a/q] * \eta_{a,U}\{\eta_{h,a_q} + \eta_{h,q}\eta_{f,e_e} - \eta_{g,a_q}[V * g_a/U]\}. \quad (4)$$

Using the results in [4], an increase in quality affects water use in three ways:

1. **Marginal effectiveness effect,** η_{h, a_q} — a tendency to increase water use as water effectiveness increases, thus, yield increased.
2. **Marginal productivity effect,** $\eta_{h,q}\eta_{f,e_e}$ — tendency to save water as marginal productivity of applied water declines (note that $\eta_{f,e_e} < 0$).
3. **Marginal drainage effect,** $V*\eta_{g,a_q}$ — a tendency to apply more water as drainage cost declines (note that $\eta_{g,a_q} < 0$).

The overall effect depends on the relative importance of these three effects. Consider the case of a simple multiplicative relationship between effective and applied water (Caswell and Zilberman, 1986). In this case $h(a, i, q, c) = a*\mathbf{h}(i, q, c)$. Then, $\eta_{h,a_q} = \eta_{\mathbf{h},q}$ and thus,

$$da/dq = -[a/q]*\eta_{a,U}\{(1 + \eta_{f,e_e})\eta_{\mathbf{h},q} - \eta_{g,a_q}[V*g_a/U]\}.$$

It is argued by Caswell and Zilberman (1986) that the elasticity of marginal productivity is greater than 1 in absolute value in most situations where water use is well inside the third phase of the production function. Therefore, with the multiplicative specification and without drainage, it is clear that less water will be applied as quality increases. The introduction of an additional drainage cost may reverse this tendency.

Consider now the impact of climatic change on drainage. Differentiation of [1] with respect to c and using the above notation yields

$$da/dc = -[a/c]*\eta_{a,U}\{\eta_{h,a_c} + \eta_{h,c}\eta_{f,e_e} + \eta_{f,e_c} - \eta_{g,a_c}[V*g_a/U]\}. \quad (5)$$

Again, based on [5], an increase in average temperature has several distinct impacts on water resource use:

1. Marginal effectiveness effect, η_{h,a_c} — a tendency to reduce water application rates as irrigation water efficiency declines with higher temperature ($\eta_{h,a_c} < 0$).
2. Marginal productivity effect of applied water, $\eta_{h,c}*\eta_{f,e_e}$ — a tendency to increase water application rates as marginal productivity of effective water increase due to decline in effective irrigation ($\eta_{h,c} < 0$, $\eta_{f,e_e} < 0$).
3. Marginal productivity effect of temperature, η_{f,e_c} — a tendency to increase water application rates as marginal productivity of effective water increases with temperature ($\eta_{f,e_c} > 0$).
4. Marginal drainage effect, $V*\eta_{g,a_c}$ — reflecting the impact of higher temperature on marginal generation of drainage, it is reasonable that higher temperatures tend to increase evaporation and reduce drainage. Therefore $\eta_{g,a_c} < 0$. In this case, increased temperature tends to reduce drainage.

The overall effect depends on the relative importance of these four effects. Consider again the case with no drainage problems, and assume that effective water is a multiplicative function of applied water. The impact of increased temperature on applied water becomes

$$da/dc = -[a/c]*\eta_{a,U}\{(1 + \eta_{f,e_e})\eta_{h,c} + \eta_{f,e_c}\}. \tag{6}$$

Since $\eta_{h,c} < 0$, and assuming $|\eta_{f,e_e}| > 1$, equation [6] suggests that higher temperatures will result in increased water application both due to high productivity associated with the higher temperature and the need to compensate for evaporation loss.

Technology Choice

Once optimal water application is decided for each technology, it determines the quasi-rent (π_i) of the technology. The technology with the highest quasi-rent value is selected given that the quasi-rent is neither negative nor smaller than the land rent value.

To better understand technology choice, consider the case where there are two technologies, a traditional one with $i = 0$ and a new technology with $i = 1$. The new technology is selected when

$$\pi_1 > \pi_0, \quad \pi_1 \geqslant r \geqslant 0$$

where r is the rental rate of land.

The quasi-rent difference between the two technologies can be written as

$$\Delta\pi = P*\Delta y - W*\Delta a - V*\Delta z - \Delta k$$

where Δy is yield difference, Δa is difference in applied water, Δz is difference in drainage volumes, and Δk is the difference in fixed cost. Selection of the new technology is likely to increase yield and reduce applied water rates and drainage volumes. If these impacts overcome the extra cost the new technology entails, it is selected.

To relate technology choice pattern to variation in quality and temperature, one needs to investigate how the quasi-rent difference changes as a result of changes in these factors.

First consider quality and note that

$$d\Delta\pi/dq = d\pi_1/dq - d\pi_0/dq. \tag{7}$$

To solve [7] differentiate quasi-rent with respect to quality which yields

$$d\pi_i/dq = P*f_e*[h_a*(da/dq) + h_q] - W*da/dq$$
$$- V*[g_a*da/dq + g_q].$$

Note that $P*f_e*h_a - W - V*g_a = 0$ from [2], and thus

$$d\pi_i/dq = P*f_e*h_q - V*g_q = U*h_q/h_a - V*g_q \geqslant 0. \tag{8}$$

The quasi-rent associated with each technology increases with quality, reflecting the values of the increase in effective water and the decline in drainage associated with a marginal increase in quality. Using [8] and introducing elasticity notations, the impact of quality on quasi-rent differential becomes

$$d\Delta\pi/dq = [1/q]*[U*a*\eta_{h,q}/\eta_{h,a}(1) - U*a*\eta_{h,q}/\eta_{h,a}(0)] - V*[g_q(1) - g_q(0)]. \quad (9)$$

When drainage is not associated with additional cost ($V = 0$), marginal cost of water under both technologies are the same [$U(1) = U(0)$], and rent differential declines with quality if

$$a(1) < a(0)*\{[\eta_{h,q}/\eta_{h,q}(0)]/[\eta_{h,q}/\eta_{h,a}(1)]\}.$$

This result is quite likely since less water is used with the new technology. It also reasonable that at the optimum the effective water with the traditional technology will be relatively more responsive to changes in quality compared to changes in applied water than effective water with the new technology. This point was proven by Caswell and Zilberman (1986) for $h(a, i, q, c) = a*\mathbf{h}(i, q, c)$.

Introduction of drainage considerations might actually operate in the same way and increase the decline in the quasi-rent differential associated with increased quality. The reason is that marginal contribution of quality to drainage declines with quality, but it declines faster with the old technology, thus $-V*[g_q(1) - g_q(0)] < 0$.

With very high land qualities there is not likely to be much difference between water use and output under different technologies. Thus, at these qualities the traditional technologies are likely to be more profitable because they require smaller investments. As $d\Delta\pi/dq < 0$ it is likely that there will be a range of high qualities where the old technology will be applied, but there might be a range of qualities below a certain critical level for which the new technology is more profitable.

Considering the impact of changes in temperature on technology choice, one can use a similar argument to show that if the new technology has a higher yield effect of temperature that dominates the water use effect, the new technology will be selected.

The main general conclusions derived from the analysis indicate that it is essential to have knowledge about the whole system in which the technology operates. Modern technologies may provide incentives to conserve resources if combined with appropriate policies of input and output prices and regulations on pollution. Heterogeneity of the production process, both in the physical conditions (e.g., weather, land, and water), and technology performances may result in different responses in different locations.

III. Data and Empirical Specifications of Variables

The San Joaquin Valley is the biggest agricultural area in California and one of the prime agricultural areas in the world. It produced nearly $4.5 billion of agricultural products which in 1987 accounted for 33 percent and 3.3 percent of the agricultural output of California and the U.S., respectively (SJVDP, 1990).

The development of irrigated agriculture in the SJV since the beginning of the century has been associated with the importation of tremendous quantities of water to the valley. This, of course, has brought prosperity to the region. But it has created new kinds of problems, including water clogging, rising water tables, salinity accumulation in soil and ground water, and elevated levels of selenium in the agricultural drainage water. These problems which are associated with the geomorphology of the westside of the region, (Gilliom, 1991) now threaten the long-term viability of agricultural production in the SJV (SJVDP, 1990).

One of a series of management options that has been suggested to resolve these problems (SJVDP, 1990) is on-farm, drainage-water source control which includes improving existing irrigation management and the adoption of new irrigation technologies.

The implementation of these prescriptions at the farm level, however, is not a simple matter. Individual farms possess different characteristics, face different environmental situations, and are operated by decision makers with potentially differing objectives and priorities. Estimates of the rates and likelihood of adoption of new technologies or improved management practices under proposed incentive programs are therefore essential to policy design aimed at control and management of drainage and drainage-related problems.

A survey of farms on the westside of the SJV was conducted during the 1989 crop year (Dinar and Campbell, 1990, 1991; Campbell and Dinar, 1991). The study area was divided into 76 "squares" identified through physical characteristics. The number of farms in the study area was estimated using County agricultural Commissioner records and a sample consisting of 6 percent (Orlich, 1978) of the full-time farms in the study area was selected. A total of 285 farms were surveyed, using a data collection instrument, during the summer of 1989.

Farm and field level data collected in the survey were for the 1988 agricultural season. The total number of field level observations was 786. The data was grouped by crops for fields that had two observations or more. Crops with one observation (field), or those missing key variables were omitted from the data set. The final analysis of field-level data included 571 observations.

FARM LEVEL VARIABLES

Variables accounting for long-term climate conditions as measured at reference weather stations in the area are $\mathbf{R^c}$ (measuring average of 1982—1989 annual rainfall value (inches)), and $\mathbf{T^c}$ (measuring average of 1982—1989 daily temperature (F°)). Variables accounting for weather during the 1988 crop year are $\mathbf{R^w}$ (measuring 1988 annual rainfall) and $\mathbf{T^w}$ (measuring 1988 average daily temperature). These variables were compiled based on data from CIMIS (various years) and NOAA (1989). For more details see Dinar and Campbell (1990).

A variable **E** is a proxy for on-farm environmental conditions. It includes on-farm information on (a) depth to water table, (b) salinity, and (c) selenium in the shallow ground water. Where on-farm data is not available, aggregated data that contains the same information at "square" level was used (Dinar and Campbell, 1990, pp. 18—20). The "square" level data is based on Geological Information Service (GIS) information provided in Appendixes 3, 4, and 5 in Dinar and Campbell (1990).

The variable **E** was calculated ($\mathbf{E} = \Sigma_i \Sigma_j \beta_{ij} \Omega_{kij}$) using a principle component analysis where β_{ij} are the estimated eigen values and Ω_{kij} are the probabilities of a farm in square $k(k = 1, \ldots, 76)$ of facing severity level $j(j \in J_i)$ of environmental condition $i(i =$ depth to water table, salinity, selenium). The estimated coefficients (eigen values) have no special interpretation, however, the variable calculated on the basis of the principal component analysis should provide meaningful interpretation. In this case, a higher value of **E** means better environmental conditions.

$\mathbf{D^0}$ is farm organization type (Campbell and Dinar, 1991). It is used here as a dummy variable with several organization categories (Unified, Primary Hierarchy, Simple Functional Hierarchy, Complex Functional Hierarchy, and Market Hierarchy). $\mathbf{D^i}$ is another dummy variable that accounts for the existence of a full time irrigator on farm. These two variables are correlated and will not be included in the same equation (see also discussion in the Empirical Models Section).

Q is total yearly quota of water from surface sources (af/yr); **S** is average cost of surface water from all source (\$/af); and **G** is average cost of groundwater from all sources (\$/af).

P is crop price index by crop category (\$/ton). Crops were grouped into the five crop categories (vegetables, orchards, cotton, field crops, pasture), and the variable **P** was weighted by each crop area.

$\mathbf{C^h}$ is farm total cost of hand move sprinkler irrigation system (\$/acre), and $\mathbf{C^l}$ is farm total cost of linear move irrigation system (\$/acre).

These variables were computed in the following way:

$$C_j^l = \left\{ \sum_{k \in K_l} X_{jtk_j}^k M_{kt}^l \right\} \Big/ A_{jt},$$

$$l = \begin{cases} \text{high} & \text{if there is full time irrigator on farm,} \\ \text{medium} & \text{if } C_j^w < 15 \text{ and there is no full time irrigator on farm,} \\ \text{low} & \text{else,} \end{cases}$$

where j is farm, t is technology, k is crop category, l is management level (low, medium, and high), X_{kt}^l is the crop area, M_{kt}^l is cost of technology ($/acre). These costs were based on CH2M HILL (1989). A_{jt} is the total area of technology t on farm j (acre), and C_j^t is the weighted cost of technology t on farm j ($/acre), and C_j^w is average cost of irrigation water.

FIELD LEVEL VARIABLES

For each field (crop), the following information is provided: **t** is a dichotomous variable indicating whether the field is owned (= 1) or rented (= 0); **a** is field area (acres); **h** is the geometrical shape of the field (1 = square; 2 = rectangle; 3 = triangle; 4 = irregular); **s** is slope of field (%); **o** is the weighted average soil texture in the field. It was calculated in the following way:

$$\mathbf{o = dist1*x1 + dist2*x2 + dist3*x3}$$

where **x1**, **x2**, **x3** are soil texture with three possible textures on one field (Each can be: sand; loamy sand; sandy loam; loam; silty loam; clay loam; clay; silty clay loam; and silty clay). Note that soil textures increase in water retention capacity. The variables **dist1**, **dist2**, **dist3** are the percent of the field with texture **x1**, **x2**, **x3**, respectively (%). Therefore, the lower the value of the soil variable, the lower its water retaining capacity.

e is portion of field having drainage or salinity problems (%). **r** is irrigation technology used on the field last year (0 = border, furrow; 1 = sprinkler, linear move, micro-sprinkler, drip). $\mathbf{p^w}$ is cost of water at field ($/AF). **y** is crop yield (ton/acre). $\mathbf{p^c}$ is crop yield gross price ($/ton). **v** is total cost of production ($/acre). Gross margin per land unit (**g**) was calculated as $\mathbf{g = y*p^c - v}$

IV. Empirical Models

Given the nature of the data and the theoretical framework, several models have been formulated and estimated. First, a farm level model is estimated, which explains the share of farm land irrigated by modern technologies, using farm level data on prices, climate, environmental conditions, human capital, and irrigation technology cost. Second, a field-level empirical model is estimated using field data and a dummy variable for crops.

For the purpose of estimating farm- and field-level adoption, two sets of dependent variables were created. Border and furrow irrigation have been compressed into one variable that represent traditional irrigation technologies. In the same way, all pressurized irrigation technologies were aggregated into a single variable representing modern irrigation technologies. In the case of farm-level analysis a continues independent variable measuring the share of farm land equipped with modern technologies is used, and in the case of the field-level analysis a dichotomous variable is used. It is assigned a value of 1 if the field is equipped with any modern technology and 0 if it is equipped with a traditional technology.

FARM-LEVEL TOBIT ESTIMATES

Previous studies have considered the adoption process at the farm level as a discrete one. This is not necessarily here since at the farm level there are heterogeneous conditions that justify adoption of modern technologies on part of the farm while other parts continue to utilize the traditional technologies.

Here we attempt to estimate the share of farm land equipped with modern irrigation technologies as affected by environmental, economic, and organization variables. The share of modern technologies on a given farm is bounded by zero and one, but may have values between zero and one. Therefore, a maximum likelihood tobit procedure was used as the estimation technique.

In the tobit model (Maddala, 1989) we have N_0 observations for which the share of modern technologies is 0, and N_1 observations for which the share >0. This provides us with the following log likelihood function to be estimated

$$\text{Log}\, L = \sum_{i \in \{0\}} \text{Log}(1 - F_i) + \sum_{i \in \{1 > I \geqslant 0\}} \text{Log}\left(\frac{1}{(2\pi\sigma^2)^{1/2}}\right) - $$

$$- \sum_{i \in \{1 > I \geqslant 0\}} \frac{1}{2\sigma^2}(y_i - \beta' x_i)^2$$

where F_i is the the density function, and σ^2 is the variance. $\beta' x_i$ is the nonstochastic portion of the model. In our application, for observation i

$$\beta' x = \beta_0 + \beta_1 \mathbf{R} + \beta_2 \mathbf{T} + \beta_3 \mathbf{E} + \beta_4 \mathbf{S} + \beta_5 \mathbf{G} + \beta_6 \mathbf{Q} +$$

$$+ \sum_k \beta_j \mathbf{P} + \sum_j \beta_j \mathbf{C} + \sum_n \beta_n \mathbf{D}$$

where variables were previously explained. Two versions (climate and weather) were used for rainfall and temperature: weather was represented by data for 1988 and climate by annual average for 1982—1988. **C** is technology ($\mathbf{C} = \mathbf{C}^h$ or $\mathbf{C}^l$) cost ($/acre), and **D** are dummy variables (for existence of irrigator on farm, $\mathbf{D}^i$, and five farm organization variables $\mathbf{D}^0$). And the β's are the estimated regression coefficients.

Our hypotheses regarding the behavior of the variables, based on the theoretical model and other relevant empirical studies that have been cited in the literature review are as follows:

We expect annual rainfall to negatively, and average temperature to positively affect on-farm share of modern technologies. In the analysis we have two sets of variables representing climate and weather. We will test each set of these variables separately and we expect that their relative values in the estimated equation will be influenced by the mean and standard deviation of the weather variables compared to that of the climate.

The principal component environmental variable was constructed such that values decrease as environmental conditions become unfavorable to the farm. Therefore, it is expected that it will be negatively related to the share of modern technologies present on the farm.

Surface and ground water cost at the farm level are expected to be positively related to the share of modern technologies. Modern technologies are characterized by better water efficiencies and might therefore conserve water, increase yields, and reduce pollution, and, thus, offset the investment costs which would be higher using traditional technologies (CH2M HILL, 1989).

Surface water quota is expected to be negatively related to share of modern irrigation technologies. Since higher values mean that water is not a constraint to decision making, there is no incentive for the farmer to switch to more efficient technologies and conserve water.

Crop price indexes were constructed for 6 main crop categories. However, in the analysis we included only the crops that may have a potential effect on cropping decisions and investment in irrigation technologies. The three groups that capture 75 percent of the study area are vegetables, cotton, and orchard crops (Dinar and Campbell, 1990, Table 23). The economic theory suggests that crop prices should be positively related to adoption. Therefore, we expect crop prices to have a positive sign in the regressions. However, based on our empirical observation regarding shares of modern technologies in cotton (Table II), it is likely that an increase in cotton price may increase cotton acreage and in turn, decrease the farm area equipped with modern technologies.

We included two sets of variables that may account for farm organization. The first variable is the existence of an on-farm, full time irrigator. The ability of a farm operation to fully utilize the potential of modern irrigation technologies is contingent upon highly trained staff, applying water precisely and timely and properly maintaining the irrigation systems. The presence of an irrigator function on-farm may achieve these outcomes, and consequently, facilitate the use of modern technologies.

Another variable is the farm organization type (Campbell and Dinar, 1990). We present several hypotheses that are only partially considered by the organization literature. Nevertheless, organizational theory and its adaptation to the case of agricultural production (Campbell and Dinar, 1990) supports our projections. The farm organizations in our analysis include the unified organization as the precursory organizational form, the functional organizations and the market organizations. In the case of the unified organization labor and management are corporally unified in one person, or organizational unified by an undifferentiated work process. We expect that for economies of smallness (Yaron *et al.*, 1992), it is more likely that modern technologies will be on larger portions of unified farm operations. Market farm organizations are more likely to face a diversity of constraints, such as crops, delivery systems, and harvest contracts. Modern technologies provide better flexibility and allow control over crop production and harvesting than do traditional technologies. Market farm organizations, as a result, are more likely to use modern technologies: Complex Functional Hierarchy farms whose production processes approximate market hierarchies are for the same reasons more likely to use modern technologies. Functional farm organizations are characterized by a moderately skilled work force directly supervised by the owner or manager(s). The Primary Hierarchy type is less likely to adopt modern irrigation technologies than the Complex, the Market, and the Functional Hierarchies because of the inefficiencies of a management and labor force that is unspecialized.

FIELD-LEVEL LOGIT ESTIMATES

Our empirical analysis is based on the assumption that only one irrigation technology is applied to any given field.[2] Therefore, for a given field, the probability that a farmer will adopt a modern technology is bounded by zero and one, and can be estimated using a maximum-likelihood logit technique.

In the logit model (Gujarati, 1988) the probability that a modern technology is in place on a given field is given by

$$P_i = E(Y = 1|X_i) = \frac{1}{1 + e^{-(\gamma_0 + \gamma' x_i)}}$$

where $\gamma' x_i$ is the nonstochastic portion of the model. In our application, for observation i

$$\gamma' x = \gamma_0 + \gamma_1 \mathbf{a} + \gamma_2 \mathbf{s} + \gamma_3 \mathbf{h} + \gamma_4 \mathbf{o} + \gamma_5 \mathbf{e} + \gamma_6 \mathbf{p}^w + \gamma_7 \mathbf{p}^c + \gamma_8 \mathbf{v} + + \sum_t \gamma_9 \mathbf{t} + \sum_n \gamma_n \mathbf{d}$$

where all variables were previously defined. **v** is a measure for profitability. In one version $\mathbf{v} = \mathbf{g}$ and in another version $\mathbf{v} = \mathbf{p}^c$. The variable **d** is a vector of dummy variables for main crop categories (pastufre, row field crops, grain crops, vegetables, cotton, and orchard crops, where the latter group serves as a bench mark).

As in the case of the farm level model, we base our hypotheses on the theory provided in Dinar and Zilberman (1991) and other relevant studies discussed in the Literature Review section.

Larger fields are more likely to be equipped with modern technologies due to economies of scale in the capital investment and the maintenance cost. Steeper fields are more likely to have modern technologies. Fields with less regular shapes (triangles, or other irregular shapes) are also more likely to have modern technologies, since modern technologies are divisible and more easily applied to nonregular field shapes. Sandy soils are more likely to be equipped with modern irrigation technologies which are assumed to be land quality augmented (Dinar and Zilberman, 1991). Fields facing salinity and drainage problems, are more likely to have modern technologies that are characterized by a better irrigation application uniformity.

Fields bearing higher water costs are more likely to be equipped with modern irrigation technologies. Higher crop prices should enhance adoption of modern irrigation technologies. In addition, the greater the margin of profit the higher the probability to adopt modern technologies. Fields owned rather than rented or leased are more likely to be equipped with modern technologies (Feder *et al.*, 1988).

The crop dummy variables provide the effects of different cultural practices on adoption of irrigation technologies. Orchard crops were used as a bench mark in the analysis, and coefficients of cotton, pasture and grain crops are expected to have negative signs.

V. Results

The shares of the main irrigation technologies, aggregated over sub-technologies, at field level are presented in Table I by crop category. The majority of the orchards plots are equipped with low volume irrigation technologies while none of the other crops are irrigated with low volume technologies. The majority of the fields planted in field crops are irrigated with either furrow irrigation or border irrigation. Zero to 15.6 percent of these fields are irrigated with sprinklers. Vegetables are irrigated with border, furrow and sprinkler irrigation. Low value vegetables are irrigated with border, furrow and sprinkler technologies, and high value vegetables are mostly irrigated with furrow and sprinklers. Pasture land is irrigated mainly by border and less frequently by furrow irrigation. Cotton is irrigated mostly by furrow.

In Table II the distribution of technologies, using farm- and field-level data, is presented (for the principle irrigation technologies). Except for dry land, which is not presented at the farm-level data, all other technologies appear to be similarly distributed. Border, furrow, sprinklers, and low volume irrigation capture 28, 49, 21, and 2 percent of the irrigated farm area based on farm-level data and 29.4, 54.4, 13.7, and 1.0, percent based on the field-level data. Other descriptive statistics for farm-level data is presented in

Table I. Percent use of main irrigation technologies by crop categories

Crop category	Nonirrigated	Percent of fields irrigated with Border	Furrow	Sprinklers	Low volume
Vegetables	0–14.3	0–44.0	0–82.6	0–100.0	0
Field crops	0–4.3	0–92.9	7.1–84.4	0–15.6	0
Pasture	0	83.3	16.7	0	0
Cotton	2.1	7.8	79.9	10.4	0
Orchards	0	0–45.5	0–35.7	12.5–40.9	7.1–66.7

Table II. Percent use of main irrigation technologies at field- and farm-level

Irrigation technology	Distribution (%) (Farm level data)	Distribution (%) (Field level data)
Dry land	—[a]	2.0
Border	28.0	29.4
Furrow	49.0	54.4
Sprinklers	21.0	13.7
Low volume	2.0	1.0

[a] Not reported on a farm level.

Table III, and field-level data for the main crops grown in the study area are presented in Table V.

FARM-LEVEL TOBIT ANALYSIS RESULTS

Table IV reports tobit maximum likelihood estimates of modern irrigation technology shares. Several equations were estimated which differed in the use of climate versus weather variables and the definitions of farm organization. The results presented for two estimated equations include log-likelihood function values and correlation between observed and expected values of shares at mean values of the independent variables. All parameter estimates are consistent with the hypotheses suggested earlier. In equation 1, 9 out of 12 estimated coefficients were found to be significant, and in equation 2, 10 out of 16 estimated coefficients were significant. Weather and climate variables did not have significant coefficients in any of the estimates nor did the surface water quota variable or cotton price index.

A full time irrigator (Eq. 1) is one of the variables contributing to high shares of modern irrigation technologies on farm. The coefficient for a full time irrigator is positive, significant, and has a high value compared to other

Table III. Descriptive statistics of main farm-level variables (284 observations[a])

Variable	Mean value	Std. Dev
Share of modern tech.	0.098	0.231
1988 average annual rainfall	12.214	6.767
1988 average daily temperature	62.351	1.321
1982—1988 annual average rainfall	8.606	4.274
1982—1988 average daily temperature	63.759	2.326
Principle component for environmental conditions	0.305	0.351
Existence of irrigator on farm	0.373	0.484
Total available annual water on farm (204 obs. excluding observations with unlimited water)	7475	28668.420
Average price of water on farm	22.068	17.061
Average price of groundwater	28.900	15.531
Vegetable price index	76.514	182.687
Field crops price index	68.838	215.631
Orchards price index	206.482	541.039
Cotton price index	930.670	714.142
Cost of hand move system	171.831	80.807
Cost of linear move system	149.641	50.982

[a] One observation was omitted due to unexplained high annual rainfall value (44 inch) that was provided by CIMIS.

variables in this equation. Also, the elasticity of this variable is relatively high (Table VII).

Farm organization dummies were also found to be significant (Eq. 2, Table IV). The coefficients imply that Unified, Complex Functional Hierarchy, Simple Functional Hierarchy, and Market Hierarchy organizations are more likely to have a larger share of modern technologies than the Primary Hierarchy. Table IV indicates also that the Unified has the highest coefficient among farm organizations, where the Simple and the Market have nearly the same coefficient.

Both surface and groundwater coefficients imply that a greater share of the farm is likely to be equipped with modern technologies as cost of these sources increases (the two variables, however, are not correlated). Vegetables and orchards price indexes have positive signs where cotton has a negative sign. The negative coefficient for cotton price is explained simply by the fact that, holding all other values constant, cotton profitability (Table III) can not provide enough margin to bear the cost of modern technologies. This means that an increase in cotton price over the range reported here will result only in an increase in cotton area equipped with traditional technologies. This in turn means a possible decrease in the share of modern technologies on such farms. It should be noted that the coefficient for cotton price is not significant. A price index for field crops was also included in the analysis but found

Table IV. Tobit analysis for share of farm land irrigated with modern technologies

Variable	Share of farm land irrigated with modern technologies Eq. 1	Eq. 2
Constant	−0.791	−0.503
	(−2.19)	(1.71)
1988 average annual rainfall	−0.00017	
	(−0.02)	
1988 average daily temperature	0.00015	
	(0.03)	
1982—1988 average annual rainfall		−0.0069
		(−0.52)
1982—1988 average daily temperature		0.00055
		(0.01)
Principal component var. for environment	−0.143	−0.76
	(−1.04)	(−0.55)
Average surface water cost	0.0066	0.0063
	(2.41)	(2.14)
Average groundwater cost	0.0071	0.0066
	(2.24)	(2.08)
Surface water quota	-1.65×10^{-6}	-1.19×10^{-6}
	(−0.84)	(−0.60)
Vegetable price index	−0.00037	−0.00032
	(1.72)	(1.43)
Orchard price index	0.00012	0.00015
	(1.66)	(2.03)
Cotton price index	-1.96×10^{-4}	-4.36×10^{-4}
	(−0.02)	(−0.05)
Cost of hand move irrigation system	−0.00076	−0.00077
	(−1.55)	(−1.15)
Full time irrigator on farm (dummy)	0.394	
	(4.26)	
Organization:		
Unified		0.331
		(2.25)
Primary hierarchy		−0.286
		(−3.27)
Complex functional hierarchy		0.044
		(0.45)
Simple functional hierarchy		0.199
		(2.06)
Market hierarchy		0.210
		(1.54)
Predicted prob. of modern tec. at aver. values	0.215	0.198
Log likelihood function	−125.70	−119.41
Squared corr. observed & expected values	0.263	0.288

In parentheses are asymptotic *t*-values.

Table V. Descriptive statistics of the main field-level variables for main crops in each crop category

Crop	Cotton	Melons	Wheat	Almonds	Alfalfa
Field area [ac]	181.34	105.06	194.22	421.07	112.85
Slope [%]	1.02	1.51	0.79	1.26	1.18
Av. Yield [ton/ac]	0.563	14.22	2.77	0.817	7.31
Crop price [$/ton]	1543.27	408.44	132.20	2345.53	93.79
No. of Irrig.	5.17	5.13	3.71	7.50	8.83
Ann Irrig. Rate [af/ac]	2.86	2.5	2.00	4.72	3.98
Tot. Var. Cost [$/ac]	467.86	455.20	331.56	968.75	307.97
Share of salinity & drainage problems [%]	15.35	28.40	14.88	9.76	11.61
Surface wtr. cst. [$/af]	36.29	38.07	32.61	22.44	52.95
Gross margin [$/ac]	400.6	4878.38	13.00	934.84	293.60
Ownership[a]	0.59	0.52	0.64	0.76	0.52

[a] This variable represents average values of 1 and 0. Therefore, it should be interpreted as percentage of the crop fields that are owned.

inconsistent and not significant. The results for equations including that variable are not presented.

Increased cost of linear move irrigation systems (representing modern technologies) adjusted to crops and management level (see data section) was expected to be negatively related to on-farm share of modern technologies. Surface water quota has a negative sign, as expected, meaning that the less water availability is constrained, the lower the share of modern technologies. The principal component variable for environmental conditions has a negative sign meaning that the more stringent environmental conditions are, the larger the share of modern technologies on farm.

FIELD-LEVEL LOGIT ESTIMATES

Field-level results are presented in Table VI for five functional specifications. The first specification includes only the physical characteristics of the field. The second specification includes dummy variables for major crop types (pasture, row crops, grain crops, vegetables, and cotton). The third specification includes, in addition to the variables in the second equation, the cost of irrigation at the field. The third specification replaces crop dummy variables by a variable for crop price and adds a variable that accounts for share of the field facing salinity and drainage problems. The fifth specification is similar to the forth except that it includes a dummy variable for owned fields.

All estimated coefficients have signs which agree with expectations presented in the theory section or provided by the literature. MacFadden R^2 value varies between 0.071 to 0.119, and log likelihood values of the esti-

Table VI. Logit estimates of technology adoption at field level (471 observations at 0 and 82 observations at 1)

Variable	Eq. 1	Eq. 2	Eq. 3	Eq. 4	Eq. 5
Constant	−1.862	−1.873	−1.971	−2.036	−2.069
	(−6.77)	(−6.45)	(−6.66)	(−6.80)	(−5.99)
Area (acres)	0.0016	0.0015	0.0014	0.0015	0.0015
	(3.11)	(2.97)	(2.79)	(3.05)	(3.05)
Slope (%)	0.182	0.189	0.184	0.174	0.175
	(2.858)	(2.83)	(2.74)	(2.70)	(2.70)
Shape	0.013	0.016	0.016	0.015	0.015
	(1.16)	(1.36)	(1.39)	(1.27)	(1.27)
Soil	−0.103	−0.106	−0.112	−0.110	−0.109
	(−1.91)	(−1.91)	(−2.00)	(−2.01)	(−2.00)
Ownership					0.051
					(0.19)
Field share with salinity and drainage				0.0043	0.0042
				(1.25)	(1.22)
Water cost at field			0.0035	0.0030	0.0030
			(2.59)	(2.31)	(2.31)
Crop price				0.000046	0.000045
				(0.27)	(0.26)
Crop type:					
Pasture		−0.372	−0.490		
		(−1.21)	(−1.52)		
Row crops		0.280	0.311		
		(0.91)	(1.00)		
Grain crops		−0.519	−0.486		
		(1.43)	(−1.33)		
Vegetables		−0.253	−0.243		
		(−0.65)	(−0.63)		
Cotton		−0.018	−0.018		
		(−0.90)	(−0.90)		
McFadden R^2	0.071	0.106	0.119	0.084	0.084
Log likelihood	−215.66	−207.34	−204.28	−212.60	−212.58
% of right prediction	76.6	77.4	77.7	76.9	76.9

In parentheses are asymptotic t-values.

mated equations vary between −204.28 to −215.66. Percent of right prediction varies around 77 percent. This means that the model predicts in 77 percent of the cases that a field will be equipped with a modern technology when it is observed to be equipped with this technology. Except for the coefficient for crop price, all other coefficients were found to be significant at a level of 5 percent or less.

Fields of larger area, steeper slope, irregular shape, and containing lower water retaining soils are more likely to be equipped with modern tech-

Table VII. Effects on share of technology at farm level of changes in level of explanatory variables (weighted aggregate elasticity)

	Eq. 1	Eq. 2
1988 average annual rainfall	−0.007	
1988 average daily temperature	0.030	
1982–1988 average annual rainfall		−0.203
1982–1988 average daily temperature		0.012
Principal component var. for environment	−0.146	−0.081
Average surface water cost	0.492	0.485
Average groundwater cost	0.683	0.666
Surface water quota	−0.036	−0.027
Vegetable price index	0.096	0.086
Orchard price index	0.087	0.112
Cotton price index	−0.006	−0.014
Cost of hand move irrigation system	−0.435	−0.462
Full time irrigator on farm (dummy)	0.490	

Table VIII. Effects on choice probabilities at field level of changes in level of explanatory variables (weighted aggregate elasticity)

	Eq. 1	Eq. 2	Eq. 3	Eq. 4	Eq. 5
Area (acres)	0.209	0.190	0.177	0.196	0.197
Slope (%)	0.195	0.189	0.182	0.183	0.185
Shape	0.033	0.039	0.039	0.036	0.036
Soil	−0.316	−0.315	−0.328	−0.331	−0.329
Ownership					0.022
Field share with salinity and drainage				0.064	0.063
Water cost at field			0.111	0.098	0.099
Crop price				0.021	0.020

nologies. Coefficients for these variables in all estimated regressions are significant at a level of 5 percent or less.

Crop dummies were included in estimates 2 and 3. They represent a shift in the regression line relative to the orchard crops that was used as a bench mark. Results indicate that pasture and grain crops are less likely to be equipped with modern irrigation technologies than are orchard crops. The coefficients for these crops are also significant at a level of 10 percent or less. Vegetables and cotton have negative coefficients and row crops have a positive coefficient, meaning that vegetables and cotton are less likely and row crops are more likely to be equipped with modern technologies compared to orchard crops. However, the coefficients for vegetables, cotton, and row crops are not significant.

Coefficients for water cost (Eqs. 3, 4 and 5) and share of field with drainage and salinity problems (Eqs. 4 and 5) are positive and significant. This indicates that the more costly the irrigation water at the field and the larger the area of the field with drainage and salinity problem, the more likely the field will be equipped with modern irrigation technologies which are more water efficient and input quality augmenting. Coefficients for crop price (Eqs. 4 and 5) and ownership (Eq. 5) are positive, which comply with the theory, but are not significant.

Elasticities of key variables are presented in Table VII and VIII for the farm- and field-level analyses, respectively. In Table VII the dependent variable is the share of farm area equipped with modern technologies. The values should be read as follows (e.g., Eq. 2, for vegetable price index). An increase of 10 percent in vegetable price index will result in an increase of 0.086 percent in the share of modern technologies on farm. The range of values for different variables is wide. Environmental and locational variables are associated with elasticities varying from $|0.007|$ to $|0.203|$. The existence of a full time irrigator on farm turns out to be very important (elasticity of 0.49); While groundwater cost appears to be the most important variable affecting on-farm share of modern technologies.[3]

Regulatory agencies with control of several institutional variables should select those with the greatest potential effect. Of the variables, surface water cost, surface water quota, crop prices, and cost of irrigation systems, two (surface water cost and irrigation system cost) appear to be the most effective. Average surface water costs have elasticity values that vary from 0.485 to 0.492, and cost of hand move sprinkler elasticities varies between −0.462 to −0.435 in estimates 1 and 2, respectively. Thus, a policy that combines taxes on water supply with subsidies for modern irrigation technologies might be the most effective (see also Dinar and Letey, 1989 and Knapp *et al.*, 1990).

Choice probabilities at field levels are presented in Table VIII. Soil type is the most important factor in determining the technology choice. Field slope and area are also important. Two other variables with relatively lower influence on probability choice are the field shape and field share with salinity and drainage problems. Of the two possible policy variables, crop price and water cost, the latter is more efficient in terms of its effects on choice probability. It varies from 0.098 to 0.111. This means that an increase of 10 percent in surface water cost will increase the probability that a given field is equipped with a modern technology by nearly 10 percent, and, therefore, should be considered in any policy aimed at field-level source control.

VI. Policy Implications and Discussion

The results of this study highlight the important factors involved in the adoption of improved irrigation and drainage reduction technologies under limiting environmental conditions, such as high soil salinity levels and high water table. A conceptual model was empirically tested on a sample of farms in the San Joaquin Valley of California and quantitative estimates were provided. Four sets of variables were found to be related to the use of modern irrigation technologies. These variables are pricing of water, irrigation technology cost sharing, farm organization type, and environmental conditions.

Information on adoption coefficients such as those in this study may assist policy makers, concerned with on-farm source control strategies, by (a) identifying the characteristics and necessary conditions for adoption – including the relationship of irrigation practices to crop patterns, field characteristics (size, shape and soil type), organizational type, presence of full time irrigators, pricing strategies, and water supply; (b) allowing policy makers and extension agencies to delimit incentive programs according to the above items. Policy agencies should be concerned with such conditions as the cost of surface and groundwater to the farm, cost of modern technologies, surface water quotas, and the more attenuating conditions of crop choice and crop price. Price support, grants, loan programs, quotas and other such institutions may be utilized to manipulate conditions which favor the use of modern technologies.

Educational agencies, on the other hand, may be more concerned with the characteristics of farms as a means of identifying a target population. The presence of a full-time irrigator, the farm's organizational structure, their crop production patterns, ownership, and the physical characteristics of farms and fields, are all factors which anticipate a favorable response to an educational program geared towards the adoption of modern technologies.

In selecting the appropriate measures to enhance on-farm source control of conservation and drainage water reduction, policy makers must acknowledge also the diversity of farms and field conditions which play a role in decision making. Growers facing significantly different conditions can not be expected to respond similarly. Policies and their programmatic counterparts must therefore be responsive to a range of locational factors such as production costs, water quotas, and environmental conditions, which themselves may vary over time. Given the variation in physical setting of environmental and locational variables, the agency should consider also the most efficient policy instruments to achieve the desired goal with a minimal cost (e.g., monitoring and enforcement costs) for the agency and the society. This may imply a flexible policy tailored to specific farm or regional conditions.

Acknowledgement

The research leading to this paper was supported by the San Joaquin Valley Drainage Program, and The Resources and Technology Division of USDA-ERS. Mark Cary compiled data for several weather stations used in the creation of the weather and climate variables. Kathy Edgington helped in organizing the data base for the analysis. Michael Moore contributed ideas and was involved in the construction of several variables.

Notes

[1] Fisheries and wildlife habitate.
[2] This is not necessarily true, since in fields with irregular shapes, the "regular" part can be irrigated with one technology and the "irregular" part with another technology. Data collected in our survey revealed that farmers used one technology per field.
[3] Note that groundwater is not available to all farms in the study area, so a cost of $0/af for ground water means that there is no ground water available for that farm.

References

Campbell, M. B. and A. Dinar (1991), 'Organizational Classes Explain Differences among Westside Farms', *California Agriculture* **46**(1), 35—39.

Caswell, M. F. (1991), 'Irrigation Technology Adoption Decisions: Empirical Evidence', in Dinar, A. and D. Zilberman (eds.), *The Economics and Management of Water and Drainage in Agriculture.* Kluwer Academic Publishers, Boston, MA, 295—312.

Caswell, M. and D. Zilberman (1985), 'The Choice of Irrigation Technologies in California', *American Journal of Agricultural Economics* **67**(2), 224—234.

Caswell, M. and D. Zilberman (1986), 'The Effects of Well Depth and Land Quality on the Choice of Irrigation Technology', *American Journal of Agricultural Economics* **68**, 798—811.

CH2M HILL (1989), *Irrigation Systems Costs and Performance in the San Joaquin Valley.* Report prepared for the Federal-State San Joaquin Valley Drainage Program, 2800 Cottage Way Sacramento, CA 95825, USA.

CIMIS (California Irrigation Management Information System) (1982—1990), Water Conservation Office, California Department of Water Resources, Sacramento, CA.

Dinar, A. and J. Letey (1989), 'Economic Analysis of Charges and Subsidies to Reduce Agricultural Drainage Pollution', *Proceeding of the 2'nd Pan-American Regional Conference on Toxic Substances in Agricultural Water Supply and Drainage — An International Environmental Perspective*, Ottawa, Canada, June 8—9.

Dinar A. and D. Yaron (1990), 'Influence of Quality and Scarcity of Inputs on the Adoption of Modern Irrigation Technologies', *Western Journal of Agricultural Economics* **15**(2), 224—233.

Dinar, A. and D. Yaron (1992), 'Adoption and Abandonment of Irrigation Technologies', *Agricultural Economics* **6**(4), 315—332.

Dinar, A. and M. B. Campbell (1990), *Adoption of Improved Irrigation and Drainage Reduction Technologies in the Westside of the San Joaquin Valley. Part I Report: Literature Review, Survey Methods and Descriptive Statistics.* U.S. Bureau of Reclamation Contract 9-PG-20-03380 and 9-PG-20-03010. San Joaquin Valley Drainage Program, Sacramento, CA, August.

Dinar, A. and M. B. Campbell (1991), *Adoption of Improved Irrigation and Drainage*

Reduction Technologies in the Westside of the San Joaquin Valley. Part II Report: Economic Analysis and Estimates of Behavior of Farm Operators under Various Environmental Conditions. U.S. Bureau of Reclamation Contract 9-PG-20-03380 and 9-PG-20-03010. San Joaquin Valley Drainage Program, Sacramento, CA, July.

Dinar, A. and D. Yaron (1992), 'Adoption and Abandonment of Irrigation Technologies', *Agricultural Economics* **6**, 315—332.

Dinar, A. and D. Zilberman (1991), 'The Economics of Resource-Conservation, Pollution-Reduction Technology Selection: The Case of Irrigation Water', *Resources and Energy* **13**, 323—348.

Feder, G., R. E. Just, and D. Zilberman (1985), 'Adoption of Agricultural Innovations in Developing Countries, A Survey', *Economic Development and Cultural Change* **33**, 255—298.

Feder, G., T. Onchan, Y. Chalamwong, and C. Hongladarom (1988), *Land Policies and Farm Productivity in Thailand.* Johns Hopkins University Press for World Bank.

Gilliom, R. J. (1991), 'Overview of Sources, Distribution and Mobility of Selenium in the San Joaquin Valley, California', in Dinar, A. and D. Zilberman (eds.), *The Economics and Management of Water and Drainage in Agriculture.* Kluwer Academic Publishers, Boston, MA, 29—47.

Gujarati, D. N. (1988), *Basic Econometrics*, Second Edition, McGraw-Hill, New York.

Just, R. E. (1991), 'Estimation of Production Systems with Emphasis on Water Productivity', in Dinar, A. and D. Zilberman (eds.), *The Economics and Management of Water and Drainage in Agriculture.* Kluwer Academic Publishers, Boston, MA, 251—274.

Knapp, K. C., A. Dinar, and P. Nash (1990), 'Economic Policies for Regulating Agricultural Drainage Water', *Water Resources Bulletin* **26**(2), 289—298.

Letey, J., A. Dinar, C. Woodring, and J. Oster (1990), 'An Economic Analysis of Irrigation Systems', *Irrigation Science* **11**, 37—43.

Lichtenberg, E. (1989), 'Land Quality, Irrigation Development, and Cropping Patterns in the Northern High Plains', *American Journal of Agricultural Economics* **71**(1), 187—194.

Lichtenberg, E., I. E. Strand Jr., and B. V. Lessley (1991), Subsidizing Agricultural Nonpoint-Source Pollution Control: Targeting Cost Sharing and Technical Assistance. Working Paper No. 91—10 Department of Agricultural and Resource Economics, University of Maryland Colledge Park MD, 20742.

Maddala, G. S. (1989), *Limited Dependent and Qualitative Variables in Econometrics.* Cambridge University Press, Cambridge.

National Research Council (1989), *Irrigation Induced Water Quality Problems.* National Academy Press, Washington D.C.

NOAA (National Oceanic Atmospheric Administration) (1977—1989), *U.S. Department of Commerce*, Climatological Data from EARTHINFO.

Negri, D. H. and D. H. Brooks (1990), 'Determinants of Irrigation Technology Choice', *Western Journal of Agricultural Economics* **15**(2), 213—223.

Orlich, D. C. (1978), *Designing Sensible Surveys.* Pleasentville, N.Y.

SJVDP (San Joaquin Valley Drainage Program) (1990), *A Management Plan for Agricultural Subsurface Drainage and Related Problems on the Westside San Joaquin Valley*, Sacramento, CA 183 pp., September.

Utton, A. E. and L. Teclaff (eds.) (1978), *Water in a Developing World.* Westview Special Studies in Natural Resources and Energy Management, Boulder, CO.

Yaron, D., A. Dinar, and H. Voat (1992), *Innovations on Family Farms: the Case of the Nazareth Region in Israel. American Journal of Agricultural Economics* **74**(2), 361—370.

119-40
[94]

[10]

JOURNAL OF ENVIRONMENTAL ECONOMICS AND MANAGEMENT 26, 66–87 (1994)

Modeling Intrastate and Interstate Markets for Colorado River Water Resources*

J. F. BOOKER[†] AND R. A. YOUNG[‡]

[†]College of Business, Alfred University, Alfred, New York 14802; and [‡]Department of Agricultural and Resource Economics, Colorado State University, Fort Collins, Colorado 80523

Received October 17, 1992; revised December 17, 1992

A river basin optimization model is presented for investigating performance of alternative market institutions for water resource allocation. We show that existing demands for Colorado River water cannot be fully satisfied given mean annual flows of 13.0 million acre-feet at Lee's Ferry. Market transfers which minimize costs to consumptive users of such shortfalls may achieve as little as 50% of the incremental benefits possible with transfers, which incorporate the economic benefits of hydropower production and reductions in river salinity. Such efficient allocations would require large transfers from existing Upper Basin consumptive users, and annual deliveries to Mexico would exceed treaty obligations.

A regional water economy can be characterized as being in either an "expansionary" or a "mature" phase [38]. In the expansionary phase, the real incremental cost of new water supplies is relatively low and constant over time, and water supplies and project sites are readily available to meet growing demands. The mature phase is distinguished by rapidly rising incremental costs of new water supplies and by increased conflicts among water users over both water supply and quality issues. In the arid regions of the southwestern United States, the focus of our study, the combination of limited local water supplies and rapidly growing population, expanding industries, and increasing affluence has brought about a transformation to a mature water economy.

Southwestern urban entities anticipating water scarcity face major obstacles in obtaining new water supplies. Cities and states have mainly continued to emphasize traditional water storage and conveyance projects, which of necessity draw from more distant and hence more capital, energy, and environmentally expensive sites. However, agriculture remains the dominant water user, accounting for over 80% of water consumption in the arid southwestern region, with much of this use (particularly that devoted to forages and feed grain production) yielding relatively low economic value [13]. Recognizing that marginal foregone benefits of transferring agricultural water rights may be less than the cost of new engineering projects as a source for meeting growing urban water demands, increasing attention is being given to nonstructural approaches, such as market transfers of agricultural water rights [21, 34, 40].

*This work was supported by Grant 14-08-0001-G1644 from the U.S. Geological Survey, and by the Agricultural Experiment Station at Colorado State University. Laurie Walters conducted the survey and tabulated the data from municipal water utilities which formed the basis of the municipal water demand estimates. R. Garth Taylor updated the model of Imperial Valley agricultural production. The authors thank two journal referees for their helpful suggestions.

0095-0696/94 $6.00

Water markets present an attractive solution to southern California water shortages. The recent completion of the Central Arizona Project and the anticipated more complete utilization of water rights by Utah, Nevada, New Mexico, and Colorado leave southern California with a shortfall compared to previous use patterns which relied heavily on unused entitlements assigned to other states. The increased economic pressures for reallocation of water are being accommodated by shifts in the legal structure of water allocation. Traditionally regarded as a resource under state jurisdiction, following the U.S. Supreme Court's decision in *Sporhase v. Nebraska* in 1982, water must now be acknowledged as an article of interstate commerce [27].

PREVIOUS RESEARCH

Economic evaluation of proposed water transfers has taken several paths, depending in part of the nature of the proposed solutions. The economic evaluation of water transfers, of course, requires consideration of cost (supply) and benefit (demand) factors and the interaction between them, so as to help determine who gains and who loses and by how much. Early approaches analyzed the costs versus benefits of transferring a fixed, predetermined supply of water. Hirshleifer *et al.* [18] proposed a transfer of Colorado River irrigation water rights from the Imperial Valley (on the California–Mexico border) to municipal and industrial uses in metropolitan southern California as a lower cost alternative than the planned Feather River Project which was eventually built to bring northern California water to southern California. Hartman and Seastone [17] assessed rural-to-urban water right transfers, giving particular attention to measures of secondary economic impacts in the area of origin. One of their case studies examined the Imperial Valley-to-Los Angeles transfer recommended by Hirshleifer *et al.* [18]. Howe and Easter [20] developed an analytic framework for evaluating large-scale structural interbasin water transfer projects. Their approach reflected the state of the art in cost–benefit analysis and emphasized the potential importance of the opportunity cost of water in both instream and offstream uses in the basin of origin as well as secondary economic impacts. Their principal empirical case study analyzed a proposal large-scale Columbia Basin-to-southern California conveyance project. Kelso *et al.* [25] reached skeptical conclusions (confirmed since then by events[1]) in their study of the economic feasibility of the Central Arizona Project, a federal Colorado River-to-Central Arizona conveyance project. As in the work of Hartman and Seastone [17] and Howe and Easter [20], input–output models were employed to measure secondary economic impacts. Houston and Whittlesey [19] studied the potential effects of introducing markets between agricultural and hydropower sectors in the Columbia–Snake River Basins, while Hamilton *et al.* [16] concluded that interruptible markets could generate hydropower benefits 10 times greater than the corresponding foregone net farm income. Small streamflow increases in the Colorado River Basin were shown by Brown *et al.* [5] to produce hydropower benefits exceeding benefits of increased

[1]Diversions were 745 thousand acre-feet (kaf) in 1990 [45], but since then have been closer to 500 kaf as a result of reduced demand in agricultural uses.

consumptive use. Related work [4] showed that an economically based priority system could greatly reduce the cost of shortages. Colby [6] has argued generally that instream values for recreation and important wildlife species may also be comparable to consumptive use values. Oamek [36] employed quadratic programming models to develop the most complete representation to date of Upper Colorado River Basin agricultural water demands. He reported that the foregone benefit of up to 0.4 million acre-feet (maf) of water presently used in Upper Basin agriculture was no more than $10 per acre-foot, with significant additional amounts of water earning only slightly more.

A more recent line of analysis, one which our research seeks to extend, has employed optimization models to estimate market allocations and to identify the optimal interregional, intersectoral allocations and prices which would emerge from hypothetical water markets. In contrast to the evaluations of fixed-quantity proposals described above, both water quantities and shadow prices are endogenous to these models. Vaux and Howitt [48] utilized a single (physical) water supply estimate for eight sources together with five economic demand regions in California to suggest that, with few exceptions, intrastate water marketing could result in significantly greater social benefits than construction of new facilities under typical water supply conditions and projected growth rates extending to year 2030. McCarl and Parandvash [31] evaluated the efficiency of two proposed irrigation projects in the Pacific Northwest, showing that interruptible irrigation to increase hydropower production in water-short years can significantly reduce costs to hydropower of irrigation development. Using a regional programming model including energy and agricultural sectors, Keith *et al.* [24] showed that with existing storage facilities and limited water rights transfers, water quantity would not significantly limit Utah economic growth. Water quality (salinity) was identified as a potential limit, however, and additional research was suggested.

Lee [28] examined salinity for the full Colorado River Basin, estimating that decreasing Lower Basin salinity through transfers of up to 78% of approximately 1.0 maf of modeled Upper Basin consumptive use is economically efficient. A developing line of research [10, 51] addressing the water quality impacts of regional water quantity markets in California's Central Valley suggests that such markets could significantly reduce drainage water from irrigated fields and thus improve water quality.

OBJECTIVES AND SCOPE

This paper reports on work which builds on the optimization models of market transfers exemplified by Vaux and Howitt [48]. We seek to extend the previous optimization models in several respects. First, on the supply side, we provide hydrologic details in both the quantity and the quality dimensions for each of 20 regions (nodes) along the river's course. Major tributary inflows, diversion points, reservoirs, and hydroelectric power plants are explicitly represented as nodes in the model. Mineral water quality (salinity), important in the Colorado River Basin, is also represented at each node. The effect of inflows, evaporation, within-basin diversions, or exports at any node on water quantity and salinity at any downstream node is captured. Second, on the demand side, both offstream (irrigation, municipal, and thermal energy) and instream (hydropower and water quality) uses are represented by empirical marginal benefit functions. The combination of increased hydrologic and demand-side detail permits this model to capture more completely who might gain and who might lose from institutional changes permitting interstate and intersector water markets in the Basin.

The specific objective of this study is to develop a modeling framework for investigating the performance of alternative market institutions for river basins and applying it to study the potential for increasing beneficial use of Colorado River water resources. We use direct economic efficiency impacts from Basin and sub-Basin perspectives as the primary performance indicators. One important empirical objective is to estimate economic benefits of moving from existing allocations to those allowing transfers from expanded intrastate and interstate consumptive use markets.

Economic values in agricultural and municipal/industrial consumptive uses do not capture the full social value of Colorado River water resources. Previous work [5, 28] suggests that the marginal value of instream flows for hydropower production and damages from degraded water quality (salinity) in Basin water are significant relative to marginal values in traditional consumptive uses. A second primary objective to the analysis is to estimate the performance of consumptive use water markets relative to efficient allocations which include additional values in nonconsumptive uses. We hypothesize that allocations and net benefits resulting from consumptive use markets alone differ significantly from "first-best" allocations based on marginal values in all consumptive and nonconsumptive uses. (This study approaches estimation of first-best allocation of basin water resources by incorporation of major nonconsumptive use values. Inclusion of additional values of instream flows and reservoir levels for recreation and habitat, particularly for endangered species, would be desirable, but was beyond the scope of this study.) If allocations which include nonconsumptive use values do significantly differ from those based on consumptive use values alone, then the physical relationships between water use, instream flows, and water quality are important from an economic perspective.

As with all models, our approach is limited in the scope and detail of its coverage. We treat Colorado River Basin water resources as a closed system, without opportunities for supply augmentation. While this is a physically reasonable approximation, water exports from the Basin link users of Basin water with alternative supplies, particularly in California and Colorado. Opportunity costs of these alternative supplies are thus exogenous variables in the analysis. To this point, we have treated the complex but important new transfer demands (the Central Arizona and Central Utah Projects), and the existing entitlements for Denver and Las Vegas, as fixed requirements, subject only to existing Basin allocation priorities. We have not attempted modeling of secondary economic impacts of potential reallocations, nor do we treat hydrologic parameters in a stochastic fashion. Management of Colorado River water resources to meet noneconomic objectives [33] is not explicitly considered.

Because Basin reservoirs store approximately 4 years of average flows, we assume that water transfers would be largely motivated by average flow levels over periods of at least 10 years. We focus on the particular case of a naturalized flow level of 13.0 maf annually at Lee's Ferry, equivalent to one estimate [41] of the long-term mean.

MODEL DESCRIPTION

A nonlinear optimization model was developed to estimate impacts of alternative institutional scenarios, river flows, and economic demand levels. Termed the Colorado River Institutional Model (CRIM), it links river flows, salinity concentrations, and demand sectors across river locations. Annual consumptive use benefits, hydropower benefits, and costs and benefits of salt discharges are incorporated as integral model components. CRIM is formulated as a two-variable[2] optimization problem with the objective of maximization of net economic surplus, defined over selected economic sectors. Model solutions provide an annual estimate of economically efficient allocations, subject to physical and institutional constraints.

Fourteen nodes include sectors with economic demand for consumptive or nonconsumptive use, while an additional 6 nodes are used for significant sources or depletions of river water and salt loads, or for important geographical or institutional features. Uses of Basin water in these latter sectors are considered quasi-exogenous; shortfalls are imposed only under the present institutional allocation priorities. At each node, economic demand, supply, and request levels can be specified. The Basin is modeled as a single mainstem with all demand and supply sectors simple tributaries or diversions. Above the Colorado–Green River confluence the Green River is chosen as the mainstem. The mathematical programming model is constructed using the GAMS higher level language [3] and is solved using the MINOS optimization program [32]. Figure 1 shows a representation of the full Basin model

The empirical focus of this study is estimating impacts of alternative institutions on Colorado River water users from a Basin and sub-Basin perspective. Of particular interest are market institutions allowing intrastate or interstate water transfers based on economic values in several alternative uses. Estimated impacts of six alternative institutional scenarios are presented; the first group includes two scenarios which approximately reflect the existing institutional environment, while the second group of four scenarios includes water allocation institutions representing interstate transfers without and with consideration of instream water benefits.

Objective Function

The objective given a typical institutional scenario is to maximize net economic benefits from selected economic demands at nodes i and j:

$$\text{maximize} \sum_i \pi_i(w_i, x_i, z_i) + \sum_j \left(\pi_j(w_j, x_j) - c_j(w_j, x_j, \sigma_j)\right), \tag{1}$$

where $\pi_i(w_i, x_i, z_i)$ is the net benefit from Upper Basin diversions w_i, water exports x_i, and salt discharges z_i; $\pi_j(w_j, x_j)$ are Lower Basin net benefits from withdrawals and exports; and $c_j(w_j, x_j, \sigma_j)$ are Lower Basin salinity damages from

[2] Water diversions, including exports and hydropower use, and salt discharges are the control variables. Water flow and salt loads at each node are distinct, but interacting, state variables.

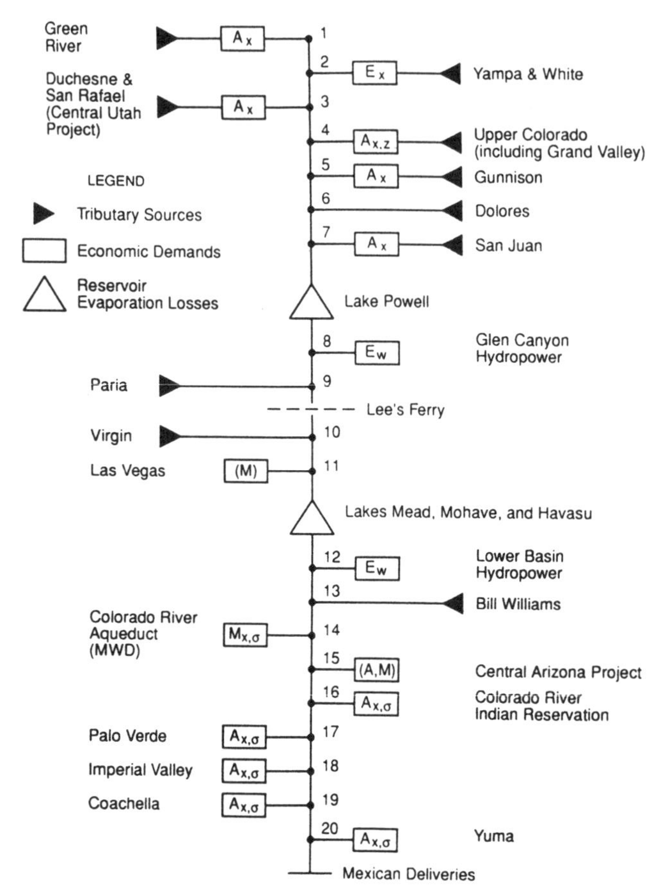

FIG. 1. Schematic representation of Colorado River Institutional Model.

salinity concentrations σ_j. For notational simplicity, only single economic demands at each node are shown in Eq. (1), although multiple demands at any node are allowed by the model. Net benefits and salinity damages in Eq. (1) are functions of withdrawals at the mainstem or its tributaries, with conveyance and treatment costs implicitly included in the objective function definition. Use of withdrawals (rather than consumptive use) simplifies incorporation of instream hydropower benefits. Net benefit functions (shown in Table I) are twice differentiable and convex in water withdrawals over the range of use considered in this study; their derivation is discussed below.

For the special case of allocations based on present institutional priorities, the objective is to minimize a weighted sum of shortages to different users, with the smallest weight given to metropolitan southern California, the most junior user of Colorado River water.

TABLE I
Net Benefit Functions and Water Use Constraints for Diversions at the Colorado–Green River Mainstem

	Existing demand			
	Consumptive use			
Region[a]	Max (kaf)	Min (kaf)	Efficiency[b] (%)	Economic benefit of withdrawals[c]
Grand Valley	163	109	0.50	$\pi = -11{,}800 + 86.3w - 0.190w^2 + 0.799z - 0.0742z^2 + 0.133zw$
Grand Valley[d]	163	109	0.50	$\pi = -10{,}400 + 76.0w - 0.110w^2$ $z = -473.5 + 7.9w - 0.0195w^2$
Lower Gunnison	396	265	0.50	$\pi = -25{,}200 + 76.0w - 0.0452w^2$
Fontenelle and Flaming Gorge	349	234	0.53	$\pi = -22{,}200 + 81.0w - 0.0584w^2$
Duchesne and San Rafael Rivers, including Central Utah Project	647	433	0.57	$\pi = -41{,}100 + 86.2w - 0.0357w^2$
Navajo Reservoir and Lower San Juan	747	500	0.63	$\pi = -47{,}500 + 96.1w - 0.0269w^2$
Thermal energy, Yampa River	13		1.00	$\pi = 2830w - 109w^2$
Upper Basin hydro			0	$\pi = 13w$
Lower Basin hydro			0	$\pi = 31w$
MWD municipal use	1300		1.00	$\pi = 1088w - 0.306w^2 - 497w(\sigma - 415)/1300$
Imperial Valley	2640	1302	1.00	$\pi_{800} = 94{,}600 + 12.7w - 3.67 \times 10^5(w - w_0)^{-0.43}$ $\pi_{1100} = 63{,}700 + 12.7w - 5.24 \times 10^5(w - w_0)^{-0.59}$
Colorado River Indian Reservation	398	187	0.60	$\pi_{800} = 24{,}400 + 7.6w - 0.53 \times 10^5(w - w_0)^{-0.43}$ $\pi_{1100} \doteq 16{,}500 + 7.6w - 0.61 \times 10^5(w - w_0)^{-0.59}$
Palo Verde	387	198	0.50	$\pi_{800} = 26{,}000 + 6.4w - 0.62 \times 10^5(w - w_0)^{-0.43}$ $\pi_{1100} = 17{,}500 + 6.4w - 0.75 \times 10^5(w - w_0)^{-0.59}$
Coachella	344	155	1.00	$\pi_{800} = 12{,}200 + 12.7w - 0.20 \times 10^5(w - w_0)^{-0.43}$ $\pi_{1100} = 8{,}200 + 12.7w - 0.20 \times 10^5(w - w_0)^{-0.59}$
Yuma	662	410	0.62	$\pi_{800} = 40{,}700 + 7.9w - 1.08 \times 10^5(w - w_0)^{-0.43}$ $\pi_{1100} = 27{,}400 + 7.9w - 1.34 \times 10^5(w - w_0)^{-0.59}$

[a] For crop irrigation unless otherwise specified.
[b] Defined as water depletion to the Basin divided by withdrawal at the mainstem.
[c] Net benefit π (in \$ thousand) as a function of withdrawals w (kaf), maximum withdrawals w_0, salinity discharge z (ktons), and salinity concentrations σ (mg/liter). Subscripts on π refer to alternative salinity concentrations in mg/liter.
[d] Net benefits and salinity discharges when salinity damages are not included in objective function.

Physical Constraints

Total flows and salinity concentrations are functions of withdrawals, exports, and salt discharges, the decision variables. Flows are found by the mass balance constraint

$$Q_i = Q_{i-1} - V_i - \eta_i w_i - x_i + Q_i^a + Q_i^f, \qquad (2)$$

where $Q_i \geq 0$ is the total flow leaving node i, $V_i \geq 0$ is the evaporation loss between i and $i - l$, $Q_i^a <> 0$ is a net flow adjustment for inflows and diversions (including reservoir releases from storage) between i and $i - 1$, $Q_i^f \geq 0$ is the tributary flow into node i, and $\eta_i w_i + x_i$ is the total consumptive use at i. Return flows from node i are thus given by $(1 - \eta_i)w_i$, where η_i is the proportion consumptively used. Technical efficiency of water exports x_i from the Basin are defined to be 1, since there are no Basin return flows. Diversions and exports are constrained by $w_i \leq \overline{w}_i$ and $x_i \leq \bar{x}_i$ with $\overline{w}_i$ and $\bar{x}_i$ determined by factors including conveyance capacities and irrigated land development. Diversions for hydropower generation are $w_i \equiv Q_i$, the river flow leaving nodes at which plants are located.

The mass balance constraint for salinity accounting is given by

$$s_i = s_{i-1} + s_i^n + z_i - s_{i-1}(x_i/Q_i), \tag{3}$$

where $s_i \geq 0$ is the total salt load leaving node i, $s_i^n <> 0$ is the net salt addition (reservoirs may act as salt sinks) from sources between $i - 1$ and i, $z_i > 0$ are the salt discharges from modeled economic activities at node i, and $s_{i-1}(x_i/Q_i) \geq 0$ are salt exports arising from out of Basin diversions. Since salinity exports are nonlinear in river flow Q_i, constraint Eq. (3) is nonlinear for water-exporting nodes. Salinity concentration at any node i is $\sigma_i = s_i/Q_i$.

Minimum instream flows $Q_i \geq \overline{Q}_i$ are required, with the level $\overline{Q}_i$ altered to simulate particular modeled institutions. $\overline{Q}_8 = 8.23$ maf, the objective annual release at Glen Canyon, is used in two scenarios to bar interbasin transfers. In all cases presented here, $\overline{Q}_{20} = 1.515$ maf, assuring full treaty deliveries to Mexico, while minimum flows at additional nodes are not specified.

Derivation of Benefit Functions

The net economic benefit functions for specific uses of Basin water resources employed in the Basin model CRIM are estimated using standard techniques (as reported by Gibbons [13]). Residual imputation using linear programming models, avoided cost, and econometric estimation of consumer demand are the primary methods employed for valuing water use in agriculture, energy production, and municipal use, respectively. Marginal valuation of water use is implicit in the use of linear programming to value water use in agriculture. In contrast, there is concern that transfers of agricultural water might result in retirement of a broad class of crops. While beyond this paper's scope, the model presented here could be reformulated in terms of average (in contrast to marginal) value products of agricultural water use.

Our approach to measuring associated costs of water use is consistent with water transfers based on average flows over periods of at least 10 years. Modeled cost of production in agriculture include annual operating plus capital costs; the resulting derived water demands are long-run estimates. Valuation of hydropower assumes that capital costs of alternative generation sources are sunk costs. This is valid at present, but may significantly underestimate avoided costs in the future. Capital costs of refurbishing dam power plants are also not considered. Water demand for fossil fuel-based energy production is derived using capital costs of water saving technologies, as these technologies have not yet been implemented in the Basin. Capital costs of municipal conveyance and treatment facilities are not considered, as they have generally already been incurred; only operations and maintenance costs of water treatment and delivery are included in municipal water-derived demand estimates. Economic surplus measures are explicitly developed for the Grand Valley (Colorado), the Imperial Valley (California), and the southern California municipalities, cooling water for fossil fuel energy resources (coal-fired electric generation, coal gasification, and oil shale) in northwest Colorado, and hydroelectric generation at Glen Canyon, Hoover, Davis, and Parker Dams. Estimated net benefit functions for 1990 (including damages from salinity) are presented in Table I. Detailed derivation of the net benefit functions used here was reported in the senior author's doctoral dissertation [2]. The level of all other Basin water demands and requests is derived from water depletion figures developed by the U.S. Bureau of Reclamation (USBR) [43] for use in their Colorado River Simulation System. All values are expressed in 1989 dollars.

Net benefit functions for Imperial Valley (California) water use were developed for a range of water deliveries and at two salinity concentrations using linear programming models updated from those reported by Gardner and Young [11]. Thirteen crops were modeled on two soil types with alternative irrigation schedules and technologies. Fifteen typical double-cropping activities were included. Double-cropping was limited to 29% of total land acreage. Crop prices and production costs from Imperial County Agricultural Commissioner [22] and Imperial County Cooperative Extension [23], respectively, were used. Continuous derived (inverse) demand functions $p(w) = a(w - w_0)^b$ were estimated by regression analysis ($r^2 = 0.95$ and 0.92 for 800 and 1100 mg/liter models, respectively) from modeled discrete profit differences in the linear programming models (w_0 is the minimum water determined by acreage constraints). Net benefits as a function of water use are of the form

$$\pi(w) = \pi_0 + \alpha(w - w_0)^{\beta}, \tag{4}$$

where $\alpha = a/(b + 1) < 0$ and $\beta = b + 1 < 0$ and π_0 is the constant of integration given by irrigator profit at $w = w_0$. Land rents included as costs in the linear programming models are interpreted as returns to water by addition of a term linear in water withdrawals (see Table I) in Eq. (4). Salinity damages $c(w, \sigma)$ are defined as the difference in profits $\pi_{800}(w)$ and $\pi_{1100}(w)$ at salinity concentrations of 800 and 1100 mg/liter, respectively. Agricultural salinity damages are linear in salinity level by assumption ($\partial c/\partial \sigma = constant$), but estimated marginal damages were found to be decreasing in water use ($\partial c/\partial w < 0$).

A linear programming model of Grand Valley (Colorado) irrigated agriculture adapted from Gardner and Young [12] was used to estimate irrigator profit as a joint function of consumptive water use and salt discharge.[3] Four irrigated crops (harvested hay, irrigated pasture, corn for grain, and pinto bean) were included with nine irrigation technology and labor combinations to model a range of alternative consumptive use levels and salt discharges. Production costs from enterprise budgets [9], yields and crop prices [7], and installed capital improvements (e.g., ditch lining) for salinity control [47] were used. Flexibility constraints limited crop acreages to the minimum reported acreage in 1976–1988 [46]. Regres-

[3]Deep percolation of irrigation water into an underlying salt-bearing shale in the Grand Valley (and numerous other irrigated regions) dissolves various salts which then enter the river with irrigation return flows. Salinity production can thus be decreased by reducing deep percolation through increased irrigation efficiency, crop switching, or irrigated land retirement.

sion analysis was used to fit a second-order polynomial,

$$\pi(w, z) = \beta_0 + \beta_1 w + \beta_2 w^2 + \beta_3 z + \beta_4 z^2 + \beta_5 wz, \tag{5}$$

($r^2 = 0.99$) to point estimates of irrigator profits at varying water diversion levels y and salt discharge z obtained from the linear program to give the continuous net benefit function $\pi(w, z)$. For the case where salt discharges are costless by assumption, a quadratic net benefit function $\pi(w)$ and a quadratic salt discharge function $z(w)$ were estimated ($r^2 = 0.99$ for each) from point solutions of the linear program with unconstrained salt discharges.

Benefits of hydropower production were defined as the costs avoided by utilities in substituting hydropower for production at existing thermal facilities. Avoided cost was calculated as the capacity-weighted average costs of the most costly 50% of total baseload capacity, calculated separately for Upper and Lower Basin states. The use of such a broad average acknowledges operational constraints imposed by transmission line capacity and other factors. Adjustment was made for operation and maintenance costs at hydropower plants, plus differences in transmission costs from hydropower sites and alternative sources to demand centers, giving avoided costs of 44.2 and 26.0 mills/kwh in Lower and Upper Basins, respectively. Peaking power values (estimated at 57 mills/kwh [5]) are not used due to uncertainties in future reservoir operation rules, particularly at Glen Canyon.[4] Model results were subsequently found to be insensitive to greater hydropower benefits. Modeled hydropower production [42], including impacts of variable heads, was used to estimate average annual Upper and Lower Basin energy production as a function of annual releases from Glen Canyon and Hoover dams, giving $E_U = 93 + 0.616Q_U$ ($r^2 = 0.99$) and $E_L = -14 + 0.724\ Q_L$ ($r^2 = 0.99$), where E_U and E_L are Upper and Lower Basin energy production in million kwh, and Q_U and Q_L are the total volume released from Glen Canyon and Hoover dams, respectively, in kaf.

Economic demand functions for cooling water by coal-fired electric generating plants, the dominant Basin consumptive use of water for energy production, are based on the cost of alternative cooling technologies. Costs of hybrid wet/dry cooling systems developed by Abbey [1] were updated to estimate net benefit functions for standard wet cooling systems.

The net benefit function for the southern California urban area was developed from econometric estimates of household water demand functions based on a direct survey of 24 regional water utilities. Benefit functions shown in Table I are net of conveyance and treatment costs.

Municipal damages in southern California from Colorado River salinity are based on our calculations adapted from data presented by the USBR [29]. Calculations in that document leading to a damage estimate of \$0.74 per mg/liter included several speculative impacts which we judged to substantially overestimate actual damages. Our annual household damage estimate of \$0.263 per mg/liter total dissolved solids is well within the range of \$0.14–0.31 per mg/liter given by Kleinman and Brown [26]. Recent evidence by Ragan [37] suggests that substitu-

[4]Avoided costs assuming a 50% mix of base and peaking power production (roughly consistent with interim operating rules for Glen Canyon Dam [44]) give Lower and Upper Basin avoided costs of 50.5 and 41.4 mills/kwh, respectively.

tion of corrosion resistant materials in piping and household appliances may cause even our estimates to overstate actual damages. We scaled total damages to zero at 415 mg/liter, the salinity of alternative California supplies.

Application to the Basin Model

The estimated derived demand functions for the Grand Valley and Imperial Valley were extrapolated to additional major agricultural sectors in the Upper and Lower Basins, respectively, by scaling of the respective net benefit functions based on water withdrawals and consumptive use reported in USBR [43]. These agricultural regions together account for 6.8 million acre-feet of the estimated total (United States) Colorado River Basin consumptive use (not including reservoir evaporation) of 11.4 maf in 1990. The high level of aggregation in estimation of agricultural water demands is undesirable, but resource and time constraints precluded more detailed study. Consumptive use requests not included as economic demand sectors, including the Central Arizona Project (0.85 maf was used for this study) and exports to Colorado's Front Range Range (0.69 maf), are met according to existing institutional priorities. The resulting allocations are then fixed as alternative institutional scenarios are considered; these quasi-exogenous requests total 2.2 maf. Nonconsumptive uses other than hydropower production and salinity reduction are not formally considered.

Hydrologic Considerations

The aggregate Basin water supply of 13.0 maf at Lee's Ferry is defined by the long-term mean flow level reported by Stockton [41]. The long-term flow estimates are reconstructed from tree-ring records reaching back to year 1560. An alternative interpretation of the flow level is a climate change [14]-induced reduction in mean flows (see, for example, Nash and Gleick [35]).[5] Spatial distribution of flows are determined from the historical flow sequence for the years 1960–1969, corresponding to the lower decile flow from the 1906–1983 historical record.[6]

During the flow period specified above, Basin reservoirs are expected to be drawn down to meet shortfalls in filling requests for Basin water. Simulations of Basin reservoir operations reported by the USBR [42] were used to estimate an additional supply of 0.2 maf drawn from storage.

EVALUATIVE APPROACH AND RESULTS

Alternative Institutional Scenarios

Six alternative institutional scenarios are evaluated. The water use sectors employed in each are identified in Table II. Scenario 1 approximates the allocation of Colorado River water under the existing interstate and intrastate priorities

[5]There is considerable uncertainty concerning long-term flow levels in the basin. The level chosen here is best considered as representative of conditions which are certain to occur, but with unknown frequency.

[6]The lower decile flow from the historic record is used, defined as that with 10% of all 10-year flows lower than this flow. For comparison, median flows at Lee's Ferry from the historic record are 14.5 maf.

TABLE II
Summary of Sectors Used in Alternative Institutional Scenarios

	Scenario number					
	Group 1		Group 2			
Demand sector	1	2	3	4	5	6
Grand Valley agriculture	P	X	X	X	X	X
Other Upper Basin agriculture	P	X	X	X	X	X
Upper Basin energy	P	X	X	X	X	X
Other Upper Basin uses	P					
Upper Basin hydropower				X		X
Lower Basin hydropower				X		X
Imperial Valley agriculture:						
Water use	P	X	X	X	X	X
Salinity damage					X	X
Other California agriculture	P		X	X	X	X
Salinity damage					X	X
MWD service area						
Water use	P	X	X	X	X	X
Salinity damage					X	X
Central Arizona Project	P					

Note. Key to inclusion of sectors in objective functions. P, water right under existing institutional priorities (1922 Colorado River Compact, 1948 Upper Basin Compact, and Central Arizona Project enabling legislation). X, Economic net benefit function included.

known as the "Law of the River." This complex institutional setting, characteristic of an expansionary water economy, includes state water law, federal legislation, two interstate compacts, a treaty with Mexico, and several court decisions. The second scenario allows unrestricted intrastate transfers based on water values in consumptive uses.[7] Present water allocation institutions of Basin states allow varying levels of intrastate water marketing [34]; the emergency California water bank [30] and recent federal legislation [50] reducing restrictions on marketing water from federal projects may expand these intrastate markets.

Scenario 3 considers unrestricted interstate (and intrastate) transfers based on consumptive use values alone. While interstate marketing is not presently a reality in the Basin, formal proposals for Upper to Lower Basin transfers [15, 49] and establishment of an interstate water bank [8] have been put forth, and marketing of Indian reserved water rights has been discussed [39]. Scenarios 2 and 3 are used to address our first empirical objective, estimating the extent of benefits and transfers under expanded intrastate and interstate use markets.

The remaining scenarios allow interstate water transfers based on alternative consumptive and nonconsumptive use values. Scenarios 4 and 5 include consumptive use values and economic values for hydropower production and salinity reduction, respectively. Scenario 6 includes all identified economic values and allows unrestricted transfers. While institutions which link Basin nonconsumptive

[7]Property rights to consumptive use are thus implicitly assumed; this was an important stumbling block in the agreement between the Metropolitan Water District (MWD) and the Imperial Irrigation District in California for transfer of water salvaged through MWD-financial efficiency improvements [39]. Third-party impacts from return flows (except as they influence salt loads) are not considered in this study.

use values with consumptive uses have yet to be proposed, economically efficient water allocations clearly require this linkage. Scenarios 4 and 5 are chosen to illustrate the significance of considering only a single nonconsumptive use value, while scenario 6 is the most comprehensive institutional scenario considered in this research and is used to provide an estimate of efficient water allocation for the Basin. Allocations modeled under scenarios 3–6 are used to test our hypothesis that allocations from consumptive use markets differ significantly from allocations with nonconsumptive use values included.

Scenario 6 is theoretically equivalent to a comprehensive water resources bank (with exclusion of free riders) accepting bids from hydropower beneficiaries and consumptive water users, including those damaged by salinity levels. Given the multiple beneficiaries of increased streamflow for hydropower and salinity dilution (and hence, difficulty in excluding free riders) a market process would encounter difficulties in fully internalizing all nonconsumptive use values. Because Basin hydropower facilities are federally owned, however, a decision to maximize hydropower rents (scenario 4) could in principle lead to water use patterns similar to those under economically efficient allocations. If such policies are not pursued because they would substantially reduce upstream irrigation, this suggests that the primary barrier in moving toward efficient allocations may lie not in the limitations of market processes, but in concerns for regional equity.

Scenario 1: *No transfers*. Using flows of 12.99 maf/year at Lee's Ferry, most present requests for Basin water can be fully satisfied. Under existing institutions, the annual Basin shortfall of 0.26 maf found with this flow level is borne by southern California municipal users served by the Metropolitan Water District (MWD), the junior rightholders in the Lower Basin. Upper Basin consumptive uses are not constrained by the Lee's Ferry delivery obligation of 8.25 maf per year. The full Basin water budget under the existing "Law of the River" is shown in Table III.

Scenario 2: *Impacts of intrastate water transfers*. Removing barriers to water marketing within California would likely result in significant transfers between southern California municipal users and California agricultural water rights holders in the Imperial, Coachella, and Palo Verde Valleys. Model results indicate that with present demand and long-term flow levels, these municipal users would benefit from water transfers up to the Colorado River Aqueduct capacity if the marginal value (i.e., foregone benefits in alternative uses) at the River is less than $300/af. Imperial Valley irrigators could benefit from such transactions at marginal values as low as $20/af. [By way of comparison, Vaux and Howitt [48] found marginal values of $210/af in southern California municipal uses and $45/af in Imperial Valley agriculture in a 1980 scenario (their values adjusted to 1989 constant dollars using the GNP deflator). For 1995 they estimated values of $360 and $60/af, respectively. Their study predicted very large transfers from the Imperial Valley (over 1 maf) in both time periods, accounting for the relatively large marginal values. Such large transfers would not be possible, we believe, without costly and environmentally controversial additions to the capacity of the Colorado River Aqueduct.]

The total benefit to southern California municipal users of eliminating the 260 thousand acre-foot shortfall found under existing priorities and the long-term flow

TABLE III
Scenario 1 (Allocation by Existing Compact and State Priorities): Water Mass Balance Summary in Thousand Acre-Feet per Year for an Annual Naturalized Flow of 14.3 Million Acre-Feet

Demand regions	Nonmodeled flow additions	Nonmodeled flow adjustments[a]	Nonmodeled consumptive use	Modeled consumptive use	Modeled evaporation losses	Modeled mainstem flow[b]
Upper Green River	1,814		92	349		1373
Yampa and White Rivers	1,932		165	13		3127
Duchesne and San Rafael Rivers	903		124	647		3259
Upper Colorado River	3,047		970	163		5173
Gunnison River	1,858	381	84	396		6933
Dolores River	740		48			7625
San Juan River	1,740			747		8618
Glen Canyon Dam		411	51		575	8404
Paria River	241					8646
Virgin River	142	345	49			9083
Las Vegas		251	152			9182
Hoover–Davis–Parker Dams		−164			1123	7895
Bill Williams River	58	160				8113
Colorado River Aqueduct		62		1037		7138
Central Arizona Project		−120		850		6168
Colorado River Indian Reservation			333	398		5437
Palo Verde Irrigation District				423		5014
Imperial Irrigation District		271		2640		2646
Coachella Irrigation District				344		2302
Yuma			125	662		1515
Delivery to Mexico			1515			(0)
Upper Basin total	13,067[c]		1534	2315		
Lower Basin total	1,005[c]		659	6354		
Full Basin total	14,072[c]		3708	8669	1698	

[a] Flow adjustments include additions from minor tributary flows, channel gains and losses, net releases from storage, and certain minor depletions.
[b] Defined as the flow leaving each node after accounting for all consumptive use, flow additions, and flow adjustments.
[c] Totals include flow additions plus net flow adjustments.

TABLE IV
Annual Consumptive Water Use, Total Benefits, and Marginal Values, by Institutional Scenario

Institutional scenario	Upper Basin consumptive use (maf)	Lower Basin consumptive use (maf)	Annual net benefits ($million) Total	Consumptive uses	Hydropower generation	Salinity damages	Difference from Scenario 1 ($million) Total	Consumptive uses	Marginal opportunity costs of water and salinity: Upper Basin water ($/af)	Lower Basin water ($/af)	Upper Basin salt discharges ($/ton)
1. Allocation by priority	2.315	6.354	1365	1076	389	−100	na	na	na	na	na
2. Use values, intrastate	2.315	6.354	1434	1169	389	−124	69	93	1	20	na
3. Use values only	2.161	6.501	1453	1170	396	−113	88	94	18	19	na
4. Use and hydropower values	1.708	6.617	1490	1158	417	−85	125	82	46	0	na
5. Use and salinity values	1.629	6.546	1500	1151	421	−72	135	75	51	53	−56
6. Use, hydropower, and salinity values	1.555	6.561	1503	1148	424	−68	138	72	96	52	−56

Note. Institutional scenarios are explicitly defined, by number, in Table II. Upper Basin marginal values are calculated at the Grand Valley, Colorado; Lower Basin marginal water values are at the Colorado River Aqueduct; and marginal water values include the benefits of salinity dilution, where applicable. na, not applicable.

level is nearly $100 million annually (see Table IV). Distributed over the estimated 1990 MWD service area population of 15 million, annual benefits (not including costs paid to Imperial Valley irrigators, or salinity damages) are about $7 per capita.

Impacts on salinity levels and hydropower generation from transfers between California users of Colorado River water would be small. The dominant salinity impact would be increased damages to southern California municipal water users from salinity levels in Colorado River water higher than those in imported supplies from northern California. In principle this impact would be included in decisions by MWD on importing Colorado River water.

Impacts of Interstate Water Transfers

Scenario 3: *Consumptive use values only.* With interstate transfers which include only consumptive use values in the allocation decision, Upper Basin use declines by 150 kaf over the full request level, with the reductions distributed across agricultural areas whose existing requests total about 2300 kaf. Lower Basin agricultural uses decrease by 110 kaf. Upper and Lower Basin marginal water values are $18 and $19/af, respectively; the difference reflects evaporation and conveyance losses. The marginal value of delivered water to southern California urban uses is much higher, but does not affect values at the river because the constraint reflecting the Colorado River Aqueduct capacity is binding.

The additional increase in net consumptive use benefits over that achieved with scenario 2 allowing intrastate transfers is only 1%. This result follows from the small difference in marginal water values in Upper and Lower Basin agricultural uses.[8] Total benefits increase by 22% over that of scenario 2, however, as a result of incidental increases in hydropower production and reduced salinity damages through dilution. Foregone agricultural income is reduced 14% compared to that of scenario 2 by distributing use reductions between Upper and Lower Basins. Total use and benefits and marginal water values are shown in Table V.

While impacts on Basin-wide net income are minimal (considering consumptive use values only), regions impacted by transfers would change considerably; the actual pattern of any transfer would likely be based on factors such as legal constraints, transaction costs, and the terms offered by existing agricultural users.

Scenarios 4–6: *Consumptive and nonconsumptive use values.* The magnitude of nonconsumptive use values in Upper to Lower Basin transfers suggests that inclusion of these values in a water market would result in significant Upper-to-Lower Basin transfers. Model results confirm that large transfers are in fact economic efficiency improvements when more complete sets of values are considered. Inclusion of hydropower values (scenario 4) results in reductions in Upper Basin use of 610 kaf, or 26% of modeled requests, while inclusion of salinity values (scenario 5) gives 30% (690 kaf) reductions in Upper Basin modeled requests. These results imply that down-river hydropower benefits and reductions in salinity damages by dilution each exceed marginal benefits in over one-quarter of existing Upper Basin irrigation uses. Because significant salt discharges arise from Upper

[8] Marginal opportunity costs for consumptive use in the Imperial and Coachella Irrigation Districts are, a priori, substantially lower than those in surrounding regions because return flows drain to the Salton Sea and cannot be reused.

TABLE V
Scenario 3 (Interstate Consumptive Use Market): Summary of River Flows, Consumptive Use Allocations, and Marginal Opportunity Costs

Demand regions	Mainstem flow (kaf)	Mainstem salinity load (ktons)	Salinity concentration (mg/liter)	Consumptive use (kaf)	Marginal opportunity cost of consumptive use ($/af)
Upper Green River	1397	1113	586	325	18
Yampa and White Rivers	3151	1859	434	13	18
Duchesne and San Rafael Rivers	3323	2619	580	608	18
Upper Colorado River	5250	3302	463	150	18
Gunnison River	7035	4779	500	371	18
San Juan River	8772	6664	559	695	18
Glen Canyon Dam	8554	6664	573		18
Hoover–Davis–Parker Dams	8042	7936	726		18
Colorado River Aqueduct	7022	6697	701	1300	19
Central Arizona Project	6052	5872	714	850	19
Colorado River Indian Reservation	5321	5872	812	398	19
Palo Verde Irrigation District	4898	5872	882	423	19
Imperial Irrigation District	2618	2974	836	2552	19
Coachella Irrigation District	2302	2616	836	316	19
Yuma	1515	2616	1270	662	19

Note. Consumptive use, flow, and salinity load are annual values. Based on model results allowing interstate transfers for consumptive uses only.

Basin irrigated agriculture outside the Grand Valley and are not included here, the estimated level of water transfer quantitatives and benefits represents an underestimate of those that could occur with an interstate market including salinity reduction values.

Scenarios 4 *and* 5. Economic surplus changes for scenario 4, measured relative to those of scenario 2 (intrastate consumptive use transfers only), give a surplus loss of $11 million for consumptive uses, but surplus gains of $28 million in hydropower production and $39 million in salinity reduction, for a net gain of $56 million. This gives average net benefits of $92/af for the transfer of 610 kaf. The marginal foregone benefits to downstream users (including hydropower) of Upper Basin consumptive use rises to $46/af, the value of hydropower production. Since Lower Basin consumptive use requests are satisfied in full and water quality is not considered in this optimization, the marginal opportunity cost of Lower Basin water is zero. Comparable figures for scenario 5 are a surplus loss of $18 million for consumptive uses and surplus gains of $32 million in hydropower production and $52 million for salinity reduction, a net gain of $66 million. The average net benefit for the 690-kaf transfer is $95/af. Marginal foregone benefits of Upper and Lower Basin consumptive use are $51 and $53/af, respectively, the value of water for dilution. The marginal damage from salt discharge is $56/ton. Grand Valley salt discharge is 0.92 tons/af of consumptive use (also $51/af), in contrast to 1.81 tons/af when water use and discharge is unconstrained.

Scenario 6. The most comprehensive scenario presented in this paper includes values for consumptive uses, hydropower production, and salinity reduction. The level of transfers is 760 kaf, or 33% of the included agricultural demand, the upper

TABLE VI
Scenario 6 (Allocations Maximizing Consumptive and Nonconsumptive Use Benefits):
Summary of River Flows, Consumptive Use Allocations, and Marginal Opportunity Costs

Demand sectors	Mainstem flow (kaf)	Mainstem salinity load (ktons)	Salinity concentration (mg/liter)	Consumptive use (kaf)	Marginal opportunity cost of consumptive use ($/af)	Marginal opportunity cost of salt discharges ($/ton)
Upper Green River	1488	1113	550	234	96	−55
Yampa and White Rivers	3243	1859	422	13	96	−56
Duchesne and San Rafael Rivers	3588	2619	537	433	96	−56
Upper Colorado River	5556	3129	414	109	96	−56
Gunnison River	7446	4606	455	265	96	−56
San Juan River	9378	6491	509	500	96	−56
Glen Canyon Dam	9144	6491	522		96	−56
Hoover–Davis–Parker Dams	8621	7763	662		83	−56
Colorado River Aqueduct	7601	6629	641	1300	52	−28
Central Arizona Project	6631	5876	652	850	29	−32
Colorado River Indian Reservation	5928	5876	729	370	32	−29
Palo Verde Irrigation District	5533	5876	781	395	30	−26
Imperial Irrigation District	3165	3204	745	2640	26	−15
Coachella Irrigation District	2821	2856	745	344	15	−12
Yuma	2034	2856	1033	662	12	

Note. Consumptive use, flow, and salinity load are annual values. Based on model results allowing interstate transfers, and including benefits of hydropower production and salinity reduction.

limit allowed by the model.[9] Net benefits increase by $69 million, or $91/af of transferred Upper Basin water over those with just intrastate transfers. The marginal opportunity cost of Upper Basin water is $96/af, of which $50/af is the value of salinity dilution. The estimated marginal damage of salt discharges is again $56/ton. Table VI shows consumptive use and marginal opportunity costs of consumptive use and salt discharges by economic demand region.

Upper Basin transfers again permit deliveries to Mexico in excess of treaty obligations. Because the Mexican delivery constraint is not binding, Lower Basin water is valued only for salinity dilution, confirming the work by Lee [28]. Basin-wide consumptive use is limited not by physical shortfalls, but by the economic benefits of downstream hydropower production and salinity reductions.

CONCLUSIONS AND POLICY IMPLICATIONS

Under one estimate [41] of long-term water supply in the Basin, present consumptive uses, now including Central Arizona Project allotments, cannot be fully satisfied. Water allocations by existing institutions under this supply shortfall do not maximize beneficial consumptive use of Basin water, as shown in Table IV, and significant efficiency gains in consumptive uses are possible under either intrastate or interstate water use markets. Because differences in marginal consumptive use values are modest between Upper and Lower Basin agricultural users, there is only a negligible (1%) difference in benefits from interstate transfers over intrastate transfers if economic values in traditional consumptive uses alone are considered. Although consumptive use efficiency gains are almost identical in either case, an external economy of interstate over intrastate markets is a 26% increase in nonconsumptive use benefits resulting from increased hydropower production and reduced salinity concentration.

Current efforts by California and Las Vegas to establish interstate markets for Colorado River water are not inconsistent with these conclusions. Indirect impacts within California of transfers from agricultural uses would be reduced if interstate transfers satisfied a portion of the Metropolitan Water District's shortfall. While Las Vegas does not presently suffer any shortfall, its future demands cannot readily be met from transfers by existing state agricultural users. These new demands, in addition to tribal interest in marketing reserved water rights and continued drought, could be catalysts for the introduction of limited interstate markets. Estimating future benefits of such markets is an area for further research.

Proposed interstate markets which exclude instream flow values achieve efficiency gains considerably smaller than those possible with allocations based on the full set of consumptive and nonconsumptive use values. This supports our hypothesis that interstate consumptive use markets alone would not efficiently allocate Colorado River Basin water resources. Although consumptive use markets are a step toward efficient resource allocation, we estimate that intrastate and interstate consumptive use markets generate only 50 and 64%, respectively, of the total efficiency gains possible with optimal allocations. The maximum economic benefit

[9]Results of this full economic optimization are similar to those found above with inclusion of either hydropower or salinity damage values because the level of transfers from Upper Basin demand sectors is constrained in the model 33% of requests. The limit is imposed because the estimated Upper Basin demand functions are not well defined past this level of reduction.

achievable with optimal allocation of Colorado River Basin water, relative to allocations by existing priorities, is estimated at $140 million annually given a Basin consumptive use shortfall of 260 kaf. This is composed of gains of $74 million for consumptive uses, $35 million for hydropower generation, and a reduction in salinity damages of $31 million. The average net benefit of the reallocation, almost entirely from reduced Upper Basin irrigation use, is $185/af. Mexican users would also benefit from increased water deliveries and salinity reductions, although quantitative estimates of these benefits are not attempted here.

Inclusion of nonconsumptive use values in Basin water allocation institutions is problematic, however, on technical grounds and for reasons of regional equity. Because benefits of increased instream flows and salinity reduction are nonrival, achieving efficient allocations through market mechanisms would require an institution such as a comprehensive water resources bank to sum bids from the common beneficiaries of reductions in Upper Basin consumptive use and salt discharges. Such a water resource bank would likely not be sufficient, however, since benefits are largely nonexclusive and free riders could not be eliminated. Finally, because optimal allocations imply very large reductions in Upper Basin consumptive use (in excess of Lower Basin consumptive use shortfalls), they are likely to be viewed as regionally inequitable.

While interstate consumptive use markets do not maximize economic benefits of Colorado River water resources, they could potentially provide important benefits. Opportunity costs faced by Upper Basin water users would rise from near zero under existing conditions to more accurately reflect the basinwide impacts of consumptive use. An external benefit of higher costs and reduced use in existing Upper Basin uses would be greater hydropower production and reduced salinity concentrations. It is not clear, however, that consumptive use benefits with interstate markets would be sufficient to overcome additional transactions costs. Because economic benefits of interstate over intrastate markets are presently almost exclusively in nonconsumptive uses, interstate transfers would likely be the result of future growth in Las Vegas or constraints on transfers within California. Apart from Las Vegas' limited future demands, and in the absence of intrastate water marketing constraints, significant private incentives for interstate over intrastate transfers might occur only if market institutions including both consumptive and nonconsumptive use values develop.

REFERENCES

1. D. Abbey, Energy production and water resources in the Colorado River Basin, *Natur. Resour. J.* **19**, 275–314 (1979).
2. J. F. Booker, Economic allocation of Colorado River water: Integrating quantity, quality, and instream use values, Unpublished Ph.D. dissertation, Department of Agricultural and Resource Economics, Colorado State University, Fort Collins (1990).
3. A. Brooke, D. Kendrick, and A. Meeraus, "GAMS: A User's Guide," Scientific Press, San Francisco (1988).
4. T. C. Brown, B. L. Harding, and W. B. Lord, Consumptive use of streamflow increases in the Colorado River Basin, *Water Resour. Bull.* **24**, 801–814 (1988).
5. T. C. Brown, B. L. Harding, and E. A. Payton, Marginal economic value of streamflow from upland watersheds in the Colorado River Basin, *Water Resour. Res.* **26**, 2845–2860 (1990).
6. B. G. Colby, Enhancing instream flow benefits in an era of water marketing. *Water Resour. Res.* **26**, 1113–1120 (1990).
7. Colorado Department of Agriculture, "Colorado Agricultural Statistics," Denver (annual, 1978–1989).

8. Colorado River Board of California, "Conceptual Approach for Reaching Basin States Agreement on the Interim Operation of Colorado System Reservoirs, California's Use of Colorado River Water Above Its Basic Apportionment, and Implementation of an Interstate Water Bank," Proposal to the Colorado River Basin States Board (1991).
9. N. L. Dalstead *et al.*, "Selected 1986 Crop Enterprise Budgets for Colorado," Department of Agriculture and Resource Economics Information Report IR: 87-6, Colorado State University (1987).
10. A. Dinar and J. Letey, Agricultural water marketing, allocative efficiency, and drainage reduction, *J. Environ. Econom. Management* **20**, 210–223 (1991).
11. R. L. Gardner and R. A. Young, Economic evaluation of the Colorado River Basin Salinity Control Program, *West. J. Agr. Econom.* **10**, 1–12 (1985).
12. R. L. Gardner and R. A. Young, Assessing strategies for control of irrigation-induced salinity in the Upper Colorado River Basin, *Amer. J. Agr. Econom.* **70**, 37–49 (1988).
13. D. C. Gibbons, "The Economic Value of Water," Resources for the Future, Washington, DC (1986).
14. P. H. Gleick, Climate change, hydrology, and water resources, *Rev. Geophys.* **27**, 329–344 (1989).
15. S. P. Gross, The Galloway Project and the Colorado River Compacts: Will the Compacts bar transbasin water diversions? *Natur. Resour. J.* **25**, 935–960 (1985).
16. J. R. Hamilton, N. K. Whittlesey, and P. Halverson, Interruptible water markets in the Pacific northwest, *Amer. J. Agr. Econom.* **71**, 63–75 (1989).
17. L. M. Hartman and D. A. Seastone, "Water Transfers: Economic Efficiency and Alternative Institutions," Johns Hopkins Univ. Press, Baltimore (1970).
18. J. Hirshleifer, J. C. DeHaven, and J. A. Milliman, "Water Supply: Economics, Technology and Policy," Univ. of Chicago Press, Chicago (1969).
19. J. E. Houston, Jr., and N. K. Whittlesey, Modeling agricultural water markets for hydropower production in the Pacific Northwest, *West. J. Agr. Econom.* **11**, 221–231 (1986).
20. C. W. Howe and K. W. Easter, "Interbasin Transfers of Water: Economic Issues and Impacts," Johns Hopkins Univ. Press, Baltimore (1971).
21. C. W. Howe, D. R. Schurmeier, and W. D. Shaw, Jr, Innovative approaches to water markets, *Water Resour. Res.* **22**, 439–445 (1986).
22. Imperial County Agricultural Commissioner, "Imperial County Agriculture," El Centro, CA (annual, 1982–1987).
23. Imperial County Cooperative Extension, "Guidelines to Production Costs and Practices," Holtville, CA (1988).
24. J. E. Keith, G. A. Martinez Gerstl, D. L. Snyder, and T. F. Glover, Energy and Agriculture in Utah: Responses to water shortages, *West. J. Agr. Econom.* **14**, 85–97 (1989).
25. M. M. Kelso, W. E. Martin, and L. Mack, "Water Supplies and Economic Growth in an Arid Environment: An Arizona Case Study," Univ. of Arizona Press, Tucson (1973).
26. A. P. Kleinman and F. B. Brown, "Colorado River Salinity: Economic Impacts on Agricultural, Municipal, and Industrial Users," Colorado River Water Quality Office, U.S. Bureau of Reclamation, Denver (1980).
27. A. V. Kneese and G. Bonem, Hypothetical shocks to water allocation institutions in the Colorado Basin, *in* "New Courses for the Colorado River: Major Issues for the Next Century" (G. D. Weatherford and F. L. Brown, Eds.), Univ. of New Mexico Press, Albuquerque (1986).
28. D. Lee, Salinity in the Colorado River Basin: A dynamic modelling approach to policy analysis, Unpublished Ph.D. dissertation, Department of Agricultural Economics, University of California, Davis (1989).
29. L. C. Lohman, J. G. Milliken, W. S. Dorn, and K. E. Tuccy, "Estimating Economic Impacts of Salinity of the Colorado River," Colorado River Water Quality Office, U.S. Bureau of Reclamation, Denver (1988).
30. J. B. Loomis, The 1991 state of California water bank: Water marketing takes a quantum leap, *Rivers* **3**, 129–140 (1992).
31. B. A. McCarl and G. H. Parandvash, Irrigation development versus hydroelectric generation: Can interruptible irrigation play a role? *West. J. Agr. Econom.* **13**, 267–276 (1988).
32. B. A. Murtagh and M. A. Saunders, "Minos-Augmented-Users Manual," Technical Report SOL 80-19, Department of Operations Research, Stanford University (1980).
33. National Academy of Sciences, "Water and Choice in the Colorado Basin," National Academy of Sciences, Washington, DC (1968).

34. National Research Council, "Water Transfers in the West: Efficiency, Equity, and the Environment," National Academy Press, Washington, DC (1992).
35. L. L. Nash and P. H. Gleick, Sensitivity of streamflow in the Colorado River Basin to climatic changes, *J. Hydrol.* **125**, 221–241 (1991).
36. G. E. Oamek, "Economic and Environmental Impacts of Interstate Water Transfers in the Colorado River Basin," Monograph 90-M3, Center for Agricultural and Rural Development, Iowa State University, (1990).
37. G. Ragan, "Economic Benefits of Salinity Control: New Models, New Techniques, and New Estimates," Unpublished Ph.D. dissertation, Department of Agricultural and Resource Economics, Colorado State University, Fort Collins (1992).
38. A. Randall, Property entitlements and pricing policies for a maturing water economy, *Australian J. Agr. Econom.* **25**, 195–212 (1981).
39. M. Reisner and S. Bates, "Overtapped Oasis: Reform or Revolution for Western Water," Island Press, Washington, DC (1990).
40. S. J. Shupe, G. D. Weatherford, and E. Checchio, Western water rights: The era of reallocation, *Natur. Resour. J.* **29**, 413–434 (1989).
41. C. W. Stockton, "Long-Term Streamflow Records Reconstructed from Tree Rings," Paper 5, Laboratory of Tree-Ring Research, Univ. of Arizona Press, Tucson (1975).
42. U.S. Bureau of Reclamation, "Colorado River Alternative Operating Strategies for Distributing Surplus Water and Avoiding Spills," Engineering and Research Center, Denver (1986).
43. U.S. Bureau of Reclamation, "Colorado River Simulation System: Inflow and Demand Input Data," Engineering and Research Center, Denver (1989).
44. U.S. Bureau of Reclamation, "Glen Canyon Dam Interim Operating Criteria," Salt Lake City (Oct. 1991).
45. U.S. Bureau of Reclamation, "Central Arizona Project Status: Quarterly Update," Phoenix (Nov. 1992).
46. U.S. Department of Commerce, "Colorado State and County Data, 1987 Census of Agriculture," Bureau of the Census (1988).
47. U.S. Department of the Interior, "Quality of Water: Colorado River Basin," Progress Report 14 (1989).
48. H. J. Vaux and R. E. Howitt, Managing water scarcity: An evaluation of interregional transfers, *Water Resour. Res.* **20**, 785–792 (1984).
49. S. J. Viscoli, The Resource Conservation Group proposal to lease Colorado River water, *Natur. Resour. J.* **31**, 887–907 (1991).
50. Water Intelligence Monthly, Congress passes H.R. 429, Stratecon, Inc., (Oct. 1992).
51. M. Weinberg and Z. Willey, Creating economic solutions to the environmental problems of irrigation and drainage, *in* "The Economics and Management of Water and Drainage in Agriculture" (A. Dinar and D. Zilberman, Eds.), Kluwer Academic, Boston (1991).

Statement of ownership, management, and circulation required by the Act of October 23, 1962, Section 4369, Title 39, United States Code: of

JOURNAL OF ENVIRONMENTAL ECONOMICS AND MANAGEMENT

Published bimonthly by Academic Press, Inc., 6277 Sea Harbor Drive, Orlando, FL 32887-4900. Number of issues published annually: 6. Editor: Dr. Ronald G. Cummings, Policy Research Center, College of Business Administration, Georgia State University, University Plaza, Atlanta, GA 30303-3083.
Owned by Academic Press, Inc., 1250 Sixth Avenue, San Diego, CA 92101. Known bondholders, mortgagees, and other security holders owning or holding 1 percent or more of total amount of bonds, mortgages, and other securities: None.
Paragraphs 2 and 3 include, in cases where the stockholder or security holder appears upon the books of the company as trustee or in any other fiduciary relation, the name of the person or corporation for whom such trustee is acting, also the statements in the two paragraphs show the affiant's full knowledge and belief as to the circumstances and conditions under which stockholders and security holders who do not appear upon the books of the company as trustees, hold stock and securities in a capacity other than that of a bona fide owner. Names and addresses of individuals who are stockholders of a corporation which itself is a stockholder or holder of bonds, mortgages, or other securities of the publishing corporation have been included in paragraphs 2 and 3 when the interests of such individuals are equivalent to 1 percent or more of the total amount of the stock or securities of the publishing corporation.
Total no. copies printed: average no. copies each issue during preceding 12 months: 1953; single issue nearest to filing date: 1998. Paid circulation (a) to term subscribers by mail, carrier delivery, or by other means: average no. copies each issue during preceding 12 months: 1394; single issue nearest to filing date: 1440. (b) Sales through agents, news dealers, or otherwise: average no. copies each issue during preceding 12 months: 0; single issue nearest to filing date: 0. Free distribution by mail, carrier delivery, or by other means: average no. copies each issue during preceding 12 months: 77; single issue nearest to filing date: 77. Total no. of copies distributed: average no. copies each issue during preceding 12 months: 1471; single issue nearest to filing date: 1517.

(Signed) Evelyn Sasmor, Senior Vice President

Q53 Q15 [11]

141-53 [96]

Modeling Regional Agricultural Production and Salinity Control Alternatives for Water Quality Policy Analysis

Donna J. Lee and Richard E. Howitt

Water development and allocation to competing uses without well-defined water quality rights contribute to water use externalities. Federal legislation to address the salinity externalities in the Colorado River Basin comprises a set of arbitrary quality standards and millions of dollars in federal projects. This study specifies economic criteria to empirically determine first- and second-best quality standards and to indicate opportunities for efficiency gains in existing policy. A basin-wide, nonlinear programming model optimizes river water quality, resource allocation, production levels, and total expenditures for control. Revealed are the economic tradeoffs between water uses, regions, and control strategies.

Key words: nonlinear programming, regional production, river basin, salinity, water quality policy.

The Colorado River Basin drains 242,000 square miles of land in Wyoming, Utah, New Mexico, Arizona, Nevada, and California. Beneath basin soils lie vast deposits of salt left behind by prehistoric seas. Water development projects and irrigated agriculture have accelerated the rate at which these naturally occurring salts are leached from the soil. Groundwater flows from irrigated agriculture and natural springs transport nine million tons of salt each year from basin soils to the Colorado River. River salinity is further concentrated by Upper Basin diversions, basin exports, and evaporation. Colorado River water is diverted many times for irrigation, municipal, and industrial uses, and becomes progressively more saline as it moves downstream. As a result of salt loading and salt concentrating, more than half of the basin's 3.1 million irrigated acres are classified as saline.[1] In the Lower Basin, river salinity is responsible for millions of dollars in annual crop losses, nonreclaimable soils, and added costs to municipal and industrial uses.

Water salinity reduces returns to irrigated production by lowering crop yields and increasing farm production costs. Applying more irrigation water, installing drainage systems, and planting salt-tolerant crops are among the alternatives available to farmers for mitigating the effects of rising water salinity levels, but when all the feasible alternatives are exhausted cropland can and has gone out of production. In municipal uses, salinity accelerates the deterioration rate of pipes, plumbing fixtures, and home appliances, and increases household expenditures for drinking water, soaps, detergents, and water treatment. In the industrial sector, water is used in food processing, chemical manufacturing, and as a coolant. The demand for water quality varies widely. In food processing, for example, the tolerance level ranges from 500 to

---

Donna J. Lee is an assistant professor in the Food and Resource Economics Department, University of Florida. Richard E. Howitt is a professor in the Department of Agricultural Economics, University of California, Davis.

The authors acknowledge Oscar Burt and Miguel Mariño for their contributions to the development of this research effort. The comments from the two journal reviewers and the senior editor of the *AJAE*, which helped improve the overall quality of the final manuscript, are appreciated. Initial funding for this research came from the Resource and Technology Division of the Economic Research Service, USDA. Additional support was provided by the Department of Economics, University of Hawaii.

[1] Soils with total dissolved solids in excess of 1,300 mg/l (2 mmhos/cm) are considered saline.

1,000 mg/l. As a coolant for a coal-fired power plant, salinity levels up to 10,000 mg/l are tolerable. More frequent equipment replacement and outlays for water treatment are the primary costs to industries supplied by salty water.

Water quality in the Colorado River Basin exhibits the classic case of the downstream externality problem. Salinity impairs 63% of the irrigated acreage in the Lower Basin, yet 72% of the salt in Lower Basin river water originates in the Upper Basin. Legally, Lower Basin water users have no recourse. Although they are entitled to a specific quantity of water, their quality rights are poorly defined. As a result, approximately three million tons of salt from agricultural return flows are loaded into the river each year, affecting hundreds of thousands of irrigated acreage and millions of households. The extent of the affected acreage, the absence of markets for transferring costs, and the conflicting goals between agricultural, and municipal and industrial uses rank salinity as the most important water quality problem in the Colorado River Basin. To assure Lower Basin users of a clean and reliable supply of water, Congress established water quality standards at three locations along the Colorado River. At Hoover Dam in Arizona, Parker Dam in Arizona, and Imperial Dam in California, the maximum allowable salinity levels are 723, 747, and 879 mg/l, respectively.[2] Average observed salinity at Imperial Dam was 753 mg/l in 1991. At the present level of development, with no additional salinity control through the year 2010, water quality at Imperial Dam is expected to vary between 639 and 1,095 mg/l (U.S. Department of the Interior 1993). Although water conservation practices could stem the load of salt from the Upper Basin, the economic incentive to use less water simply does not exist. In the Upper Basin, water is plentiful, and at $8 per acre-foot, water-intensive farming is a common practice. To assure compliance with the standards, federal funds have been allocated through the year 2015 to stem annual salt loads. To date, federal salinity control projects have reduced total annual salt loading by 270,000 tons. An additional 1.26 million tons removal is proposed at a cost of approximately $700 million (U.S. Department of the Interior 1993).

This research develops a model to evaluate the efficiency of current water quality policy in the Colorado River Basin based on the value of water and water quality in agricultural, municipal, and industrial water uses and to determine the level of water quality that maximizes aggregate net returns to the river basin. For evaluating and comparing a wide range of water quality control alternatives, the model developed is the most comprehensive to date.

Previous Work

Three previous works estimated the marginal value of water quality to uses in the Imperial Valley. Moore, Snyder, and Sun developed farm-level models for nine crops grown on three soil types to quantify the value of clean water in irrigated production. The selected crops, soil types, and farm sizes were representative of the Imperial Valley Irrigation District. Inputs to production were irrigation treatment, water quality, and leaching fraction. Four water-quality levels between 490 and 1,950 mg/l[3] were simulated. Their results showed that the value of Imperial Dam water quality to Imperial Valley agriculture was between $2,000 and $4,300 per mg/l.[4] The value of water quality was quantified by Kleinman and Brown in a study to analyze the effects of salinity at Imperial Dam on agricultural, municipal, and industrial water uses. For salinity levels between 875 and 1,225 mg/l, they estimated that a 1 mg/l increase in salinity would result in $418,500 worth of damages to municipal uses, $17,200 in lost productivity to agricultural uses, and $89,400 in indirect costs to regional uses. The value of water quality at the margin was estimated to be $525,100 per 1 mg/l change in salinity.[5] Using linear programming, Gardner and Young compared the cost of salt load reduction in the Grand Valley with the benefits to Imperial Valley agriculture. They found that a rise in input water salinity from 800 mg/l to 1,100 mg/l would cost Imperial Valley producers $13.88 million. Allowing for fluctuations in crop prices,

---

[2] The concentration of salt in river water is expressed in units of milligrams per liter which is abbreviated mg/l.

[3] Results were reported in units of mmhos/cm; 650 mg/l = 1 mmhos/cm.

[4] In 1986 dollars. Dollar amounts reported by Moore, Snyder, and Sun were inflated to allow for a 54% increase in crop prices between 1970 and 1986.

[5] Kleinman and Brown's agricultural figures were inflated by 4% to account for the increase in crop prices between 1976 and 1986. Similarly, the municipal and industrial figures were increased 74% to capture the rise in GNP between 1976 and 1986.

the marginal value of water quality to Imperial Valley growers was estimated to be $39,000 per 1 mg/l salinity.[6]

This research expands the scope of previous work by evaluating agricultural production in five Upper Basin regions responsible for 80% of the annual agriculturally induced salinity and assessing the downstream effects on agricultural, municipal, and industrial uses at both Grand Junction and Imperial Dam. A set of regional production models were estimated which allowed us to (*a*) capture the cost of managerial alternatives to salinity control (e.g., input use reduction, shifting to less water-intensive crops) and (*b*) determine the effect of changes in water quality on downstream productivity. All six of the estimated production models were evaluated simultaneously within a basin-wide model that included production activities in multiple regions, a physical model of river basin hydrology, and a benefit function to account for agricultural returns, salinity control costs, and municipal and industrial water quality values.

Regional Production Models

Nonlinear regional production functions in land, capital, water, and water quality for Colorado River Basin crops have not been estimated in previous work, and the limitations of existing data precluded estimation by econometric methods. Traditional calibration approaches require calibration constraints that restrict the range of alternative scenarios and for that reason are limiting in policy analysis. For this work, a new procedure was employed. Flexible agricultural production models were calibrated using a new two-step procedure developed by Howitt. The calibrated models were augmented in a third step to include the effects of water quality on crop productivity.

Step 1: Obtain Resource Shadow Values

In the first step, we used linear programming to estimate the shadow value of resource inputs in regional production. We assumed efficient allocation of regional resource inputs and then obtained the shadow value of each resource in production as revealed by its marginal contribution to production and net returns.

The shadow value λ_j for each resource input $j \in J$ was defined to be the marginal profitability of the input in the lowest valued crop $i = 1, \ldots, I$. In equation (1), p is crop price, Q is total production, X is input quantity, r is input cost, and a is the linear rate of input use.

$$(1) \quad \lambda_j = \min \left[\frac{\left(p_1 \frac{\partial Q_1}{\partial X_1} - r_j\right)}{a_{1j}}, \frac{\left(p_2 \frac{\partial Q_2}{\partial X_2} - r_j\right)}{a_{2j}}, \ldots, \frac{\left(p_I \frac{\partial Q_I}{\partial X_I} - r_j\right)}{a_{Ij}} \right]$$

We estimated the shadow value of each input in each region.

Step 2: Calibrate Nonlinear Production Parameters

Using the estimated resource shadow values from step 1, the production parameters were calibrated to satisfy the first- and second-order conditions for profit maximization, observed output levels, existing cropping patterns, and constant returns to scale in the least profitable crops.

A total of $I + IJ$ Cobb-Douglas production parameters were obtained by minimizing the sum of squared errors (E), where I is the number of crops and J is the number of inputs. The objective was to choose the parameters (α) that minimize E,

$$(2) \quad \min_{\alpha} E = \varepsilon'\varepsilon$$

subject to the definition of the error terms in equations (3), (5), (6), and (7), and the inequality constraint in equation (4). The error vector, ε, is defined as $\varepsilon = [\varepsilon 1' \ \varepsilon 2' \ \varepsilon 3' \ \varepsilon 4']'$ and is of dimension $(1 + 2I + IJ \times 1)$.[7]

Equations (3) and (4) constrain the parameters to satisfy the first- and second-order conditions for profit maximization. In equation (3), $\hat{\lambda}_j$ is the estimated resource shadow value from step 1.

[6] Gardner and Young (GY) attribute the difference in their estimates and those of Kleinman and Brown (KB) to be primarily due to the way crop sensitivity to salinity and soil quality were handled within the region-wide models. GY allowed for higher yields from crops planted on the best soil class. KB used regional average yields to represent yields on the best soils. Further, KB assumed that salt-sensitive crops would only be planted on well-drained soils (where increases in salinity are more easily mitigated). Thus, KB's estimated farm returns were relatively less sensitive to changes in salinity than were GY's, so KB estimated a smaller marginal value of water salinity than did GY.

[7] $\varepsilon 1$ is $(IJ \times 1)$, $\varepsilon 2$ and $\varepsilon 3$ are $(I \times 1)$, and $\varepsilon 4$ is (1×1).

$$(3)\quad p\frac{\partial Q}{\partial X_j} = r_j + \hat{\lambda}_j + \varepsilon 1_j$$

$$(4)\quad \frac{\partial^2 Q(\alpha, X)}{\partial X_j^2} < 0.$$

Equation (5) requires that the calibrated functions $Q(\hat{\alpha}, X)$ reproduce observed output levels Q^0 at the base level of resource allocation X^0:

$$(5)\quad Q^0 + Q(\alpha, X^0) + \varepsilon 2.$$

Equation (6) requires the difference in value between average and marginal yields to equal the shadow value of land ($\hat{\lambda}_L$), where X_L^0 is the observed allocation of land:

$$(6)\quad p\left(\frac{Q^0}{X_L^0} - \frac{\partial Q(\cdot)}{\partial X_L}\right) = \hat{\lambda}_L + \varepsilon 3.$$

Equation (7) imposes constant returns to scale on the lowest-valued crop in the region. For Cobb-Douglas production the constraint is

$$(7)\quad \sum_j \alpha_j = 1 + \varepsilon 4.$$

Empirical Results

To calibrate the production functions, yield, price, cost, and resource use data were obtained from various sources, including the Colorado Department of Agriculture, the U.S. Department of Commerce (1984, 1986), and the Utah Department of Agriculture. For yields and output price by crop and region, county-level prices received by farmers were used. Yield and price data are published annually. Unit costs of resource inputs for each crop are available by state (USDA-ERS 1985). The production parameters were calibrated using data representing 70% of the irrigated acreage in the Lower Gunnison Basin and the Grand Valley, approximately 80% of the irrigated acreage in Uinta, Price, and San Rafael, and 69% of the harvested irrigated acreage in the Imperial Valley.

Results show that output elasticities in land are large relative to the elasticities of water and capital. The output elasticity of land is largest for alfalfa production in the Grand Valley (0.888) and in Lower Gunnison (0.884); it is lowest for corn silage production in Price-San Rafael (0.214). The estimated output elasticity of water ranges from 0.368 in Imperial Valley alfalfa production to 0.039 in Price-San Rafael and Uinta corn silage production. The elasticity of capital is small, ranging from 0.007 for cotton and wheat grown in the Imperial Valley up to 0.033 for alfalfa in Lower Gunnison and other hay in Price-San Rafael and Uinta. Calibrated parameters are displayed in table 1.

The calibrated results compare favorably to econometric results obtained by Just, Zilberman, and Hochman. For vegetable crops in Israel, output elasticities for water ranged from 0.004 to 0.0788. The output elasticities of water from the calibrated model for field crops ranged from 0.039 to 0.368. Dinar and Knapp estimated output elasticities of water for cotton and alfalfa in California to be 0.335 and 1.005. The calibrated elasticity of water was 0.258 for cotton and 0.368 for alfalfa grown in the Imperial Valley.

Step 3: Determine Production Response to a Change in Irrigation Water Salinity

The third step introduces crop yield response information in the calibrated Cobb-Douglas models to predict the effect of a change in irrigation water salinity on agricultural production. To this end the Cobb-Douglas production models were augmented with agronomically derived salinity response coefficients. Agronomic estimates from Letey and Dinar were obtained for this purpose and used to derive a production shift function which when multiplied by the calibrated Cobb-Douglas functions yielded regional production estimates in land, capital, water, and water quality.

Letey and Dinar brought together linear yield response to applied water (from various sources), yield response to soil salinity (from Maas and Hoffman 1977), and functions representing the relationship between soil and applied water salinity, plant consumptive use of water, leaching fraction, and root zone depth in order to generate relative yield[8] data for a range of applied water and water salinity levels. The generated data were estimated using linear, log-log, and quadratic functional forms. The best fit for explaining yield decline with increasing salinity and decreasing water application was the quadratic function

[8] Relative crop yield (y) as a proportion of maximum yield when salinity is zero, $0 \leq y \leq 1$.

(8) $y = \beta_0 + \beta_1 C + \beta_2 C^2 + \beta_3 CW + \beta_4 W + \beta_5 W^2.$

From equation (8), a change in relative crop yield due to a small change in water salinity can be expressed as

$$(9) \quad \frac{\partial y}{\partial C} = \beta_1 + 2\beta_2 C + \beta_3 W.$$

Empirical estimates of the coefficients from equation (9), using data from Letey and Dinar, are shown in table 2.

Using Letey and Dinar's coefficient estimates, observed regional yields, and existing regional irrigation water quality, a "yield shift" function, $\Psi(\cdot)$, was derived. In equation (10), y^0 is proportion of maximum yields under base-year irrigation salinity and C^0 is base-year salinity:

$$(10) \quad \psi(y^0, C^0, \hat{\beta}, W, C) = \frac{1}{y^0} \int_{C^0}^{C^0+\Delta C} \frac{\partial y(\hat{\beta}, C, W)}{\partial C} dC.$$

When salinity is constant, $\Psi(\cdot) = 0$. Regional

Table 1. Calibrated Cobb-Douglas Production Parameters

	Intercept	Land	Water	Capital
		Grand Valley		
Alfalfa	3.62	0.888	0.080	0.032
Barley	145.53	0.734	0.058	0.018
Corn grain	327.0	0.539	0.042	0.015
Corn silage	44.4	0.516	0.054	0.027
Wheat	66.3	0.651	0.050	0.014
		Imperial Valley		
Alfalfa	5.32	0.580	0.368	0.020
Cotton	4.23	0.421	0.258	0.007
Wheat	80.2	0.784	0.209	0.007
		Lower Gunnison		
Alfalfa	3.36	0.884	0.083	0.033
Barley	105.0	0.806	0.066	0.020
Corn grain	431.0	0.535	0.043	0.029
Corn silage	35.3	0.687	0.074	0.019
Wheat	148.0	0.705	0.056	0.015
		Price and San Rafael		
Alfalfa	16.1	0.357	0.076	0.023
Other hay	4.6	0.403	0.087	0.033
Wheat	63.8	0.357	0.097	0.020
Corn grain	66.8	0.275	0.051	0.013
Corn silage	24.2	0.214	0.039	0.021
Barley	61.9	0.513	0.127	0.024
Oats	93.7	0.282	0.096	0.029
Pasture	3.8	0.586	0.295	0.119
		Uinta Basin		
Alfalfa	19.0	0.522	0.065	0.020
Other hay	2.19	0.877	0.090	0.033
Wheat	86.5	0.583	0.082	0.017
Corn grain	149.0	0.440	0.050	0.013
Corn silage	47.5	0.354	0.039	0.021
Barley	105.0	0.716	0.093	0.018
Oats	79.3	0.797	0.100	0.031

production as a function of salinity is

$$\tilde{Q}(\cdot) = Q(\hat{\alpha}, L, W, K)[1 + \psi(y^0, C^0, \hat{\beta}, C, W)]. \quad (11)$$

From equation (11), the marginal product of water is

$$\frac{\partial \tilde{Q}(\cdot)}{\partial W} = \frac{\partial Q(\cdot)}{\partial W} + Q(\cdot)\frac{\partial \psi(\cdot)}{\partial W}. \quad (12)$$

The augmented production function, $\tilde{Q}(\cdot)$, in equation (11) allows both total product and marginal product of water to vary with water salinity. Total production decreases with increasing water salinity for all crops. In table 2, $\beta_3 < 0$ for both alfalfa and corn, revealing from equation (12) that the marginal product of water decreases as water salinity rises. For barley, cotton, and wheat, $\beta_3 > 0$, indicating that for those crops the marginal product of irrigation water increases as irrigation water becomes more saline.

Optimization Model

To determine the level of water salinity that maximizes net returns to the river basin, an optimization model was specified. The optimization model includes three components: the downstream benefits of improved water quality, the cost of upstream salinity reduction, and the hydrologic relationship between upstream salinity reduction and downstream water quality.

Table 2. Quadratic Crop Yield Coefficients

Crop[b]	β_1	Coefficient[a] β_2	β_3
	C	C^2	$C\,W$
Alfalfa	−0.022	0.000068	−0.018
Barley	−0.002	−0.001	0.008
Corn	−0.035	−0.001	−0.015
Cotton	−0.020	−0.009	0.020
Wheat	−0.002	−0.001	0.008

Source: Letey and Dinar.

[a] Coefficients that appear in the partial derivative of the quadratic crop response function [equation (9)].

[b] Crop yield in percent of maximum, salinity (C) in ds/m, and water (W) in cm-ha.

Downstream Benefits

Within the model context, the beneficiaries of a reduction in Upper Basin salt loads are the agricultural users in the Grand Valley, and the agricultural, municipal, and industrial users at Imperial Dam.

Agricultural profit. Agricultural profits vary by region and crop. Under existing salinity conditions, profits for field crop production in Grand Valley range from \$100 to \$249/ac, and in Imperial Valley from \$249 to \$567/ac. As salinity levels fall, yields and hence profits improve, input use shifts, and cropping patterns change.

Agricultural profit is defined as

$$\pi_g = \mathbf{p}_g'\tilde{\mathbf{Q}}_g - r_{Lg}'\mathbf{L}_g - r_{Wg}'\mathbf{W}_g - r_{Kg}'\mathbf{K}_g \quad \text{for regions } g = 2, 6 \quad (13)$$

where π_g is agricultural profit for regions g, $\mathbf{p}_g$ is the $(I \times 1)$ vector of crop prices, $\tilde{\mathbf{Q}}_g$ is the $(I \times 1)$ vector of crop production, $\mathbf{r}_{jg}$ is the $(I \times 1)$ vector of costs, and $\mathbf{L}_g$, $\mathbf{W}_g$, and $\mathbf{K}_g$ are the $(I \times 1)$ vectors of input use for land, water, and capital. Production for crop i in region g is Cobb-Douglas in land, water, and capital, and concave in increasing water salinity, as defined in equation (11). Vector $\tilde{\mathbf{Q}}_g$ in equations (13) and (14) is a $(I \times 1)$ vector of crop production in region g, and $\Psi_{ig}(\cdot)$ [equations (14) and (10)] is the proportionate change in crop yield due to a change in irrigation water salinity:

$$\tilde{\mathbf{Q}}_g = \begin{bmatrix} \tilde{Q}_{1g} \\ \tilde{Q}_{2g} \\ \vdots \\ \tilde{Q}_{Ig} \end{bmatrix}, \quad (14)$$

$$\tilde{Q}_{ig} = Q_{ig}^{*}(\alpha_{ig}, L_{ig}, W_{ig}, K_{ig})[1 + \psi_{ig}(W_{ig}, C_g)].$$

Resource use in each region is limited to $\bar{L}_g$, $\bar{W}_g$, and $\bar{K}_g$, and water use is additionally constrained by WD_g, the volume of water withdrawn from the river and R_{idg} return flows from sources indexed by d, which includes deep percolation from fields, and on-farm and off-farm conveyance systems. These restrictions are expressed as follows:

$$\sum_i L_{ig} \le \bar{L}_g, \sum_i K_{ig} \le \bar{K}_g, \quad (15)$$

$$\sum_i W_{ig} \leq WD_g - \sum_i \sum_d R_{idg} \leq \overline{W}_g .$$

Municipal and industrial benefits. The values of municipal and industrial water uses in region 6 (Imperial Dam) were modeled as a linear function of the change in water salinity (dC). Defining m to be the average combined municipal and industrial value of water quality per mg/l, total municipal and industrial benefits from a change in salinity are

$$(16) \qquad B_g = m_g dC_g \qquad \text{for } g = 6.$$

Municipal and industrial uses in the Imperial Valley gain \$607,000 for each 1 mg/l reduction in water salinity, as estimated by Lohman, Milliken, and Dorn.

Control Costs

Water quality control alternatives included in the model were all proposed federal salinity control projects and an unconstrained reallocation of Upper Basin agricultural resources.

Federal projects. A series of federal projects have been proposed to reduce deep percolation from irrigated fields and on-farm and off-farm water conveyance systems. The projects were represented as a vector of discrete choice variables $\mathbf{F}$ $(k \times 1)$ whose elements take on values of one if chosen and zero otherwise.[9] Project costs are annualized and represented by r_F $(k \times 1)$, so total annual investment (Z) in federal salinity control projects ($\mathbf{F}$) is

$$(17) \qquad Z = r_F' \mathbf{F}$$

Federal projects to reduce salt loading ranged from \$4.44/ton to \$300.46/ton for improving on-farm and off-farm irrigation water use and water conveyance efficiency.

Managerial alternatives. Salt loading can also be reduced by reallocating agricultural inputs on upstream farms. For example, switching to crops that are less water intensive, reducing water application rates, and removing land from irrigated production can effectively reduce Upper Basin salt loads. The cost of these managerial alternatives is measured in terms of foregone agricultural profits:

$$(18) \qquad \text{ag loss} \equiv \pi_g(Q^0, X^0) - \pi_g[Q(X), X]$$

where

$$\pi_g = p_g' Q_g - r_{Lg}' L_g - r_{Wg}' W_g - r_{Kg}' K_g \qquad \text{for } g = 1, \ldots, 5.$$

Profits for field crop production in these Upper Basin regions range from \$11/ac to \$359/ac. In these predominantly agricultural regions the secondary effects of acreage reduction likely would be significant. Due to the nature of this study, however, the secondary economic effects were not included in the analysis.

Hydrology Model

To link upstream agricultural activities with downstream salinity, the Colorado River Basin hydrology model developed by Lee, Howitt, and Mariño was used. The change in downstream salinity (dC_g) in regions 2 and 6 is modeled as a function of the change in upstream salt loading (ΔS) and the change in upstream water use (ΔU). Region 1 (Lower Gunnison) is located upstream from region 2 (Grand Valley); regions 3 (Price), 4 (San Rafael), and 5 (Uinta) are situated upstream of region 6 (Imperial Dam). So, within the context of the model, water quality in regions 2 and 6 are influenced by activities that reduce salt loading. Regions 1, 3, 4, and 5 are hydraulically "upstream" or parallel to all other regions in the model; so for purposes of this analysis, their irrigation water quality is assumed to be in steady-state.

Let ρ represent the rate of decay, η the rate of evaporation, and T_g the rate of salt flow at g; V_g is river flow volume at g. River water salinity for regions 2 and 6 is represented by equations (19) and (20), as follows:

$$(19) \qquad dC_2 = \frac{[\rho_1 V_2 \Delta S_1 - \eta_1 T_2 \Delta U_1] dt}{V_2^2}$$

$$(20) \qquad dC_6 = \frac{\sum_{g=2}^{5} [\rho_g V_6 \Delta S_g - \eta_g T_6 \Delta U_g] dt}{V_6^2} .$$

[9] To illustrate, let $\mathbf{f}_R$ represent the vector of proposed federal projects in terms of volume of return flow reduced. Then $\mathbf{f}_R' F$ would indicate total reduction in return flows from federal projects. Multiplying $\mathbf{f}_R$ by salinity of the return flows c_R and then by F would yield $(\mathbf{f}_R c_R)' F$ or total salt load reduction from federal projects. So $r_F' F$ is total annualized cost of chosen federal salinity control projects.

Let $\mathbf{R}_g$ denote the vector of return flow volumes from region g to the river. Return flow volume is defined to be a function of water application rate (W), acreage planted (L), federal salinity control projects implemented (F), water diverted to the region (WD_g), and a series of crop- and soil-specific parameters (γ) (e.g., crop consumptive use, leaching fraction):

$$(21)\quad \mathbf{R}_g = R(W_g, L_g, F, WD_g, \gamma_g).$$

In equation(22), $\mathbf{c}_g$ is the vector of the salinities of the return flows. The change in salt load from region g is expressed as

$$(22)\quad \underset{\text{for } g=1 \dots 5}{\Delta S_g} = -\frac{\partial WD_g}{\partial t} C_g + (V_g - WD_g)\frac{\partial C_g}{\partial t} + \mathbf{c}_g' \frac{\partial R_g}{\partial t}.$$

The change in river flow below region g is

$$(23)\quad \underset{\text{for } g=1 \dots 5}{\Delta U_g} = -\frac{\partial WD_g}{\partial t} + \frac{\partial R_g}{\partial t}.$$

Objective Function

The objective is to choose land, water application rate, capital, and investment in federal projects (F) to maximize net returns (N) over all uses. Because of the dams and large reservoirs in the river basin, several years will elapse before the effect of a change in Upper Basin water can be detected in Lower Basin water quality. To account for this lag, δ, in equation (24) discounts net benefits from an improvement in water quality. In equation (24), π_g is regional profit, ∇D is the gradient of discounted net returns to downstream uses from a small change in salinity, and dC is the total change in salinity.[10]

$$(24)\quad \max_{L,W,K,F} N = \sum_{g=1}^{6} \pi_g - Z + \int_{C^0}^{C^0+\Delta C} \nabla D' dC$$

where

$$D = \begin{bmatrix} \pi_2(\delta_2 - 1) \\ (\pi_6 + M)(\delta_6 - 1) \end{bmatrix}, \quad dC = \begin{bmatrix} dC_2 \\ dC_6 \end{bmatrix}.$$

Empirical Model

Currently, regional resources are allocated to irrigated agriculture without regard to the quantity of salt loaded in irrigation return flows or the consequences downstream. To determine if an alternative allocation of resources would net a higher level of aggregate returns than currently exists, the model was parameterized with existing published and unpublished data on Colorado River Basin production and hydrology. The values of water quality in downstream municipal and industrial uses were taken from estimates reported in Lohman, Milliken, and Dorn. Federal irrigation projects included on-farm conservation measures (e.g., laser leveling, shortening irrigation runs), water conveyance improvements (e.g., pipe laterals, canal lining), and soil conserving practices (e.g., gully plugging, contour furrowing). Federal projects and costs were obtained from the U.S. Department of Agriculture, Soil Conservation Service (1981) and the U.S. Department of the Interior, Bureau of Reclamation (1982, 1986, 1989). Annual dollars to fund federal projects were assumed to be unconstrained. Nonstructural (managerial) alternatives to salinity control—such as shifting to crops that are less water intensive, reducing water application rate, and removing land from irrigated production—were captured through the nonlinear production functions for the Upper Basin regions [equation (14)], and the cost inferred by the corresponding reduction in farm profits [equation (18)]. Land, water, and capital are constrained regionally for agricultural irrigation, but within the model context, water may be reallocated downstream for purposes of dilution. Data from the U.S. Department of the Interior (1993) were used to update the hydrology model from Lee, Howitt, and Mariño. The optimization model was run using GAMS/Minos software (Kendrick and Meeraus).

Optimization Model Results

The optimal reduction of salt from the river was estimated to be 1.268 million tons per year; this would cause river salinity in the Grand Valley to fall 71 mg/l from a base level of 500 mg/

---

[10] $\int_{C^0}^{C^0+\Delta C} \nabla D' dC \equiv \int_{C_2}^{C_2+\Delta C_2} \frac{\partial \pi_2}{\partial C_2}(\delta_2 - 1)dC_2 + \int_{C_6}^{C_6+\Delta C_6} \frac{\partial \pi_6}{\partial C_6}(\delta_6 - 1)dC_6.$

Table 3. Water Quality Results

	Base Year	Optimal	Second Best: Upper Basin Agriculture — Status Quo[a]	Second Best: Upper Basin Agriculture — Constant Profit[b]
Change in Annual Water Use (%)				
Lower Gunnison	0	−89	0	−0.046
Grand Valley	0	−89	0	−0.257
Price and San Rafael	0	−84	0	−0.041
Uinta Basin	0	−87	0	−0.045
Imperial Valley	0	+14	+6.3	+0.068
Salt Load (10^3 tons)				
Lower Gunnison	730	202	449	444
Grand Valley	709	173	441	419
Price and San Rafael	150	42	42	42
Uinta Basin	202	105	179	175
Δ Net salt load	0	−1,268	−680	−710
Water Quality (mg/L)				
Lower Gunnison	260	260	260	260
Grand Valley	500	429	466	465
Price and San Rafael	300	300	300	300
Uinta Basin	300	300	300	300
Imperial Valley	753	600	688	683

[a] Resource allocation constrained to base year.
[b] Profit loss limited to less than 0.01%.

l salinity to 429 mg/l, and Imperial Dam salinity to drop 153 mg/l from current levels of 753 mg/l to 600 mg/l, as shown in table 3. To achieve this level of salinity reduction at the lowest cost, Upper Basin agriculture in the Lower Gunnison and the Grand Valley in Colorado shifted largely out of production. In Utah, the Price and San Rafael regions' crops shifted slightly out of alfalfa and pasture and into other crops, and in Uinta, crops shifted out of hay and into alfalfa. Land use results appear in table 4. As a result of the large reduction in acreage, profits in the Upper Basin regions fall and imputed land values drop. The net cost to Upper Basin agriculture from removing land from irrigated production, shifting crops, and reallocating resources was estimated to be $16.27 million. Salt load reduction from investment in proposed federal projects would cost $22.59 million. Gains to Lower Basin agriculture in the Imperial Valley were $1.61 million, and savings to municipal and industrial water uses were estimated to be $92.84 million. The net social benefit of this outcome would be $55.45 million per year. The results are displayed in tables 3 through 5.

Alternatively, we can consider the current situation, in which Upper Basin farmers are expected to continue farming, as usual, and all salinity reduction is expected to come about as a result of federally funded projects. In this scenario, labeled the "status quo," salinity reduction is more costly, so equilibrium water salinity levels at Imperial Dam are higher than the unrestricted optimum, total expenditures for salinity control are greater, and net welfare gains are smaller. Results show that in this second-best scenario "optimal" salt load reduction is 680,000 tons, water salinity at Imperial Dam is 688 mg/l, and annualized expenditures for federal salinity control projects are $37.5 million. Net welfare gain is $4.3 million per year. Results appear in tables 3, 4, and 5.

A slightly less restrictive alternative allows

50 *February 1996* *Amer. J. Agr. Econ.*

Table 4. Agricultural Land Use Results

	Base Year	Optimal	Second Best: Upper Basin Agriculture — Status Quo[b]	Second Best: Upper Basin Agriculture — Constant Profit[c]
	Proportion[a]			
Lower Gunnison				
Alfalfa	0.493	0.049	0.493	0.490
Barley	0.092	0.025	0.092	0.093
Corn grain	0.147	0.082	0.147	0.148
Corn silage	0.083	0.025	0.083	0.083
Wheat	0.184	0.055	0.184	0.186
Grand Valley				
Alfalfa	0.597	0.060	0.597	0.585
Barley	0.105	0.010	0.105	0.109
Corn grain	0.168	0.031	0.168	0.174
Corn silage	0.119	0.022	0.119	0.122
Wheat	0.010	0.001	0.010	0.011
Price and San Rafael				
Alfalfa	0.536	0.528	0.536	0.538
Other hay	0.137	0.145	0.137	0.109
Wheat	0.034	0.035	0.034	0.034
Corn grain	0.012	0.013	0.012	0.013
Corn silage	0.037	0.040	0.037	0.038
Barley	0.034	0.029	0.034	0.034
Oats	0.011	0.055	0.011	0.044
Pasture	0.165	0.108	0.165	0.161
Uinta Basin				
Alfalfa	0.603	0.630	0.603	0.604
Other hay	0.233	0.192	0.233	0.230
Wheat	0.021	0.022	0.021	0.021
Corn grain	0.016	0.018	0.016	0.016
Corn silage	0.045	0.050	0.045	0.045
Barley	0.052	0.050	0.052	0.052
Oats	0.030	0.038	0.030	0.031
Imperial Valley				
Alfalfa	0.681	0.907	0.781	0.789
Cotton	0.068	0.068	0.068	0.068
Wheat	0.252	0.025	0.143	0.143

[a] Proportion of land allocated to base-year acreage.
[b] Resource allocation constrained to base year.
[c] Profit loss limited to less than 0.01%.

for some reallocation of Upper Basin resource use to reduce salt loading, provided that Upper Basin agricultural profits are maintained. This scenario, referred to as "constant profits," yields a slight reduction in federal expenses for salinity control than the "status quo," while at the same time achieves a 5 mg/l lower salinity level at Imperial Dam. In this scenario, equilibrium salt loading is reduced 710,000 tons, water salinity at Imperial Dam is 683 mg/l, federal salinity control expenditures are $36.4 million per year, and net welfare gains are $8.3 million annually. Results are displayed in tables 3, 4, and 5.

Remarks

Due to the high value of water quality to Lower Basin users and the high marginal cost of some federal projects, these results indicate that the least-cost means of improving Colorado River water quality, mitigating the downstream externality problem, and moving toward a more efficient allocation of basin resources includes a combination of water conservation, acreage reduction, and federal salinity control projects. Although the same level of salinity reduction can be achieved with federal projects alone, the cost would be $700 million (U.S. Department of the Interior 1993), eighteen times more than a policy that includes acreage reduction. These results support the findings of Gardner and Young, Howe and Orr, and Booker and Young.

Retention of current Upper Basin agricultural farming practices and agricultural profits would require greater expenditure for more costly federal projects to control salinity. Results indicate that federal expenditures up to $37.5 million are warranted by downstream benefits from improved water quality if Upper Basin farming remains unchanged.

Table 5. Economic Results

			Second Best	
			Upper Basin Agriculture	
	Base Year	Optimal	Status Quo[a]	Constant Profit[b]
	$ million			
Agricultural Profit				
Lower Gunnison	14.80	5.60	14.80	14.80
Grand Valley	6.96	1.93	6.96	6.96
Price and San Rafael	4.03	3.58	4.03	4.03
Uinta Basin	15.07	13.48	15.07	15.07
Imperial Valley	80.98	82.59	81.63	81.68
Imputed Land Value				
Lower Gunnison	0.094	0	0	0
Grand Valley	0.100	0	0	0
Price and San Rafael	0.011	0	0	0
Uinta Basin	0.080	0.057	0	0
Imperial Valley	0.226	0.226	0.226	0.226
Aggregate Annual (ized) Change				
Lower Basin M&I		+92.84	+41.12	+44.09
Agricultural profits		–14.66	+0.673	+0.695
(Upper Basin)		(–16.27)	(+0.023)	(–0.004)
(Lower Basin)		(+1.61)	(+0.650)	(+0.699)
Federal project expenses		–22.59	–37.548	–36.483
Net change in welfare (million)		$55.45	$4.324	$8.302

[a] Resource allocation constrained to base year.
[b] Profit loss limited to less than 0.01%.

Irrigation projects and resources are allocated based on their relative effectiveness at improving downstream water quality in the absence of transactions costs and equity considerations. The additional costs incurred from water quality monitoring, enforcement, and litigation would likely offset the net benefits, suggesting that when transaction, monitoring, and enforcement costs are considered, a smaller reduction in salt loading may be optimal.

Conclusion

In this paper we present an empirical approach for analyzing the economics of agriculturally induced externalities. Irrigated agriculture for six regions in the Colorado River Basin were modeled with Cobb-Douglas production functions in land, capital, water, and water quality. These nonlinear regional production models serve two purposes: (*i*) to include input use reduction and cropping pattern shifts as a means of upstream salinity control, and (*ii*) to assess the downstream agricultural benefits of upstream salinity control. Since the production models were constrained only by fixed resources, a range of policy options beyond the scope of previous studies could be evaluated and compared. The hydrology model provided the physical linkage between Upper Basin water use and salt loading and Lower Basin water salinity, and it allowed the equations of the basin-wide model to be solved simultaneously. Nonlinear programming was used to evaluate the trade-off between upstream production and the value of downstream water quality in the absence of transaction costs and constraints to water transfers. Results showed that on-farm water conservation and acreage reduction offers huge potential savings over more costly structural projects for improving water quality and meeting water quality standards.

To obtain a more comprehensive model of basin agriculture, to control more of the agriculturally induced river salinity, and to fully value the benefits to water quality improvement, additional information on acreage planted, resource use, and cropping patterns is needed. Future studies might also consider evaluating advances in irrigation technology and management to conserve water, dispose of salts, and reuse return flows. New developments to cope with rising water salinity through irrigation scheduling, salt-tolerant crop varieties, and water and soil treatment can be studied in an economic context to better model downstream response, and to quantify the value of water quality improvement.

[Received May 1994; final revision received September 1995.]

References

Booker, J.F., and R.A. Young. "Modelling Intrastate and Interstate Markets for Colorado River Water Resources." *JEEM* 26(1994):66–87.

Colorado Department of Agriculture, Crop and Livestock Reporting Service. "Colorado Agricultural Statistics," 1985.

Gardner, R., and R.A. Young. "Assessing Strategies for the Control of Irrigation Induced Salinity in the Upper Colorado River Basin." *Amer. J. Agr. Econ.* 70(February 1988):37–49.

Howe, C.W., and D.V. Orr. "Effects of Agricultural Acreage Reduction on Water Availability and Salinity in the Upper Colorado River Basin." *Water Resour. Res.* 10(October 1974):893–97.

Howitt, R.E. "Calibration Methods for Agricultural Economic Production Models." *J. Agr. Econ.*, in press, 1995.

Just, R.E., D. Zilberman, and E. Hochman. "Estimation of Multicrop Production Functions." *Amer. J. Agr. Econ.* 65(November 1983):770–80.

Kendrick, D., and A. Meeraus. *GAMS: An Introduction.* Washington DC: Development Research Department, The World Bank, 1985.

Kleinman, A.P., and B. Brown. "Colorado River Salinity Economic Impacts on Agricultural, Municipal, and Industrial Users." Colorado River Water Quality Office, Engineering and Research Center, Department of the Interior, December 1980.

Lee, D.J. "A Dynamic Modelling Approach to Policy Analysis: Salinity in the Colorado River Basin." PhD dissertation, University of California, Davis, 1989.

Lee, D.J., R.E. Howitt, and M.A. Mariño. "A Stochastic Model of Surface Water Quality: Application to Salinity in the Colorado River Basin." *Water Resour. Res.* 29(1993):3917–23.

Letey, J., and A. Dinar. "Simulated Crop Water Production Functions for Several Crops when Irrigated with Saline Waters." *Hilgardia* 54(1986):1–32.

Lohman, L.C., J.G. Milliken, and W.S. Dorn. "Estimating the Economic Impacts of Salinity of the Colorado River Final Report." With K.E. Tuccy. Milliken Chapman Research Group, Inc. Prepared for the U.S. Department of the Interior, Bureau of Reclamation, February 1988.

Maas, E.V., and G.J. Hoffman. "Crop Salt Tolerance—A Current Assessment," *J. Irrig. Drain. Div.*

Amer. Soc. Civil. Eng. 103(June 1977):115–34

Moore, C.V., J.H. Snyder, and P. Sun. "Effects of Colorado River Quality and Supply on Irrigated Agriculture." *Water Resour. Res.* 10(April 1974):137–44.

U.S. Department of Agriculture, Economic Research Service, "U.S. Mathematical Programming Agricultural Sector Model." Firm Enterprise Data System, 1985.

U.S. Department of Agriculture (USDA), Soil Conservation Service. "On-Farm Program for Salinity Control." *Final Report for the Grand Valley Salinity Study*. December 1977.

___. "Potential On-Farm Irrigation Improvements." Final Report of the Lower Gunnison Salinity Control Study. September 1981.

___. "Preliminary Cost Figures for the Price-San Rafael Unit." As per phone conversation with Lee Page, SCS economist, December 1988.

U.S. Department of Commerce, Bureau of the Census. "Colorado State and County Data." *1982 Census of Agriculture vol. 6*, no. 1, 1984.

___. 1984 Farm and Ranch Irrigation Survey. June 1986.

U.S. Department of the Interior. "Quality of Water in the Colorado River Basin." Progress Report No. 14, 1989.

U.S. Department of the Interior, Bureau of Reclamation. "Grand Valley Unit Stage Two Development." *Supplement to the Definite Plan Report*. June 1983, revised (May 1986).

___. "Lower Gunnison Basin Unit." *Feasibility Report/Final Environmental Impact Statement*. November 1982.

U.S. D.O.I. "Quality of Water in the Colorado River Basin." Progress Report No. 16, 1993.

___. "Uinta Basin Unit." Planning Report/Final Environmental Impact Statement. April 1986.

Utah State Department of Agriculture, Agricultural Statistics Service. "Utah Agricultural Statistics" Annual Report, 1988.

Appendix

Variable Definitions

Table 1. Notation Used

Variable	Definition
α	Cobb-Douglas production elasticity
β	Quadratic yield response coefficient
δ	Discount factor
λ	Resource shadow value
π	Profit
ρ, η, γ	Hydrology model parameters
$\Psi(\cdot)$	Yield shift function
a	Input use per acre
B	Benefits to municipal and industrial uses
C	Water salinity
dC	Change in water salinity
E, ε	Sum of squared calibration errors, calibration error
$\mathbf{F}, \mathbf{f}$	Vector of federal salinity control projects undertaken (0, 1), vector of proposed projects
i, j, g	Crop index ($1 \ldots I$), resource index ($1 \ldots J$), region index ($1 \ldots G$)
L, W, K	Land use, water use, capital use
m	Marginal benefits to municipal and industrial uses from an improvement in water quality
p, r	Crop price, resource input cost
Q	Production
R	Return flow volume
$\mathbf{r_F}$	Cost vector of federal salinity control projects
ΔS	Change in river salt load
ΔU	Change in river flow volume
t	Time
T	Rate of salt flow
V	River flow volume
WD	Water diverted to region
X	Resource input use
y	Proportion of maximum crop yield
Z	Total cost of federal salinity control projects

154-67
(96)

U.S.

458

[12]

Uncoordinated Agricultural and Environmental Policy Making: An Application to Irrigated Agriculture in the West

Marca Weinberg and Catherine L. Kling

Agriculture and the environment are linked by a mutual reliance on scarce resources and prevalent market distortions. In this paper we examine the efficiency costs of policies that correct distortions in one sector while ignoring those in another. The dual distortions of an environmental externality and water subsidies are studied in the context of irrigated agriculture in California. The welfare costs associated with independent action by either a water agency or an environmental authority when that policy maker attempts to correct the respective market distortion without consideration of the distortion in the other market are estimated. Policies to correct the two distortions are found to be complementary. Under these conditions, the independent correction of either distortion improves welfare by at least $118 per acre. In most cases, the simultaneous adoption of various second-best policies further reduces welfare losses associated with these distortions.

Key words: environmental policy, irrigated agriculture, second best, water quality.

The theory of the second best (Lipsey and Lancaster) sends a strong warning to economists and policy makers that the correction of a single market distortion without simultaneously correcting other sources of market failure can lead to pareto-inferior resource allocations. Even if the independent correction of one distortion generates gains in welfare, there may be large opportunity costs to correcting only one of the sources of market failure. However, it is political reality that a policy maker with jurisdiction over one distortion often does not have jurisdiction over another, even when both distortions occur in the same sector, and policy making is rarely coordinated.

Marca Weinberg is assistant professor in the Division of Environmental Studies at the University of California, Davis. Catherine L. Kling is associate professor of economics at Iowa State University.

This is Journal Paper No. J-16415 of the Iowa Agriculture and Home Economics Experiment Station, Ames, Iowa; Project No. 3246. Weinberg was at the Economic Research Service, USDA, and at the Congressional Budget Office during the course of this research. The opinions expressed in this paper are the authors' and do not necessarily represent those of the authors' current or previous employers. The authors thank Jim Wilen for early contributions, and Arun Malik, Paul Sabatier, Michael R. Moore, and Bruce Larson, and three anonymous referees for helpful comments and insights.

Agriculture is one industry where the counsel of Lipsey and Lancaster should be seriously heeded. Nearly all agricultural production decisions are affected in some form by policies that distort input or output markets, and many agricultural practices generate pollution externalities that are the focus of environmental legislation. Examples of the pervasive influence government programs have had on agricultural markets are readily enumerated. Many crop prices are supported through commodity programs, input prices are often distorted, and input use is sometimes controlled through quantity restrictions and use regulations. Subsidization of irrigation water, electricity, and farm credit, cost sharing for soil conservation practices, seed standards, labor laws, and pesticide regulations provide a sample of policies that distort relative input prices. The Clean Water Act and the Endangered Species Act are examples of environmental legislation with potential implications for farm practices.

Just et al. peruse the array of agricultural and environmental policies, noting that such policies "may at best be uncoordinated and at worst conflicting and counterproductive." They fur-

Amer. J. Agr. Econ. 78 (February 1996): 65–78

ther find that while the agricultural-environmental interface is much discussed, few empirical studies systematically analyze these problems. Despite the strong presence of government programs in agricultural markets, analysis of the regulation of agricultural pollutants has generally ignored the simultaneous market distortions. For example, most models and analyses of nonpoint source pollution control implicitly assume that the pollution externality represents the only source of market failure in the sector being analyzed. More generally, regulatory analysis of nonagricultural policies also typically ignores multiple market distortions.

Previous empirical studies examine the relationship between agricultural policy and the effectiveness of environmental regulations (e.g., Lichtenberg and Zilberman; Kopp and Krupnick; Hrubovcak, LeBlanc, and Miranowski; Howitt; Just, Lichtenberg, and Zilberman; Miranowski, Hrubovcak, and Sutton). Just, Buss, and Donoso (p. 375) note that such studies "serve to better understand the interface of agricultural and resource policy but do not emphasize the opportunities to better coordinate the two types of policies. What is needed is a total welfare or policy criterion that can consider agricultural and resource issues simultaneously in the development and design of joint policies. This should be a primary goal of future research."

In this paper we analyze a particular agricultural-environmental policy interface—the interaction between irrigation water policies in the West and policies for the control of drainage from irrigated agriculture—from the same perspective proposed by Just, Buss, and Donoso. The implications of water policies for drainage production has been explored by Caswell, Lichtenberg, and Zilberman, by Dinar and Letey, and by Weinberg, Kling, and Wilen. In contrast to the one-way linkage examined in previous papers, we consider the water and drainage policy distortions simultaneously. We ask, What are the welfare losses of each authority acting independently? or, conversely, What is the magnitude of the welfare gains if the authorities act in coordination?

Policy Setting

Large-scale irrigated agriculture in the San Joaquin Valley is made possible by the importation of large quantities of surface water. The Bureau of Reclamation supplies water at subsidized prices and in fixed quantities specified in long-term contracts. The terms of these contracts vary by district so that one farmer may receive a generous allotment at a relatively low price, while a farmer in a neighboring district may receive limited quantities of water and pay a higher price.

Irrigation water is applied in excess of crop water requirements to leach accumulated salts out of the root zone and to provide the minimum amount of water required by plants in all portions of nonuniform fields. This excess applied water may contribute to high water tables that are highly saline, which can cause yields to decline. Artificial drain systems are installed in many areas to maintain sufficient depth to the high water table. Much of the water collected in subsurface drain systems installed on the west side of California's San Joaquin Valley is high in dissolved solids and contains naturally occurring selenium, molybdenum, boron, and other elements that can be toxic in high concentrations. Discharge of these waters historically has been through the rivers, sloughs, and wetlands in the area.

This problem is characterized by a water pollution externality with distortions in the market for an input (water). Policy debates in California regarding the two distortions have, for the most part, been independent. Both water markets and reform of federal reclamation policies have been proposed as efficiency-enhancing measures to correct the distortions from the water price subsidies and institutionally set quantities (e.g., Howe, Schurmeier, and Shaw; Moore). Concurrently, budgetary and environmental concerns have lent momentum to congressional actions to reduce federal water subsidies. Congress provided new direction for pricing and management of Central Valley Project water with the Central Valley Project Improvement Act, signed into law 31 October 1992 (U.S. Congress 1992). However, the exact manner in which many of the provisions of the act will be implemented remains to be determined.

Meanwhile, the State of California has established water quality standards for selenium, molybdenum, boron, and salts in the San Joaquin River (State of California). Independent discussions by state and local lawmakers concerning policies for reducing drain water have been ongoing since the water quality standards were set by the State Water Resources Control Board. The uncoordinated manner in which the irrigation water market distortions and the agricultural pollution problems have been considered suggests a strong potential for the adoption of inefficient policies by the respective authorities. This study examines the possible welfare losses from such uncoordinated policies.

Conceptual Framework

Profit-maximizing farmers are assumed to jointly produce crops and effluent through the use of two inputs, a polluting input (x) and an abating one (z). For purposes of discussion, input x may be thought of as irrigation water, while z may represent irrigation technology or management. To reflect federal water allocation policies, farms face a maximum constraint ($\bar{x}$) on the quantity of the polluting input used and receive that quantity at a distorted price (w_i). The optimization problem for a representative farm with no regulation of the pollution externality is to choose input levels to

$$(1) \quad \max_{x_i, z_i} \pi_i = pf^i(x_i, z_i) - w_i x_i - rz_i - E_i \quad \text{subject to } x_i \leq \bar{x}_i$$

where π_i represents net returns to land and management for the ith farm; p is the price received for crop output; $f^i(\cdot)$ describes crop production opportunities for the ith farm and is assumed to be increasing in the relevant range for both inputs, twice differentiable and concave; r represents the marginal cost of the abating input (z); and E_i is expenditures on other inputs.

In contrast to the problem faced by the individual farm, the first-best resource allocation can be found by maximizing social welfare (S), defined as the sum of farm-level net returns less the social damages incurred as a result of the production externality:

$$(2) \quad \max_{x_1,\ldots x_I, z_1, \ldots z_I} S = \sum_{i=1}^{I} [pf^i(x_i, z_i) - w^* x_i - rz_i - E_i] - p_d D[g^1(x_1, z_1), \ldots, g^I(x_I, z_I)]$$

where w^* represents the marginal cost of supplying the input x, $D(\cdot)$ represents total effluent levels, p_d represents the marginal social cost of damages from emissions, and $g^i(\cdot)$ is a production function describing emissions from the ith farm, $i = 1, \ldots, I$. The pollution production function is assumed to be twice differentiable and convex, and to increase with input x and decrease with levels of z ($g_x > 0$, $g_z < 0$, $g_{xx} \geq 0$, $g_{zz} \geq 0$, and $g_{xx}g_{zz} - g_{xz} \geq 0$). For tractability, the total damages are assumed to be an additive function of farm-level emissions; i.e., $D = \Sigma_i g^i(x_i, z_i)$.[1]

Note that the first-best problem incorporates the true supply costs of water, w^*, does not constrain water consumption by individual farms, and internalizes the pollution externality. The appropriate policy recommendation involves the simultaneous correction of the water market distortions and the externality. However, the best policy recommendation is not readily apparent unless both distortions are corrected. The following example illustrates the potential implications of uncoordinated policy making. Two firms are assumed to contribute to water pollution with emissions that arise from use of input x. The firms face different institutional parameters related to input x. Firm 1 receives a large price subsidy while firm 2 pays the true supply price; i.e., $w^* = w_2 > w_1$.

In figure 1, firm 1 and firm 2 are depicted on the right and left sides of a back-to-back diagram. The net private marginal benefits ($PMB_j = pf_x - w_j$, $j = 1, 2$) and the net social marginal benefits resulting from use of input x are illustrated for each firm.[2] Social marginal benefits (SMB_j) equal net private marginal benefits when firms are charged w^* for water, minus the marginal social cost of using input x ($SMB_j = pf_j - w^* - p_d g_x$, $j = 1, 2$). Curve PMB_1 represents the net private marginal benefits enjoyed by firm 1 at the subsidized prices. Firm 2 pays the true cost of water, but its net private marginal benefits do not equal the social marginal benefit of water use due to the externality. As drawn, firm 2's private and social net marginal benefit curves are flatter than those of firm 1. Curve PMB_1' is the net private marginal benefit curve firm 1 would face if it also paid full cost for water.

The socially optimal level of input use by each firm occurs at input levels x_1^* and x_2^*. However, firm 2 will choose the input level that sets net private marginal benefits equal to zero in the absence of policy intervention ($x_2 = x_2^o$). Firm 1 does the same ($x_1 = x_1^o$). The welfare losses associated with this base case, relative to

[1] As noted by a reviewer, the constant marginal social cost of damages and additive total damages may not be typical in many real world situations.

[2] Two assumptions in the diagram merit notice. There is a one-to-one relationship between input use and effluent production ($g_x = 1$), and the inputs x and z are independent ($f_{xz} = g_{xz} = 0$). These assumptions are made for diagrammatic purposes, but are relaxed in the empirical model.

68 *February 1996* *Amer. J. Agr. Econ.*

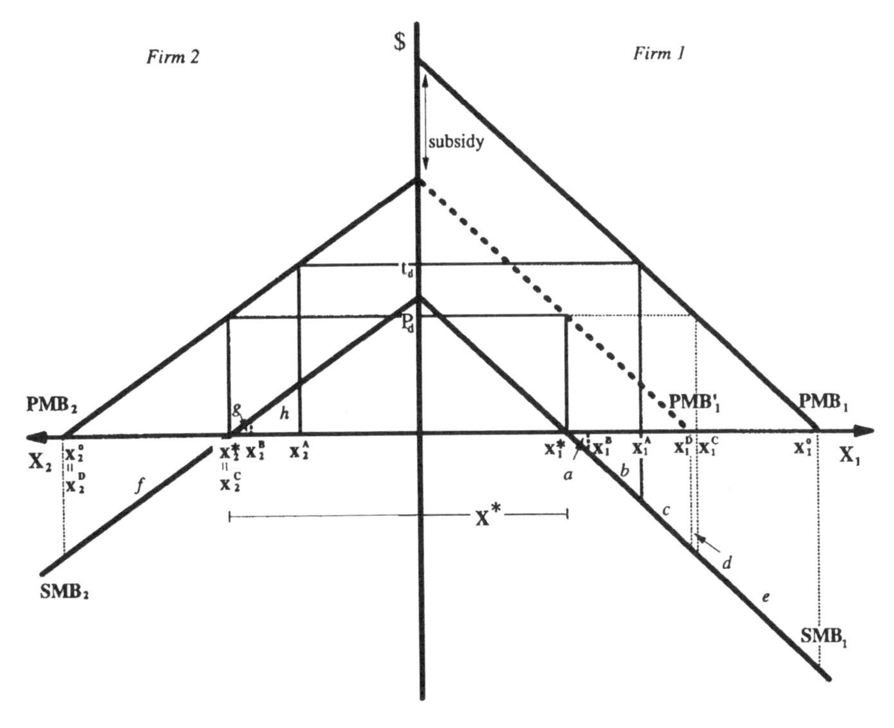

Summary of Welfare Consequences of Alternative Policies

Policy	Water Use	Welfare Loss
Current Position (Base Case)	$x_1^0,\ x_2^0$	$(a+b+c+d+e)\ +\ f$
Correct Both Distortions (First-Best)	$x_1^*,\ x_2^*$	0
Externality Policies		
A. Marketable Permits (tax=t_d)	$x_1^A,\ x_2^A$	$(a+b)\ +\ (g+h)$
B. Uniform Standard	$x_1^B,\ x_2^B$	$a\ +\ g$
C. Pigouvian tax=P_d	$x_1^C,\ x_2^C$	$(a+b+c+d)\ +\ 0$
Correct Water Price	$x_1^D,\ x_2^D$	$(a+b+c)\ +\ f$

Note: $X^* = x_1^* + x_2^* = x_1^A + x_2^A = x_1^B + x_2^B < x_1^C + x_2^C$

Figure 1. Welfare effects of independent application of policy alternatives for correcting input price and environmental externality distortions

the first best, are $(a + b + c + d + e) + f$. To achieve a first-best solution, both the subsidized water price and the pollution externality must be corrected. Now suppose separate, uncoordinated decision makers are charged with addressing these two problems. What are the likely policies that might be employed by each agency independently and what are the welfare consequences if a policy that corrects only one of the distortions is adopted?

Start first with an environmental authority interested in correcting the pollution externality.

Suppose that the socially optimal level of water use, given optimal levels for all inputs, is $X^* = x_1^* + x_2^*$. One policy that could be adopted is a marketable permit program where X^* marketable permits are issued.[3] This market would clear at a price of t_d. The input allocations that would arise with marketable permits are denoted x_1^A and x_2^A ($x_1^A + x_2^A = X^*$).

An environmental authority mindless of the input market distortion might logically argue that this scheme would efficiently achieve the environmental objective since the net marginal benefits will be equated across firms. Unfortunately, equating the distorted net private marginal benefits of firm 1 with those of firm 2 would not be efficient because of the water price distortion. To see this, note that the welfare loss relative to a first best is measured as area $[(a + b) + (g + h)]$.

A second method the environmental authority might consider is to set a uniform standard for all firms, denoted x_1^B and x_2^B. This policy would not be expected to be efficient since the net marginal benefits would not be equated. As drawn in figure 1, the uniform reduction scheme causes firm 1 to use too much of the input and firm 2 to use too little, relative to optimal levels, and the welfare loss is $a + g$. These input allocations are closer to optimal levels than under the marketable permit scheme, but this result depends on the parameters of the problem and need not be true in general.

A third policy, less likely to be adopted by the environmental authority due to its information requirements, but interesting from an economic efficiency point of view, is a tax equal to the true marginal social damage, p_d. This second-best policy will motivate firm 2 to select $x_2^C (= x_2^*)$, but firm 1 will reduce input use only to x_1^C because its input price is subsidized. Though this move is welfare enhancing, the welfare loss of this policy relative to the first best is area $(a + b + c + d)$.

Turn now to a separate authority interested in removing the input market distortion. Such an agency might choose to eliminate the price subsidy. A policy that eliminates the input price subsidy will cause firm 1 to treat the true supply cost of water as its opportunity cost, so that its effective net private marginal benefit curve is identical to firm 2's (PMB_1' in the diagram). In this case, water use by firm 2 will be $x_2^D = x_2^o$ and use by firm 1 will be x_1^D. The welfare loss from this second-best policy relative to the first best is areas $(a + b + c) + f$.

In summary, independent authorities charged with correcting input market distortions or pollution externalities may undertake second-best policies if they do not adequately account for other distortions in the relevant markets. Three conclusions deserve emphasis. First, none of these policies achieves a first-best solution. Second, quantity (effluent permits) and price (Pigouvian tax) policies do not produce the same solution in the presence of an input subsidy. Baumol and Oates hint at this problem, noting that the cost-minimizing properties of the charges and standards approach (or, equivalently, marketable effluent permits) require that "input prices approximate their opportunity costs" (p. 165). Third, a uniform standard may improve welfare relative to a marketable permits policy that "appears" to efficiently allocate the effluent reduction. The size of the opportunity costs associated with these second-best policies and their relative efficiencies are empirical questions. In the next section, we quantify these efficiency costs and rank the policies accordingly for the case of regulating drainage effluent from irrigated agriculture in California's San Joaquin Valley.

Modeling Framework

The implications of alternative policies for addressing the dual distortions of federal water subsidies and drainage effluent from irrigated agriculture are examined with a simulation model describing agricultural conditions in the western San Joaquin Valley. The simulation model is described first. Policy options for addressing each distortion are then specified.

Simulation Model

The model for this analysis is an agricultural production model designed to simulate farmer decision making in a drainage problem area in the San Joaquin Valley. The model describes economic, agronomic, and hydrologic characteristics pertaining to the area and predicts changes in agricultural production decisions and drainage volumes in response to policy alternatives.

Prior to the recent drought, siphon tube furrow irrigation systems with half-mile runs were

[3] For example, assume that the effluent level is mandated by the legislature and provided to the environmental authority.

typical for irrigating cotton, tomatoes, sugar beets, and melons (cantaloupes) in the area, while wheat fields were generally irrigated with border check (flood) systems. These crops represent 80% to 90% of irrigated acreage in the study area. Changes in irrigation practices can conserve water and help to reduce drainage production, but will necessarily increase costs. To incorporate this aspect of the problem, crop-specific irrigation technology cost-efficiency functions are estimated and included in the model. Irrigation efficiency is defined as the ratio of the depth of water beneficially used (plant needs plus minimum leaching fractions) to the average depth of water applied to a field.

Production of the principal crops is modeled with crop-water production functions. Water applications (x) are multiplied by irrigation efficiency (z) in the production functions so that yield is a function of effective applied water, i.e., the amount of water available for plant growth.

The objective in this problem is to choose cropping patterns and crop-specific water applications and irrigation efficiency levels to maximize net returns to land and management from crop production, subject to water allotments, land availability, and constraints on acreage allocations for selected crops, and given the technological relationships specified for crop production and irrigation technology costs. Drainage production is predicted with a mass balance equation adapted from the Westside Agricultural Drainage Economics model (San Joaquin Valley Drainage Program). The region modeled includes 68,000 acres and is divided into sixteen decision units, each approximately homogenous with respect to economic, agronomic, and hydrologic conditions. The water supply parameters of all nine water districts in the area are represented in this formulation. The model is specified as a nonlinear programming problem; details are available from the senior author.

Policy Alternatives

It has been estimated that the San Joaquin River water quality standard for selenium could be met by decreasing drain water volumes to an average of 0.45 acre-feet per drained acre discharged from a 94,000 acre drainage study area on the west side of the San Joaquin Valley, and that this objective could be achieved through water conservation and improved irrigation practices (Hanemann et al.). We assume that this drainage discharge objective, denoted D^s, was established under second-best conditions; i.e., that the policy maker ignored the water market distortions when establishing the standard.[4] We assume, further, that the true marginal social damages are constant and were known to the policy maker.[5] The policy options available to the environmental regulator are, thus, a marketable drainage discharge permit program based on the second-best drainage objective ($D = D^s$), a uniform drainage discharge standard, also based on the second-best objective, and a Pigouvian effluent tax.[6] We also model a marketable permit system in which total permitted quantities are set at the first-best drainage objective ($D = D^*$).[7]

Prices charged to farmers in the study area for irrigation water range from $0 to $36 per acre-foot. Correcting the water distortion is straightforward in principal. The water policy maker must simply ensure that farmers face the "true" value of water. However, this value is unknown. Therefore, three water price options are considered to incorporate the types of options likely to be pursued by a water policy maker. First, "full-cost" prices, w_i^f, are considered. Bureau of Reclamation estimates of the value of irrigation water are based on full-cost water prices, defined in the Reclamation Reform Act as the annual rate that amortizes, with interest, the outstanding (nonreimbursed) portion of expenditures allocated to irrigation facilities (U.S. Congress 1982). Full-cost prices vary by age of the service canal and location of the district; these prices range from $26 to $56 per acre-foot (U.S. Department of Interior).

The opportunity cost of water provides another basis for estimating the implicit water

[4] Indeed, the analysis conducted by Hanemann et al., upon which the San Joaquin River selenium standard appears to be based, contains a model of agricultural decision making that uses existing water prices and allocations.

[5] A second set of simulations was conducted in which marginal social damages were assumed to be increasing in total effluent. In general, results were nearly identical qualitatively and quantitatively to the constant marginal cost assumption; differences that do occur are noted in later footnotes.

[6] The Pigouvian effluent tax is found by first setting the drainage constraint such that the objective for regional average discharge (0.45 acre-feet of drain water per drained acre) is achieved at least cost. The shadow value on this constraint then represents marginal benefits of drainage production (marginal abatement costs). Given the assumptions that marginal social damage is constant and was known by the regulator, this value also must represent marginal social damages in this problem.

[7] The first-best drainage objective is found by simultaneously imposing the Pigouvian effluent tax and eliminating the water price subsidy.

Table 1. Returns to Crop Production, Returns Net of Environmental Damages, and Social Returns Under Alternative Policies

Policy:	Crop Returns ($/acre)	Drain Water (af/da)	Returns Net of Damages[a] ($/acre)	Water Subsidy[b] ($/acre)	Welfare Loss ($/acre)	Social Rank
Base case	418	1.44	187	292	217	8
First-best	167	0.35	112	0	0	1
Externality policies						
Permits ($D = D^S$)						
Pigouvian tax	366	0.45	293 (1)	203	21	3
Uniform standard ($D = D^S$)	237	0.32	185 (3)	139	65	6
Permits ($D = D^*$)	344	0.35	289 (2)	192	15	2
Water policies						
Full-cost ($w_i = w_i^f$)	312	0.99	153	140 (3)	99	7
Opportunity cost ($w_i = w^*$)	187	0.68	78	0 (1)	34	5
Environmental surcharge ($w_i =$ $125/af)	58	0.45	–15	–95 (2)	32	4

Note: Numbers may not add to totals due to rounding.
[a] Numbers in parentheses are ranks of policies by an environmental regulator that maximizes crop returns (evaluated at the water price faced by the farmer) net of environmental damages.
[b] Numbers in parentheses are ranks of policies by a water regulator that minimizes the absolute value of the water subsidy.

subsidy. The 1991 California Drought Water Bank purchased water for $125 per acre-foot and sold 389,770 acre-feet of water at a base price of $175 per acre-foot (Howitt, Moore, and Smith). In 1992, the Water Bank purchased water for $50 per acre-foot and sold it for $75 per acre-foot. For the purposes of this analysis, we assume that the true opportunity cost of water in the drainage study area is $75 per acre-foot.

Third, a price of $125 per acre-foot is considered to reflect the possibility that regulators might arbitrarily impose a surcharge on water use to account for environmental costs associated with water development. This is the motivation behind provisions such as the $25 surcharge on marketed water included in the Central Valley Project Improvement Act (U.S. Congress 1992).

Results: Efficiency Costs of Alternative Policies

Results are presented as regional average per acre values. The simulations analyzed include a base case in which no policies are undertaken by either authority. Of the simulations analyzed, the base case results include the highest levels of water use, the highest private crop returns (crop revenues less input costs) ($418/acre), the greatest volume of drain water (1.44 acre-feet per drained acre), and, thus, the greatest environmental damages and the lowest level of net social welfare (–$105/acre) (table 1). Net social welfare equals private crop returns less social damages from drain water production and the income transfer implied by federal water subsidies. These base case results correspond to historic resource allocations. The marginal social cost of drain water is estimated to be $160 per acre-foot.

A first-best policy, in which the water subsidy is eliminated and a Pigouvian effluent tax is imposed, yields a level of drainage that is lower than the state's 0.45 objective (table 1). The 0.35 acre-feet per drained acre predicted to be generated in this scenario represents the first-best drainage objective. Net social welfare with the first-best policy is predicted to be $112 per acre. The welfare loss associated with the base case is, thus, $217 per acre. This value corresponds to the area $[(a + b + c + d + e) + f]$ in figure 1.

The potential implications for policy choice

when correcting only one distortion are examined next. Turn first to the policy options available to the environmental regulator. Results for the marketable permit scheme implemented at second-best effluent levels include a level of crop returns net of damages from drainage production of $293/acre and a welfare loss (relative to the first-best solution) of $21/acre; these results are identical to those for the Pigouvian tax policy (table 1).

The uniform standard performs significantly worse than other environmental policy options. The drainage constraint of 0.45 acre-feet per drained acre is infeasible for four of the sixteen simulated farms in the model. The drains on these farms collect drainage from on-farm irrigation, as well as significant quantities of drainage from neighboring, undrained farms. As a result, these farms must stop producing in the uniform standard scenario. This reduces regional average crop returns, water use (and total subsidies), and drainage production. The welfare cost of this policy is $65/acre (table 1). In addition to the efficiency costs, the distributional consequences of this policy, which are a function of heterogenous institutional and physical parameters, are quite large.

The marketable permits scheme implemented at a first-best effluent level generates lower private returns and lower returns net of environmental damages, but higher social welfare than when a second-best level of permits is issued or a Pigouvian effluent tax is imposed. Increased irrigation technology costs and crop acreage allocation adjustments required to achieve the more stringent drainage reductions explain the lower private crop returns. The welfare loss associated with this policy is $15/acre (table 1). Thus, it can be seen that the Pigouvian tax (price instrument) and marketable permits for first-best drainage levels (quantity instrument) diverge in the presence of an input price distortion. That is, the solution under the correct price (defined by P_d and D^s) is not the same as the solution under the correct quantity (defined by t_d and D^*).[8]

This result, which was suggested by the conceptual analysis, indicates that policy makers should use caution in applying the basic tenets of environmental economics under these circumstances. Clearly there are quantity instruments that are dual to the price instrument and vice versa. The point here is that the dual price or quantity may not represent what we commonly consider it to represent. For example, drainage production is nearly 30% higher than optimal with the Pigouvian tax policy in this case. Further, the price that would clear a permit market based on first-best quantities is $271 per acre-foot of drain water, more than $100/acre-foot higher than the Pigouvian tax.

Turning to the policy options for the water regulator, it is readily apparent that the alternative water prices generate significantly different results. The welfare loss associated with full-cost water prices, relative to the social optimum, is $99/acre (table 1). The water subsidy is less than half that in the base case, but still averages $140/acre. The welfare loss associated with a policy that charges the opportunity cost of water, thus eliminating the water subsidy, is $34/acre and reflects a relatively high level of drainage produced with this policy. A policy of charging $125 per acre-foot of water (the environmental surcharge) results in crop returns and returns net of environmental damages that are lower than with any other policy. Rather than receiving a subsidy, farmers would pay a tax with this policy that averages $95/acre. The welfare loss associated with this policy is $32/acre.

The environmental surcharge water policy, which is conceptually equivalent to an environmental policy of imposing an input tax, achieves the second-best drainage objective, although an efficient solution to this objective is not achieved. The difference in net social welfare under the Pigouvian tax policy and the environmental surcharge water policy indicates the inefficiency of selecting this water price as an environmental policy.

Regulator Policy Choice

Analysis of the choices that the two policy makers would make given these options requires assumptions regarding the objective of each. One possible objective for the environmental regulator is to maximize the sum of private returns net of environmental damages, S^e. This objective, which is common throughout the environmental economics literature, is specified as

[8] This result does not hold when marginal social costs are increasing. Increasing marginal social costs imply that the "correct price," i.e., marginal social costs, will vary with the effluent level. Thus, the distinctions between Pigouvian and second-best taxes blur with increasing marginal social costs. A Pigouvian tax set equal to marginal social costs evaluated at the first-best effluent level would motivate D^*. But the essential point that dual prices and quantities are not obvious still holds.

(3)

$$\max_{x_1,\ldots,x_I,z_1,\ldots,z_I} \; S^e = \sum_{i=1}^{I}[pf^i(x_i,z_i) - w_i x_i - rz_i - E_i] - p_d D[g^1(x_1, z_1), \ldots, g^I(x_I, z_I)].$$

Equation (3) differs from equation (2) in that crop returns are evaluated at the private cost for water rather than at the social cost. This, then, is the objective of an environmental policy maker who seeks an economically efficient solution, but fails to recognize the existence of the price subsidy, or, equivalently, recognizes the distortion, but believes it is immutable. Correspondingly, a reasonable objective for the water policy maker would simply be to minimize the water price distortion.

Under these assumptions, the rankings of individual regulators do not match the social orderings for the policy options available to that regulator. The environmental policy maker interested in solving equation (3) would choose the Pigouvian tax (or second-best effluent permits) as the optimal policy (column 3, table 1). However, the marketable permit scheme implemented at a first-best quantity generates higher net social welfare (welfare losses are $6/acre lower), and thus ranks higher socially than the Pigouvian tax policy (column 6, table 1). Likewise, of the three water policies, the opportunity-cost water policy would be preferred by the water regulator, but society would prefer the environmental surcharge. The welfare cost of making the "wrong" decision in this case is $2/acre.

Alternative Agency Objectives

The agency objective specified in equation (3) is a rather simple formulation in several regards: all factors are implicitly assumed to be deterministic and costlessly observable. Perfect compliance is also assumed. What are the implications for policy choice of these rather unrealistic assumptions? In this section, we briefly consider the range of objectives the agencies might have and the policy choices that would be consistent with those objectives.

Griffin and Bromley examine the case that emissions are not observable, but an effluent production function is known and input use is observed. The policy maker facing this problem is found to be indifferent between four policies: taxes on input use, input use standards, a tax on estimated effluent, and a standard for estimated effluent production. The objective specified in equation (3) is equivalent to this approach. Shortle and Dunn expand on the approach of Griffin and Bromley by introducing asymmetric information in the producer's profit function—the policy maker has incomplete information regarding producer's costs—and stochastic weather and uncertain processes into the effluent production function. In this case, a tax on input use is found to be the best policy choice.

Segerson examines a problem in which stochastic ambient water quality is observable but firms' emissions are not, thus addressing free-rider issues in the control of nonpoint source pollution. The policy option suggested in this case is a variant of the Pigouvian effluent tax: each polluter pays a charge based on the marginal benefit of reduced ambient pollution levels (which is a function of the contributions of all polluters). The Shortle and Dunn and Segerson papers, among others, suggest that the environmental authority may wish to account for uncertainty and choose as its objective to maximize the difference between expected benefits and costs. The choice of a Pigouvian tax is consistent with this objective, although the target of the instrument (input use in the case of Shortle and Dunn, and ambient quality in Segerson's analysis) may differ. Empirical evidence suggests that the Environmental Protection Agency (EPA) does attempt to balance costs and benefits of regulations, although special interest input is also a factor in pesticide regulations (Cropper et al.).

Numerous studies have examined the implications of noncompliance and enforcement costs on environmental regulations. These studies typically assume that the agency objective is to minimize enforcement costs while complying with legal mandates to address the environmental problem. For example, Malik considered this issue, concluding that an agency attempting to minimize enforcement costs subject to an environmental objective could not a priori choose marketable permits (or Pigouvian taxes) over a uniform standard on efficiency grounds.

These strands of the environmental economics literature suggest a number of possible objectives for the environmental authority. Outside of the environmental economics literature, additional theories of agency decision making exist that can be applied to either the environmental or water authority.

One alternative is the "agency capture" school of thought, which presumes that agencies will select policies that appease special interests (Buchanan and Tullock). This objective would manifest itself in an attempt to minimize the cost to the industry while satisfying legal obligations to address the issue. The full-cost water rate would satisfy this criterion for the water regulator. The environmental regulator would select either marketable permits (because they minimize compliance costs to the industry) or the uniform standard (because it might be most politically palatable to the industry) (see, e.g., Hahn 1990). Another model predicts that agencies will select regulations that most closely protect the status quo while satisfying their legal obligations. Merrifield notes a strong inertia in policy preferences by polluters, regulators, and environmentalists. This inertia may bias policy choices toward the more well-known command and control policies. Environmental and water agencies biased toward the status quo likely would select uniform standards and full-cost water prices, respectively. However, Merrifield also argues that cost effectiveness is becoming an increasingly important factor in policy making. The implication is that incentive-based policies, such as taxes and marketable permits, likely will evolve as add-ons to, but will not replace, command and control policies. Hahn (1989) also provides evidence supporting this proposition.

Yet another paradigm assumes that agencies attempt to maximize their budgets, or equivalently, maximize a bureaucrat's income and nonmonetary perquisites (Niskanen 1975). Subsequent modifications of the model consider the possibility agencies aim to maximize their discretionary budget, rather than their total budget (Niskanen 1991). This objective might imply a selection of effluent charges by the environmental regulator and the environmental surcharge for the water regulator, if revenues from charges remain with the regulator. Additionally, many researchers believe that enforcement costs are higher with incentive-based policy instruments. Thus, an agency wishing to enhance its role might rank marketable permit and tax policies above uniform standards.

Finally, the role of Congress in providing direction to regulatory agencies is considered. Congressional direction with respect to economic efficiency in regulations has been mixed. For example, the Toxic Substances Control Act (TSCA) and the Federal Insecticide Fungicide and Rodenticide Act (FIFRA) require consideration of benefits and costs by regulators, while the Clean Air and Clean Water acts do not (Portney). Congress did, however, include citizen suit provisions in much of the major environmental legislation. These provisions, which give interested citizens the right to sue agencies for actions (or inaction) inconsistent with the intent of the law, may be effective in moderating the effect of potential distortions due to alternative agency objectives.

The seven policies considered—four for the environmental regulator and three for the water regulator—appear to be sufficient to capture the range of policies likely to be selected, despite this broad range of agency objectives. Casual empiricism supports this view as well; these are the types of policies actually implemented or under discussion by agencies such as EPA and the Bureau of Reclamation. In the next section, we examine the welfare consequences of the four environmental and three water policies. Which of the combinations arises depends ultimately on a variety of political factors and agency objectives and we do not attempt to predict the outcome. Rather, we note that these are the policies under consideration and we provide a positive analysis of the welfare consequences of each combination.

Results: Efficiency Costs of Policy Combinations

The analysis, thus far, has focused on the implications of addressing a single distortion at a time. The welfare losses associated with each possible combination of the environmental and water policies are presented in table 2. The first row and first column repeat results discussed above for the water policies and environmental policies, respectively. The lowest value in the first row, a welfare loss of $15/acre, represents the best the environmental regulator on its own can do for society. Likewise, the best policy available to the water authority if no action is taken by the environmental regulator, the environmental surcharge, results in a welfare loss of $32/acre. The interior cells present welfare losses associated with combinations of policies. For example, if the environmental policy maker selects a Pigouvian tax and the water policy maker simultaneously implements full-cost water prices (second cell in the second row), the social loss is predicted to be $5/acre.

None of the policies, individually or in combination with others, reduces welfare below that attained in the base case; in each case the welfare loss is significantly lower than the base-case losses of $217/acre. Moreover, a range of options exist that attain solutions with minimal losses. Welfare losses associated with

Table 2. Welfare Losses Associated with Alternative Policy Combinations ($/acre)

	Environmental Policy				
	No Action	Pigouvian Tax	Effluent Permit $(D = D^s)$[a]	Effluent Permit $(D = D^*)$[b]	Uniform Standard $(D = D^s)$[a]
Water policy					
No action	217	21	21	15	65
Full-cost prices: $w_i = w_i^f$	99	5	10	4	52
Opportunity cost: $w_i = w^* = \$75/\text{af}$[c]	34	0	5	0	45
Environmental surcharge: $w_i = \$125/\text{af}$	32	26	31	27	65

[a] Policy attains the second-best drainage objective.
[b] Policy attains the first-best drainage objective.
[c] Policy eliminates the water price distortion.

the six outcomes from combinations of either full-cost or opportunity-cost water prices and a Pigouvian tax or effluent permit scheme range from $0 to $10/acre, and are $5/acre or less with only one exception. In addition, two combinations yield the first-best solution. Thus, in this case, the costs of piecemeal policy making appear in the form of opportunity costs of not attaining the first-best solution. In addition, the opportunity cost associated with a given policy depends on the policy choice of other policy makers.

While it is true that for a range of agency objective functions each policy maker's first ranked policy is not the social first rank when implemented in isolation, they combine here to attain the first-best. From table 2, it can be seen that if the environmental regulator selects the Pigouvian tax and the water policy maker selects a water price of w^*, the result is a welfare loss of zero; the first-best solution is attained. More importantly, this result is encouraging because it illustrates that, in this case, the social optimum can be attained without coordination. One regulator need not have information nor expectations regarding the behavior of the other, and need not even recognize that they are setting policy under second-best conditions. This result is particular to the distortions examined here and results from the complementary nature of the agricultural and environmental policies examined.[9]

Conversely, policy options for one authority that are welfare-enhancing when introduced in isolation may be welfare-reducing when combined with the policy selection of the other authority. This result can be seen by noting that the welfare losses are the same or higher for the combination of the environmental surcharge and any environmental policy than with the environmental policy alone (table 2). Thus, if the environmental authority chooses any of its four policy options, it would be preferable for the water authority to take no action rather than to impose the environmental surcharge water price.

Likewise, the social cost of implementing the uniform standard can be higher than that associated with no action by the environmental authority. For two of the three policy options considered for the water authority (opportunity cost or environmental surcharge water prices), the imposition of the uniform standard by the environmental authority results in welfare levels $11/acre and $33/acre lower than with no environmental policy (table 2). Thus, if the water policy maker chooses either opportunity cost or environmental surcharge water price policies, society is better off if the environmental regulator does nothing than if the uniform standard is established. This observation is particularly striking in light of a strong tradition of evaluating the efficiency costs of implementing "command and control" types of policies, such as a uniform standard, in terms of the gains that could be attained if a policy with efficiency properties is implemented (e.g., McGartland and Oates).

In considering alternative regulator objec-

[9] The result that the policy combination that includes the effluent tax and opportunity cost of water attains the first-best solution also depends on the assumption of constant marginal social costs. In the increasing marginal social cost specification examined, this combination does not attain the first-best solution. However, welfare losses in this case are estimated to be less than $1/acre.

tives, it is clear that a likely policy combination is the choice of uniform standards at the second-best quantity by the environmental regulator, and the full-cost water prices by the water regulator. These policies satisfy agency desires to maintain the status quo, given legal obligations. They may also be consistent with an objective of minimizing enforcement costs and the agency capture theory. In this case, the efficiency cost relative to the first-best set of policies is $52/acre. However, the efficiency gain relative to the base case is $165/acre.

Before completing the discussion of the empirical results, it is worth discussing an important qualification. As a reviewer noted, if there exists a third distortion in a related market, then there is no guarantee that the coordination of agricultural and environmental policy actually will improve social welfare. The most likely candidate for such a distortion in this case is crop price supports. To consider the possible effects of this third distortion, we ran our simulation model with a third regulator. An agricultural policy maker could eliminate policies that support the price of cotton, wheat, and sugar beets. The results changed only marginally quantitatively, and did not change at all qualitatively.

Final Remarks

A diverse array of agricultural programs have become ingrained in U.S. farm policy; these programs relate to nearly all aspects of agricultural markets. With the recent growth of environmental concerns has come increased environmental regulation of the agricultural sector. Thus, there is increasing potential for conflicting and counteracting regulatory actions to be undertaken by agricultural and environmental policy makers. In this paper we have extended the inquiry into the implications of uncoordinated policy making by moving beyond simple examination of the implications of agricultural policy for the effectiveness of environmental policy, which has been the approach in previous empirical studies. A social objective is specified to examine, both conceptually and empirically, the welfare losses associated with uncoordinated policy making. The empirical application is the interface between western irrigation water policies and policies for the control of drainage from irrigated agriculture.

Several important empirical findings resulted from the analysis. The rankings of individual regulators do not necessarily match the social orderings for the policy options available to the regulators, even when these regulators seek economically efficient policies. However, in the case presented here, if both regulators independently choose their top ranked policy, the social first-best ranking results. More generally, independent policy making by both authorities tends to improve social welfare relative to cases when only one regulator acts, regardless of the agency objective. However, there are important exceptions. Given any action by the environmental regulator, the environmental surcharge option for the water regulator reduces welfare. Likewise, if the water regulator has selected either the opportunity cost or environmental surcharge water prices, then the choice of the uniform standard by the environmental authority leaves society worse off than when pollution is left unregulated.

Antle and Just (p. 127) note that agricultural and environmental policies can be complementary or in conflict depending on the characteristics of the problem and the types of policy instruments. In the particular interface studied here, the environmental and agricultural (water) policies are complementary. Irrigation water use is a primary input to drainage production. As a consequence, the volume of irrigation water used is reduced with the three externality policies considered. Likewise, drainage production is reduced with the water policy options. Because of the direct link in these two forms of market distortions, as well as the direction of the distortion, the independent correction of either distortion improves social welfare. Thus, the cost of uncoordinated policy making arises in the form of opportunity costs of not achieving greater welfare gains. The magnitude of this cost ranges from $15/acre to nearly $100/acre when each authority acts alone, depending on the policy chosen. If both authorities act simultaneously, then cost estimates range from zero to $65/acre, again depending on the choice of policies.

The nature of the relationship between the "direct link" from water applications to effluent production and the policy complementarity in our application is informative with respect to generalizing the results found here. In our case, the complementarity derives from two conditions: first, a direct (though not one-to-one) relationship exists between the instrument of one distortion (water price subsides) and the creation of the other (drainage effluent). Also, both distortions—subsidized water and underpriced effluent—create an incentive to overuse the input. If the water price distortion had been too high of a price for water, correcting that distortion would have made the environmental problem worse, not better. Those two conditions, the direct link in the production technology and the equivalent directions of the distortions, are sufficient but not necessary for complementary policies.

Given the complementary nature of the market failures discussed here, it is not completely surprising that reducing one distortion tends to improve social welfare even when the other externality persists. However, even in the case of complementary policies, the magnitude of the welfare changes from piecemeal policy making deserve careful analysis as the size of the welfare gains from coordinating action may vary substantially depending on the particulars of the case. Additionally, although it does not occur in our application, it is possible that one agency may overcorrect when ignoring distortions in other markets and thus generate actual welfare losses.

Analysis of the type presented in this paper yields insights into the need for and gains from unified policy actions. The pervasiveness of government intervention in agricultural input and output markets motivates incorporation of these results when designing policies to address the environmental problems associated with agriculture.

[Received July 1993; final revision received November 1995.]

References

Antle, J.M., and R.E. Just. "Effects of Commodity Program Structure on Resource Use and the Environment." *Commodity and Resource Policies in Agricultural Systems*. R.E. Just and N. Bockstael, eds. Berlin: Springer-Verlag, 1991.

Baumol, W.J., and W.E. Oates. *The Theory of Environmental Policy*. Cambridge: Cambridge University Press, 1988.

Buchanan, J.M., and G. Tullock. "Polluters' Profits and Political Response: Direct Controls Versus Taxes." *Amer. Econ. Rev.* 65(March 1975):139–47.

California, State of, Regional Water Quality Control Board, Central Valley Region. "Staff Report on the Program of Implementation for the Control of Agricultural Subsurface Drainage Discharges in the San Joaquin Basin (5C)." Sacramento, California, August 1988.

Caswell, M., E. Lichtenberg, and D. Zilberman. "The Effects of Pricing Policies on Water Conservation and Drainage." *Amer. J. Agr. Econ.* 72(November 1990):883–90.

Cropper, M.L., W.N. Evans, S.J. Berardi, M.M. Ducla-Soares, and P.R. Portney. "The Determinants of Pesticide Regulation: A Statistical Analysis of EPA Decision Making." *J. Polit. Econ.* 100(February 1992):175–97.

Dinar, A., and J. Letey. "Agricultural Water Marketing, Allocative Efficiency and Drainage Reduction." *J. Environ. Econ. and Manage.* 20(May 1991):210–23.

Griffin, R.C., and D.W. Bromley. "Agricultural Runoff as a Non-point Externality: A Theoretical Development." *Amer. J. Agr. Econ.* 64(August 1982):547–52.

Hahn, R.W. "Economic Prescriptions for Environmental Problems: How the Patient Followed the Doctor's Orders." *J. Econ. Perspect.* 3(Spring 1989):95–114.

___. "The Political Economy of Environmental Regulation: Towards a Unifying Framework." *Public Choice* 65(March 1990):21–47.

Hanemann, M., E. Lichtenberg, D. Zilberman, D. Chapman, L. Dixon, G. Ellis, and J. Hukkinen. "Economic Implications of Regulating Agricultural Drainage to the San Joaquin River." Report to the California State Water Resources Control Board, Sacramento, California, August 1987.

Howe, C.W., D.R. Schurmeier, and W.D. Shaw, Jr. "Innovative Approaches to Water Allocation: The Potential for Water Markets." *Water Resour. Res.* 22(April 1986):439–45.

Howitt, R.E. "Water Policy Effects on Crop Production and Vice Versa: An Empirical Approach." *Commodity and Resource Policies in Agricultural Systems*. R.E. Just and N. Bockstael, eds. Berlin: Springer-Verlag, 1991.

Howitt, R.E., N. Moore, and R.T. Smith. *A Retrospective on California's 1991 Emergency Drought Water Bank*. A report prepared for the California Department of Water Resources, March 1992.

Hrubovcak, J., M. LeBlanc, and J. Miranowski. "Limitations in Evaluating Environmental and Agricultural Policy Coordination Benefits." *Amer. Econ. Rev.* 80(May 1990):208–12.

Just, R.E., N. Bockstael, R.G. Cummings, J. Miranowski, and D. Zilberman. "Problems Confronting the Joint Formulation of Commercial Agricultural and Resource Policies." *Commodity and Resource Policies in Agricultural Systems*. R.E. Just and N. Bockstael, eds. Berlin: Springer-Verlag, 1991.

Just, R.E., A. Buss, and G. Donoso. "The Significance of the Interface of Agricultural and Resource Policy: Conclusions and Directions for Further Research." *Commodity and Resource Policies in Agricultural Systems*. R.E. Just and N. Bockstael, eds. Berlin: Springer-Verlag, 1991.

Just, R.E., E. Lichtenberg, and D. Zilberman. "The Effects of Feed Grain and Wheat Programs on Irrigation and Groundwater Depletion in Nebraska." *Commodity and Resource Policies in Agricultural Systems*. R.E. Just and N. Bockstael, eds. Berlin: Springer-Verlag, 1991.

Kopp R.J., and A.J. Krupnick. "Agricultural Policy and the Benefits of Ozone Control." *Amer. J. Agr. Econ.* 69(December 1987):956–62.

Lichtenberg, E., and D. Zilberman. "The Welfare Economics of Price Supports in U.S. Agriculture." *Amer. Econ. Rev.* 76(December 1986):1135–41.

Lipsey, R.G., and K. Lancaster. "The General Theory of Second Best." *Rev. Econ. Stud.* 24(1956):11–32.

McGartland, A.M., and W.E. Oates. "Marketable Permits for the Prevention of Environmental Deterioration." *J. Environ. Econ. and Manage.* 12(September 1985):207–28.

Malik, A.S. "Enforcement Costs and the Choice of Policy Instruments for Controlling Pollution." *Econ. Inquiry* 30(October 1992):714–21.

Merrifield, J. "A Critical Overview of the Evolutionary Approach to Air Pollution Abatement Policy." *J. Policy Anal. and Manage.* 9(Summer 1990):367–80.

Miranowski, J.A., J. Hrubovcak, and J. Sutton. "The Effects of Commodity Programs on Resource Use." *Commodity and Resource Policies in Agricultural Systems*. R.E. Just and N. Bockstael, eds. Berlin: Springer-Verlag, 1991.

Moore, M.R. "The Bureau of Reclamation's New Mandate for Irrigation Water Conservation: Purposes and Policy Alternatives." *Water Resour. Res.* 27(February 1991):145–55.

Niskanen, W.A. "Bureaucrats and Politicians." *J. Law and Econ.* 18(December 1975):617–44.

___. "A Reflection on Bureaucracy and Representative Government." *The Budget Maximizing Bureaucrat: Appraisals and Evidence*. A. Blais and S. Dion, eds. Pittsburgh PA: University of Pittsburgh Press, 1991.

Portney, P. *Public Policies for Environmental Protection*. Washington DC: Resources for the Future, 1990.

San Joaquin Valley Drainage Program, Interagency Study Team. "Overview of the Westside Agricultural Drainage Economics Model (WADE) for Plan Evaluation." Technical Information Record. Sacramento, California, October 1989.

Segerson, K. "Uncertainty and Incentives for Nonpoint Pollution Control." *J. Environ. Econ. and Manage.* 15(March 1988):87–98

Shortle, J.S., and J.W. Dunn. "The Relative Efficiency of Agricultural Source Water Pollution Control Policies." *Amer. J. Agr. Econ.* 68(August 1986):668–77.

U.S. Congress. Reclamation Projects Authorization and Adjustment Act, Pub. L. 102–575, 30 October 1992.

___. Reclamation Reform Act, Pub. L. 97-293, 12 October 1982.

U.S. Department of Interior, Office of the Secretary. *Irrigation Subsidy Legislation: Questions from the Subcommittee on Water and Power Resources of the Committee on Interior and Insular Affairs*. H.R. 1443, 24 February 1988.

Weinberg, M. *Economic Incentives for the Control of Agricultural Non-point Source Water Pollution*. Unpublished PhD dissertation, University of California, Davis, 1991.

Weinberg, M., C.L. Kling, and J.E. Wilen. "Water Markets and Water Quality." *Amer. J. Agr. Econ.* 75(May 1993):278–91.

Part III
Water Reallocation and the Environment

[13]

WATER RESOURCES RESEARCH, VOL. 26, NO. 6, PAGES 1113–1120, JUNE 1990

Enhancing Instream Flow Benefits in an Era of Water Marketing

BONNIE G. COLBY

Department of Agricultural Economics, University of Arizona, Tucson

Growing populations in the western United States demand water not only for residential use and to support urban development but also for recreation, water quality enhancement, improvement of fish and wildlife habitat and to preserve the aesthetics of riparian areas. Instream flows contribute substantial economic benefits, and emerging pressure to reserve water instream comes at a time when markets are evolving to reallocate water among offstream uses such as agriculture, industry and municipal expansion. This article examines current instream flow policies in the western states and outlines the economic values generated by stream flows. The author argues that instream values are high enough to compete in the market for water rights with offstream uses when important recreation sites and wildlife species are involved. The paper suggests how western state policies might be altered to accommodate instream flow protection within the context of water marketing, with the objective of improving the efficiency of water allocation among instream and consumptive uses.

INTRODUCTION

During the last decade, nearly every western state has grappled with the trade-offs between providing water for offstream uses and retaining water instream to support recreation, power generation, water quality, fish and wildlife. The issue has arisen in state and federal courts, state legislatures and federal agencies (see Water Market Update, 1987–1989 and Water Strategist 1987–1989 for numerous examples). Policymakers' interest has been stimulated for a number of reasons. First, many western cities are growing rapidly and new city dwellers demand boating, fishing and other outdoor recreation opportunities that rely on adequate stream and lake levels. Second, as diversions of water for offstream irrigation, industrial and residential deliveries have increased, flow levels on many stream systems have decreased, sometimes to the detriment of instream water uses. Finally, increased income and opportunities for leisure activities among urban westerners brings increased appreciation of the intrinsic and aesthetic value of free-flowing water and of its economic value in enhancing recreation and wildlife habitat, thus contributing to the western tourism industry.

This article gives an overview of western states' instream flow policies, reviews recent economic evidence on instream flow values in the West, analyzes the relationship between the water marketing phenomenon and instream flow protection and concludes with suggestions for policies that enhance the economic contributions of free-flowing waters.

Adequate instream flows are vital to preserving fish and wildlife habitat in the arid West. Water-based recreation is an important part of many westerners' leisure activities and water-related recreation opportunities draw visitors and tourism dollars to the West. Since there is little direct market evidence on willingness to pay for recreational opportunities and for wildlife preservation, a variety of nonmarket valuation approaches have been applied to estimate the value of water for these purposes (see *Water Resources Research*, volume 23, issue 5, 1987, for a review of various approaches). Studies which focus on the contribution of streamflows to recreation, water quality, community development and wildlife values are reviewed later in this article.

Scarcity is a key theme in discussing economic trade-offs between instream and offstream uses. Economists define a resource as scarce when there is not enough of it available to satisfy existing and potential demand, and, consequently, some allocative decisions must be made regarding who will have access to the resource and under what conditions. The evolution of the doctrine of prior appropriation in the West is evidence of economic scarcity as water users realized an allocative process was needed to settle conflicts and facilitate orderly use of water resources. Implicit in economic scarcity is the notion that individuals are willing to pay for access to or use of a scarce resource based on the value they attribute to the resource. As new values emerge there is often pressure to modify allocative processes in order to recognize and accommodate new demands. The proliferation of legislation and case law related to instream flow protection during the last decade is one indication that the benefits generated by instream flows have become important enough to warrant changes in the legal framework originally designed to promote offstream water uses.

OVERVIEW OF STATE INSTREAM FLOW POLICIES

While there are strong economic arguments for preserving flows to support recreation and wildlife, the western states differ a great deal in their approaches to instream flow protection. Differences are notable both in the statutory basis for establishing water rights to maintain flow levels and the extent of state agency programs directed toward protecting free-flowing waters. This section briefly summarizes the policies of selected western states. Readers should refer to *MacDonnell* [1989] for a more comprehensive discussion of state policies and their impacts.

While Arizona statutes do not explicitly recognize appropriations for instream flow maintenance, a 1976 court case held that surface water may be appropriated for instream recreation and fishing [McClellan versus Jantzen, 1976]. The Arizona Department of Water Resources (ADWR) issued two permits to the Nature Conservancy in 1983 and one to the Bureau of Land Management in 1989, authorizing these entities to appropriate water for instream flows. As of mid 1989, over 40 minimum instream flow permit applications

Paper number 89WR03617.
0043-1397/90/89WR-03617$05.00

were pending before ADWR and an instream flow task force had been appointed to assist ADWR in formulating new criteria and procedures for granting permits (State of Arizona, Department of Water Resources, 1989).

In California, case law has ruled against appropriation where there is no diversion or other physical control over the water [Fullerton versus State Water Resources Control Board, 1979; California Trout, Inc. versus State Water Resources Control Board, 1979]. However, instream uses are declared to be reasonable and beneficial and the state board must consider impacts on instream uses in approving new appropriations and transfers (State of California, Water Code Sec. 1243). The Mono Lake case [National Audubon Society versus Superior Court of Alpine County, 1983] resulted in explicit recognition by the California Supreme Court of the Public Trust Doctrine's application to water resources and this provides further basis for instream flow protection. Pursuant to this decision, the Supreme Court ordered Los Angeles to curtail its diversions from the Mono Lake area in order to avoid further damage to wildlife resources (Water Market Update, volume 3, number 9, 1989).

In Colorado, the Colorado Water Conservation Board (CWCB) may appropriate water for instream flow and lake level maintenance. Private entities are not authorized to appropriate water for instream flow protection but may dedicate water rights to the CWCB for instream flow maintenance. The CWCB is also responsible for filing objections to water transfers which may impair instream flow rights (State of Colorado, Rev. Stat. Sec. 37-92-102 (3) and Sec. 37-92-103 (4)).

Montana passed legislation in 1973 authorizing the state to reserve quantities of flows as needed for the public interest (Section 84-2-316, MCA). The state has established such flow reservations in the Yellowstone River basin and is evaluating reservations for stream flow maintenance on other key river systems [*McKinney et al.*, 1988]. Montana has also established a pilot program of water leases to maintain and enhance streamflows for fisheries (Water Market Update, volume 3, number 6).

Appropriations for instream flow and storage in lakes without a physical diversion have been granted in Nevada in specific instances. Instream flow appropriations must be acquired through the same process as any other appropriation. A 1988 Supreme Court decision determined that the federal government, representing public interests, can hold instream rights under Nevada law. Nevada versus Morros ruled that recreation constitutes a beneficial use of water, even though physical diversion of flows does not occur [Nevada versus Morros, 1988].

New Mexico has no statutes pertaining to appropriation of water for instream flow maintenance, though recognition of instream flow rights has been considered in recent legislative sessions. Case law and decisions by the state engineer imply that diversion structures are necessary for water right appropriations (Reynolds versus Miranda, 1972). While case law and administrative procedures do not require that impacts on instream flow levels (other than those which affect vested water rights) be considered in evaluating proposed water use changes, a little known statute passed in 1912 declares that no diversion shall lessen streamflows to the detriment of fish populations (New Mexico Statutes, Ann. Sec. 17-4-15). This statute could provide a framework for instream flow protection in New Mexico.

Oregon passed legislation in 1987 authorizing appropriation of water rights for instream flow protection and providing that a proportion of irrigation water salvaged through conservation efforts be dedicated to instream flow maintenance (State of Oregon, Senate Bill 24, 1987). One application for transfer of conserved water had been filed by the end of 1989.

A Utah statute enacted in 1986 allows the State Division of Wildlife Resources to acquire established water rights to maintain flows for fish habitat. The division must have legislative approval to acquire a right for instream flows (State of Utah, Code Ann. Sec. 73-3-3, 1986).

Economic Benefits Generated by Instream Flows

State policies to protect instream flows are a response to growing recognition of the benefits generated by free-flowing water. In order to understand the magnitude of these benefits and to provide policymakers with information on the relative value of water in various instream and offstream uses, economists have estimated benefits attributable to streamflows. Several concepts of value have been used in these studies. Marginal values measure the benefits generated by an additional increment of water added to a specified base flow level. Marginal values are useful in understanding the economic impacts of policies that provide higher streamflows and of policies that reduce flow levels. Total value refers to the economic benefits generated by a particular stream setting or recreation site, rather than to increments or decrements in flow levels. Instream flow benefits are often measured using travel cost and contingent valuation techniques and usually rely on willingness-to-pay concepts of value. Travel cost methods use data on costs associated with visiting a site to infer the benefits to the recreationist. Contingent valuation relies on survey methodology to elicit values from a respondent and can be used to estimate both user and nonuser values.

There are several kinds of economic benefits generated by instream flows.

Water Quality Enhancement

A stream's dilution capacity provides economic benefits related to the costs of treatment that would otherwise be incurred by dischargers and by downstream water users. Higher streamflows dilute wastes and benefit offstream industrial and agricultural water users who bear the higher operating and maintenance costs and lower crop yields associated with low water quality. As high-quality water is diverted for offstream uses, streamflows become depleted, water quality standards are more likely to be violated and municipal sewage treatment plants and industrial discharges have to incur additional expenses to assure compliance with state and federal water quality standards. Very few studies have estimated the monetary benefits of dilution capacity. However, *Young and Gray* [1972] did estimate values for assimilating biological oxygen demand. *Greenley et al.* [1982] indicate substantial benefits to recreationists from maintaining and improving surface water quality in Colorado's South Platte River Basin along the populated Front Range.

Recreation Values

Outdoor recreation in the West concentrates around lakes, rivers and streams. Adequate streamflows are essential to boating and fishing and also, because of their importance to wildlife, to hunting, bird watching and other wildlife-related recreation.

Daubert and Young [1979] examined the contribution of streamflows to recreation benefits on Colorado's Cache la Poudre River. They found the value of an additional Mm^3 of flow for fishing to be $17 during low-flow periods and the value of an additional Mm^3 for shoreline recreation to be $12 during low river flows. Values for an additional unit of flow dropped to zero at higher flow levels suggesting that minimum flow maintenance is of value to recreationists rather than additional increments to already adequate flows. *Walsh et al.* [1980*a*] investigated flow values at nine sites along Colorado mountain streams and found that flow levels of 35% of maximum streamflow were optimal for recreation. The value of an additional Mm^3 of flow beyond the 35% flow level was estimated to be $17 per Mm^3 for fishing, $4 for kayaking and $3 for rafting. *Walsh et al.* [1980*a*] estimate that leaving an additional increment of water in high mountain Colorado reservoirs for an additional 2 weeks in August is worth $39 per Mm^3 in additional recreation benefits during that peak recreation period. *Amirfathi et al.* [1984], analyzing recreation on a river in northern Utah, found that the value of additional flows is zero until flows dropped to 50% of peak levels. Marginal values for streamflows reached a maximum of $65 per Mm^3 when flows were 20–25% of peak levels. *Ward* [1987] examined the relationship between streamflow levels, recreation use levels and travel costs incurred by recreationists on New Mexico's Rio Chama to infer a value of $13 to $22 per Mm^3 of reservoir releases in the summer recreation season, assuming optimal augmentation of streamflows during low-flow periods. Consistent with other studies, Ward found that marginal values fall dramatically for high-flow periods and when stored water is available to augment natural flow levels. These results suggest a significant economic payoff in augmenting streamflows in low-flow years, even though augmentation would reduce water availability for other users.

Loomis [1987*a*] provides an overview of the various methods that have been applied to measure the economic value of instream flows, citing studies relying on the travel cost method and on contingent valuation. He argues convincingly, based on the studies cited, that dollar values of instream flows can be reasonably estimated so that they can be compared to the value of water in offstream uses, such as irrigation.

Local Economic Development

Many small towns and Native-American tribal reservations rely on seasonal tourism as a significant source of livelihood for local residents. Spending on boating, fishing, hunting and camping equipment supports recreation-related businesses and stimulates local and tribal economies. While few studies have estimated the dollars flowing into local economies from water-related tourism, *Boyle and Bishop* [1984] found that boaters on a 12-km stretch of the Wisconsin River generate over $800,000 of sales by local businesses during the summer boating season. Half of this remains in the local economy as income to local households, and the other half goes to pay for supplies and services purchased from other areas by local businesses. These figures underestimate total recreation-linked economic activity since they do not include spending by summer hunters, fisherman, hikers and fall, winter and spring recreationists who visit the river. Tribes, towns and counties have a substantial economic interest in preserving the free-flowing streams upon which local tourism depends. They also have an incentive to monitor and enforce streamflow levels they are able to legally protect under state law or, in the case of Indian tribes, under federal reserved rights and other legal avenues available to tribes.

Nonuser Values

Nonuser values, benefits to individuals who do not directly use streamflows for recreation, are of several types. Benefits associated with preserving a riparian area so that one has the option to enjoy it in the future are termed "option values." Option value is relevant when choices must be made between an irreversible alternative (or one that is costly to reverse), such as drying up a stream environment or flooding a canyon, and the alternative of leaving the river system in its current state, which is reversible since new diversions or water development can later be approved. Willingness to pay for preservation so that one's heirs can enjoy the source is termed "bequest value," and benefits generated by simply knowing a unique site will continue to exist are termed "existence values." Nonuser benefits are relevant in valuing instream flows where there are wildlife species whose survival is dependent upon streamflows and also where there are areas whose aesthetic and retreational characteristics are dependent on free-flowing water. Since nonuser values are not associated with an actual visit to a site, they are particularly difficult to estimate. However, recent studies indicate that nonuser values can be sizable, especially for unique recreation sites and for endangered species. Existence, bequest and option values ranging from $40 to $80 per year per nonuser household have been estimated for stream systems in Wyoming, Colorado and Alaska [*Greenley et al.*, 1982; *Madariago and McConnell*, 1987].

Studies on nonuser values for wildlife habitat and specific species suggest that nonuser values can be sizable, especially for unique sites and endangered species [*Walsh et al.*, 1984]. *Loomis* [1987*b*] found that individuals' willingness to pay to preserve the level of California's Mono Lake, while based partly on the enjoyment stemming from an actual site visit, was largely based on the satisfaction of knowing the lake would be preserved (existence value), assuring the opportunity for future visits (option value) and guaranteeing site availability for the next generation (bequest value). These nonuser values accounted for over 80% of total willingness to pay.

Significant but hard-to-measure nonuser values are associated with water in lakes and streams. Local economic benefits and nonuser values are difficult to translate into dollars per Mm^3 so as to be comparable with the marginal value of water in offstream uses. However, recognition of these benefits suggests that marginal values for water in recreation uses and for water quality enhancement should be regarded as a lower bound, a minimum estimate, of the actual benefits generated by maintaining instream flows and lake levels.

Policy Relevance of Economic Values

While the studies cited are not an exhaustive review of the literature, they do demonstrate that the economic value of instream flows can be estimated and compared to the value of water in offstream uses. The benefits generated by keeping another Mm^3 of water instream can be equal to or greater than the marginal value of water in competing offstream uses, especially in important recreation and wildlife areas.

Failure to incorporate instream flow values into water management decisions can result in water use patterns that do not maximize the economic benefits that are potentially available from regional water resources. *Daubert and Young*'s [1979] research on instream values in northern Colorado suggests that benefits generated by area stream systems could be enhanced by altering the timing of water storage and releases to increase instream flows during the fall recreation season. Recreational benefits associated with instream flows could be increased without decreasing water availability for irrigation, implying that payments to persuade irrigation right holders to alter water management practices in favor of recreation need not be large in this particular study area. *Loomis* [1987*b*] estimates total visitor and nonvisitor benefits from preservation of Mono Lake levels to be about $40 annually per California household, well above the cost of $2 to $3 per household per year to preserve lake levels by replacing Los Angeles diversions from streams feeding the lake with water from other sources. These figures suggest that the benefits of preservation greatly outweigh the costs and have clear implications for California and Los Angeles area water policy decisions.

Attention to the benefits generated by instream flows in other parts of the West will help to identify economically beneficial alterations in diversions for offstream uses. Without information on instream values, water policy decisions will continue to emphasize offstream diversions for uses with more easily documented values, such as irrigation, mining, energy development, manufacturing and urban growth. Recent evidence on the economic value of water instream suggests that instream benefits can often exceed the benefits generated by offstream uses and that economic development in the western states could be enhanced by more attention to instream flow protection for recreation and wildlife.

INSTREAM FLOW PROTECTION AND WATER MARKETING

As economic benefits generated by free-flowing waters become increasingly recognized, instream flow considerations are playing a greater role in western water market activity. In many areas of the West, market acquisition of water rights plays an important role in providing water for offstream uses [*Saliba and Bush*, 1987]. Acquisition of senior rights can give water users secure access to water that would not be provided by a new appropriation. Groups interested in protecting instream flows seek to purchase or lease senior appropriative rights, recognizing that new appropriations often will not be sufficient to guarantee flow levels adequate for recreation, fish and wildlife, given extensive senior rights to divert water for consumptive uses. Dry-year options, in which arrangements are made ahead of time for access to water during drought, are an attractive alternative to conventional purchases and leases. Such options can protect fish, wildlife and recreation during dry years but do not tie up water during years when streamflows are adequate. Where instream flow maintenance is recognized as a beneficial use so that water rights may be held for that purpose, market transfers could become an important means of accomplishing instream flow protection.

What Characterizes Water Markets?

A "water market" consists of the interactions between buyers and sellers of rights to use water resources either for a limited period of time or into perpetuity. Negotiated transactions generate prices and conditions of sale and lease. The term "market" generally refers to a set of transactions taking place continuously over a period of time. When relatively few transactions take place, the market is considered "thin," and a key market function, the establishment of a "going" price, may be lacking. Market transfers are only one of many processes by which water is reallocated in the western United States. Other reallocative processes include at-cost administrative transfers (in which water is leased for a rate that covers costs associated with the transfer but does not confer economic gain on the lessor); forfeiture and abandonment proceedings under state law; public agency exercise of eminent domain powers; litigation challenging existing water allocations; congressional and state legislative settlements of conflicting claims; and water project redesign to alter initial project allocations among alternative water uses.

The following characteristics distinguish market transfers from other transfer processes and from transfers of other property rights:

1. Water's value is recognized as distinct from the value of land and improvements. Water is bought or leased for its own sake, not merely as an incidental part of a land transfer.
2. Buyers and sellers agree to transfer rights to use water voluntarily, believing it is in their own best interest given the alternative opportunities available to them.
3. Price and other terms of transfer are negotiable by the buyer and the seller and are not constrained to be "not for profit" or "at cost."

The motivating force behind water markets is mutual perception by potential buyers and sellers that economic gains may be captured by transferring water to a place or purpose of use in which it generates higher net returns than under the existing use patterns. The economic returns to buyers must be large enough (or be perceived as large enough) to outweigh the costs of obtaining water through the market process. Even if the net benefits to prospective buyers of transferring water are positive, a second criterion must be satisfied for a market transfer to occur. A market transaction must be attractive relative to other processes by which buyers could achieve their water supply objectives. The costs of a market acquisition, including political and legal costs, must be less than the costs of alternative means of obtaining water, such as hooking up to an existing water utility or contracting for water deliveries from a public water project.

While prospects for increasing the economic returns of water use are the driving force behind markets, laws and policies affect the cost of market transactions and the attractiveness of market transfers relative to other means of transferring water. The legal and political setting determines the transactions costs associated with market transfers. Transactions costs are incurred in identifying legal and hydrologic characteristics of water rights (priority date, return flow obligations, etc.); in negotiating price, financing, and other terms of transfer; and in satisfying state laws and transfer approval procedures. State laws impose transactions costs on market participants in the form of approval requirements for changing the purpose and place of use of a water right. These may include court hearings, title searches consumptive use studies and other hydrologic studies to determine transfer impacts. Ambiguity in state law can also impose costs by creating uncertainty regarding how much water can be transferred and for what purposes. This ambiguity is particularly prevalent when instream flows are involved, since many states have not yet developed criteria and procedures for evaluating instream flow applications. Transactions costs influence the profitability of a given transfer and can therefore affect the level of market activity.

Role of Market Acquisitions in Protecting Instream Flows

Although purchase or lease of senior appropriative rights appear to be a logical approach for assuring adequate streamflows, there are only a few examples of market acquisition of water rights to maintain streamflows or lake levels for recreation and wildlife purposes. In 1987, Lander County, in north central Nevada, purchased 3690 Mm^3 of senior irrigation rights for $176 per Mm^3 in order to maintain a stable shoreline for fishing and boating on a new county reservoir (Water Market Update, volume 1, number 5, 1987). In 1987 the Montana Fish, Wildlife and Parks department purchased 12,336 Mm^3 of water to be released from a reservoir on the Bitteroot River in the western part of the state. Concern over the survival of trout fisheries during the unusually dry summer prompted the purchase of one-time rights to summer reservoir releases for $1.6 per Mm^3 (Water Market Update, volume 1, number 8, August 1987).

In 1989 the Nature Conservancy purchased 2467 Mm^3 of water to support flows on a stretch of Colorado's North Poudre River. The arrangement involved the cooperation of an irrigation district that stores and releases water from a reservoir above the Conservancy's Phantom Canyon Preserve. The district had agreed to schedule 1988 winter releases so as to maintain a specific flow through the preserve, and the Conservancy acquired water for the district to make up for the winter releases when irrigators faced a short fall during the 1989 summer irrigation season (Water Market Update, volume 3, number 8, September 1989).

Diverse interests cooperated in the purchase of 1230 Mm^3 to support riparian habitat near Sacramento, California, in 1989. Several years of drought threatened a rare bird species, other wildlife and fisheries along Putah Creek. County and city governments, along with several water districts, contributed money and water to assure adequate releases from an upstream reservoir and to cover the costs of substituting groundwater for diversions from the creek (Water Market Update, volume 3, number 8, September 1989).

In 1989, California's Department of Fish and Game and Grasslands Water District purchased 36,900 Mm^3 to maintain wildlife and fish in the drought-stricken San Joaquin River Basin (Water Market Update, volume 3, number 9, October 1989). The Upper Snake Water Bank in Idaho, in cooperation with the Nature Conservancy, leased water from irrigators to protect rare trumpeter swans. While the initial agreement was for 1989, negotiations are underway for longer-term arrangements to provide adequate winter flows for the swan flock (Water Market Update, volume 3, number 9, October 1989).

There are a number of other market acquisitions for instream flows being contemplated around the West. The Nature Conservancy is actively pursuing purchase of water rights on the Yampa River in western Colorado to facilitate endangered fish recovery (Water Market Update, volume 1, number 3, March 1987). Congress approved, as a part of the 1988 federal budget, 1 million dollars for acquiring water rights to protect endangered fish species in the upper Colorado River basin of Wyoming, Colorado and Utah (Water Market Update, volume 2, number 2, February 1988). The Bonneville Power Administration Department of Fish and Wildlife is considering purchase of senior irrigation rights in the Sawtooth area of Idaho to protect salmon fisheries (Water Market Update, volume 1, number 6, June 1987). Nevada waterfowl interests have raised money that is earmarked for water rights purchases to enhance the Stillwater wetlands in western Nevada (Water Market Update, volume 3, number 9, October 1989). The California legislature, in 1989, established a special fund for water management to provide environmental benefits, including purchases of water to preserve riparian areas and to improve water quality (Water Market Update, volume 3, number 9, October 1989).

While instream flow protection can be enhanced through market transactions, active water markets can make instream flow protection more controversial since instream flow rights could make water transfers more complicated and costly to implement. Currently, water rights for instream flow maintenance are few in number relative to rights for consumptive uses, and most instream flow rights are recent appropriations and have low priority relative to other water rights. Those free-flowing waters not protected by a water right have no legal recognition and thus create no legal basis for protesting transfers which will have adverse impacts on flow levels. However, as public interest and environmental concerns play a greater role in water policy, instream flow impacts may eventually be considered routinely in approval of transfers between consumptive users. *Livingston and Miller* [1986] characterize conflicts of interest between consumptive water users desiring to transfer water rights and interest groups seeking to protect instream flows as stark, unavoidable and pervasive. *Shupe* [1986] notes that since instream flow rights typically are year-round rather than seasonal, and since they often extend along a stretch of a stream rather than being diverted at a single point, they are particularly constraining for new water development and for water transfers. Consequently, establishment and enforcement of instream flow rights will continue to generate controversy among proponents of water marketing.

Why So Few Market Transactions?

If instream flow values are high enough to compete with offstream water uses, why are not purchases and leases of water rights to maintain instream flows more common? Instream flow interests are not well represented in western water markets for several reasons. First, those wishing to protect streamflow levels do not have legal access to water rights on the same terms as farmers, cities and industry. Some western states do not recognize instream flow maintenance as a beneficial use and so water rights may not be held for instream purposes. Only Alaska and Arizona, of the western states, allow a private party to hold a water right for the purpose of maintaining instream flows (Water Market Update, volume 2, number 3, March 1988). Markets could better incorporate instream flow values if state laws permitted appropriation, purchase and seasonal leasing of water rights for instream flow maintenance by both public and private organizations.

A second reason why there have not been more market transactions is that the transactions costs for instream flow acquisition are likely to be higher than for water rights purchased for offstream uses. Organizations wishing to use water rights to maintain streamflows often face opposition by neighboring water users who fear the flexibility of their own rights will be constrained and incur high costs in overcoming objections to the new instream use of the water rights. Further, many state agencies have little experience in handling applications for change in purpose of use of a water right from irrigation, for instance, to instream flow maintenance. New procedures and criteria often have to be developed, creating delays, uncertainty and additional costs for the instream use applicant.

Even if obstacles to acquisition of water rights for maintaining flow levels were abolished, instream flows have public good characteristics which make it difficult to translate collective values for instream flows into dollars to bid for water rights in the market place. Those who benefit from free-flowing waters are a large, but largely unorganized, constituency. The term "public good" refers to resources characterized by nonexcludability, meaning it is difficult or impossible to exclude those who do not pay from enjoying the benefits of the resource. Many individuals who do place a positive value on a public good may be "free riders," enjoying the resource but making no payments, since payments are not required. Funds raised to purchase water for instream flow maintenance will not represent total willingness to pay by all potential beneficiaries due to the free ridership phenomenon, the difficulty of collecting contributions from all who will benefit, and the lack of an incentive to voluntarily contribute, since those who do not contribute cannot easily be prevented from enjoying the resource.

In spite of these obstacles to raising money for instream flow acquisitions, environmental groups have successfully organized fund raising and donations to acquire water rights. For instance, the Nature Conservancy has received donations of water rights which they intend to use for fish and wildlife enhancement on the Gunnison River in Colorado and on Aravaipa Creek in southeastern Arizona (Water Market Update, volume 1, number 3, March 1988). Donations of water and funds were also crucial in implementing some of the market acquisitions described earlier. In those states which allow private organizations to acquire rights for instream flow maintenance, such donations along with market transfers, could become an important means of accomplishing instream flow protection.

The public sector is becoming more active in acquiring water for instream flows as illustrated by the Montana Fish, Wildlife and Parks Department acquisition, the acquisition by a county government in Nevada and the 1988 Congressional appropriation previously described. In 1987 the Bureau of Reclamation announced it was altering Shasta Reservoir releases into the Sacramento River of northern California in order to enhance the Chinook salmon fishery, at the expense of over 1 million dollars in foregone hydropower revenues (Water Market Update, volume 1, number 9, September 1987). The Colorado Water Conservation Board has appropriated water for junior instream flow rights on over 10,483 km of streams and numerous lakes. These rights do not guarantee specific flow levels, however, because preexisting senior rights for consumptive uses may deplete the stream below levels desirable for recreation and wildlife. CWCB has also participated in the acquisition of some senior rights which provide more reliable protection of streamflows [*Shupe*, 1989]. Further examples of public sector involvement can be found in the work by *MacDonnell* [1989], which describes federal and state efforts in support of instream flows.

Role of Litigation

The role of litigation in prompting instream flow protection is significant. The 1987 Mono Lake case discussed earlier is an important landmark. A lawsuit filed by the Environmental Defense Fund and other parties in the 1970s resulted in a 1989 California ruling requiring a large Bay Area water provider to divert less than its full entitlement from the American River during dry years to provide flows for salmon and recreation. The court also ruled that state wildlife officials could request up to 74,016 Mm^3 per year in releases from an upriver reservoir if needed for fish and wildlife in the American River [Environmental Defense Fund versus East Bay Municipal Utilities District, 1989].

In Texas the Guadalupe-Blanco River Authority has begun proceedings based on the Endangered Species Act to curtail groundwater pumping that it asserts are damaging free-flowing springs and endangered species that rely on the springs. The River Authority would benefit from reduced pumping from the Edwards Aquifer because it relies on surface water supplies affected by the groundwater withdrawals (Water Market Update, volume 3, number 7, July–August 1989).

Litigation based on the Endangered Species Act has been instrumental in the Pyramid Lake Paiute Tribe's efforts to regain adequate flows into Pyramid Lake to preserve the endangered cui-ui fish and to prevent further deterioration of the ecosystem around Pyramid Lake [*U.S. Department of Interior*, 1987].

Litigation will undoubtedly continue to be an important tool for environmental, tribal and wildlife organizations seeking to protect and enhance streamflows. Litigation has the drawbacks of being costly, protracted, and not under the control of the parties involved with regard to outcome. Further, litigation evokes a sense of antagonism rather than of cooperation among those whose interests are at stake. On the other hand, the threat of an unfavorable and arbitrary outcome is a powerful incentive for successful negotiations and resolution of conflicts. In this light, litigation for the purpose of stimulating negotiations can be complimentary to a market-based approach to protecting streamflows. The political and financial costliness of litigation can prompt interests to consider market transactions as a lower cost alternative.

POLICY IMPLICATIONS

Existing studies suggest that regional economic development can be facilitated by cooperative public and private efforts to maintain flows on critical stream systems. There are strong economic arguments for assuring flow levels that enhance recreation and wildlife habitat and that preserve the aesthetic and intrinsic appeal that free-flowing water has in the arid West. Western state water law is evolving in recognition of the values generated by water in streams and lakes. Policies for protecting and enhancing streamflows are emerging based on case law, new legislation and changes in state administration of water rights.

The economic contributions of instream flows to recreation can be substantial but are only a small portion of the total value generated by instream flows. Other benefits stem from contributions of recreation-linked spending to local economies, from the role of instream flows in maintaining fish and wildlife habitat, from the aesthetic appeal of free-flowing water, from water quality enhancement and treatment costs avoided due to dilution of pollutants by higher flows, and from nonuser values. Instream values can be higher than benefits generated by offstream uses, particularly field-crop irrigation, especially where unique recreation areas, fish and wildlife and the quality of municipal water sources depend upon streamflow levels. Western state economies continue to urbanize, shifting from agriculture, mining and energy development to construction, services, industry and tourism. Amenities generated by instream flows will become more highly valued as growing urban populations demand opportunities to hunt, fish, view wildlife and recreate on free-flowing waters. The value of water quality enhancement and assimilative capacity provided by streams is also likely to increase as water providers and wastewater treatment facilities face more stringent water quality standards.

Water market transactions are becoming common in many parts of the West. Water markets can present opportunities for instream flow protection in states where public and private entities can acquire water rights for such purposes. Instream flow values are high enough to compete with offstream uses in the market for water rights, though the beneficiaries of free-flowing water may not contribute to purchase of rights at levels consistent with benefits received, due to free rider problems. Active water markets can accentuate controversy regarding instream flow protection if market proponents perceive instream flow rights as constraining transfers for offstream uses.

Western states that allow both public and private sector acquisition of water rights for instream flows can expand the economic contributions generated by free-flowing waters. Private parties should be allowed to appropriate, lease and purchase water rights for the purpose of maintaining instream flows, giving them an "equal opportunity" to compete in water markets and in the water right appropriation process with municipal, industrial and agricultural interests for scarce water resources. State agencies and local governments concerned with recreation, wildlife and tourism should have authority and funding to acquire water rights for instream flow maintenance. Public sector involvement is necessary because acquisition of instream flows by the private sector is handicapped by the "public good" nature of instream flow benefits. Active participation by state and local governments, in cooperation with federal agencies, can complement private efforts to protect flows at adequate levels and in desirable locations. Policies that encourage participation by county, municipal and tribal governments take advantage of the fact that these entities have an incentive to identify streams deserving protection, to provide funds to protect flow levels and to monitor and enforce flow standards. Their incentive stems from the fact that local areas experience tourism, recreation and aesthetic benefits generated by local stream systems and suffer losses when nearby streams are depleted. Local areas will also have a unique perspective on the trade-offs between retaining water instream and making it available for offstream uses such as urban growth and irrigation. Many quality of life decisions, funding for parks, police protection and schools, are primarily made at the city and county level, and some local role in setting stream flow priorities seems appropriate.

Western states could enhance public and private efforts to protect free-flowing water by clarifying the criteria that must be satisfied to change the purpose and place of use of a water right from a consumptive use to instream flow maintenance. They could also define the circumstances under which instream flow interests have legal standing to object to a change in water use which they believe impairs established flow rights and riparian areas not yet protected. Clear criteria in state change of use approval procedures would reduce transactions costs and uncertainties for both instream and offstream water users.

Another opportunity presents itself in those areas of the West where municipal and industrial water users are acquiring water rights historically used for irrigation and converting them to nonagricultural uses. Where the irrigation water was diverted upstream of the new point of diversion, additional streamflows occur along the stretch between the old and new diversion points. Riparian habitat and recreation opportunities may be enhanced and, in some regions, ephemeral streams may begin to have year-round flows. Organizations participating in transfers involving a movement of a diversion point from upstream to downstream should consider how they can manage and protect the new instream flow patterns so as to enhance water quality, environmental and recreation benefits.

REFERENCES

Amirfathi, P., R. Narayanan, B. Bishop, and D. Larson, *A Methodology for Estimating Instream Flow Uses For Recreation*, Logan, Utah Water Research Laboratory at Utah State University, 1984.

Boyle, K., and R. C. Bishop, Lower Wisconsin River recreation: Economic impacts and scenic values, *Univ. Wis. Agric. Econ. Staff Pap. Ser. 216*, Univ. of Wis., Madison, January 1984.

California Trout Inc. versus State Water Resources Control Board, 207 Cal. App. 3d 585, 1979.

Daubert, J. T., and R. A. Young, Economic benefits from instream flow in a Colorado mountain stream, *Colo. Water Resour. Res. Inst. Completion Rep. 91*, Colo. State Univ., Fort Collins, June 1979.

Environmental Defense Fund versus East Bay Municipal Utilities District, no. 425955, 1989.

Fullerton versus State Water Resources Control Board, 90 Cal. App. 3d. 590, 153, 1979.

Greenley, E., R. Walsh, and R. Young, *Economic Benefits of Improved Water Quality*, Westview Press, Boulder, Colo., 1982.

Livingston, M., and T. Miller, A framework for analyzing the impact of western instream water rights on choice domains, *Land Econ.*, *62*, 306–312, 1986.

Loomis, J., The economic value of instream flow: Methodology and benefit estimates for optimum flows, *J. Environ. Manage.*, *24*, 169–179, 1987*a*.

Loomis, J., Economic evaluation of public trust resources of Mono Lake, *Inst. Ecol. Rep. 30*, Univ. of Cal., Davis, 1987*b*.

MacDonnell, L. J. (Ed)., *Instream Flow Protection: Law and Policy*, University of Colorado Press, Boulder, 1989.
Madariago, B., and K. McConnell, Exploring existence value, *Water Resour. Res.*, *23*, 936–942, 1987.
McClennan versus Jantzen, 26 Ariz. App. 723 547 p. 2d 494, 1976.
McKinney, M. J., G. Fritz, P. Graham, and D. Schmidt, The protection of instream flows in Montana: A legal-institution perspective, report, Mont. Dep. of Nat. Resour. and Conserv., Dep. of Fish, Wildlife and Parks, and Environ. Qual. Counc., Helena, 1988.
National Audubon Society versus Superior Court of Alpine County, 33 Cal. 32 419, 658 p. 2d, 1983.
Nevada versus Morros, no. 18105, 1988.
Saliba, B. C., and D. B. Bush, *Water Markets in Theory and Practice: Market Transfers, Water Values and Public Policy*, Westview Press, Boulder, Colo., 1987.
Shupe, S. J., Colorado's Instream Flow Program Protecting Free-Flowing Streams in a Water-Consumptive State, in *Instream Flow Protection in the Western States*, edited by L. MacDonnell, University of Colorado Press, Boulder, 1989.
Shupe, S. J., Emerging forces in western water law, *Resour. Law Notes Newsl. 2-8*, Nat. Resour. Law Cent., Univ. of Colo., Boulder, 1986.
U.S. Department of Interior, *Final Environmental Impact Statement for Newlands Project Proposed Operating Criteria and Procedures*, Bureau of Reclamation, Sacramento, Calif., 1987.
Walsh, R. G., R. Auckerman, and R. Milton, Measuring benefits and the economic value of water in recreation on high country reservoirs, *Colo. Water Resour. Res. Inst. Completion Rep. 101*, Colo. State Univ., Fort Collins, September 1980*a*.
Walsh, R. G., R. K. Ericson, D. J. Arosteguy, and M. P. Hansen, An empirical application of a model for estimating the recreation value of instream flow, *Colo. Water Resour. Res. Inst. Completion Rep. 101*, Colo. State Univ., Fort Collins, October 1980*b*.
Walsh, R. G., J. B. Loomis, and R. A. Gilman, Valuing Option, existence and bequest demands for wilderness, *Land Econ.*, *60*, 14–29, 1984.
Ward, F. A., Economics of water allocation to instream uses in a fully appropriated river basin: Evidence from a New Mexico wild river, *Water Resour. Res.*, *23*, 381–392, 1987.
Young, R. A., and S. L. Gray, Economic value of water: Concepts and empirical estimates, *Tech. Rep. PB210356*, U.S. Natl. Water Comm., Washington, D. C., 1972.

B. G. Colby, Department of Agricultural Economics, Economics Building, #23, Room 208, University of Arizona, Tucson, AZ 85721.

(Received December 8, 1988;
revised November 8, 1989;
accepted November 20, 1989.)

179–217 [96]

[14]

MICHAEL R. MOORE, AIMEE MULVILLE, AND MARCIA WEINBERG

Water Allocation in the American West: Endangered Fish Versus Irrigated Agriculture

ABSTRACT

This research analyzes potential water allocation conflicts between endangered fish species and irrigated agriculture in western river systems. Through geographic and statistical analyses of county-level data sets for all 17 western states, we describe a pattern of mutual dependence on limited water resources. The numbers that characterize the conflict appear large when totaled across the West: 50 fish species listed under the Endangered Species Act are linked to agricultural activity, and 235 counties contain irrigated production that relies on water from rivers with ESA-listed fish. Statistical results show that the number of ESA-listed species in a county correlates positively with the level of agriculture reliant on surface water in the county. Three features of the Reclamation program—its pervasive presence throughout the West, substantial water deliveries to agriculture, and federal-agency responsibilities under the ESA—make it possible to develop a leadership role for the Bureau of Reclamation in species recovery.

I. INTRODUCTION

Implementation of the federal Endangered Species Act (ESA) may prove to be the litmus test of the extent to which existing water-use patterns in the American West can accommodate contemporary environmental values. Many commentators have written of the need to balance traditional western water uses and new demands for instream water for ecosystem protection, river-based recreation, and aesthetic appreciation.[1]

---

1. Moore, School of Natural Resources and Environment, University of Michigan; Mulville, Yale Law School, Yale University; Weinberg, Division of Environmental Studies, University of California, Davis. The authors thank Pamela P. Eaton, Ralph E. Heimlich, Jan Lewandrowski, and, especially, Noel R. Gollehon for useful ideas and helpful comments on an earlier draft of the paper. They also thank Peter M. Feather and Daniel M. Hellerstein for technical assistance with the statistical estimation procedures.
See Deborah Moore & Zach Willey, *Water in the American West: Institutional Evolution and Environmental Restoration in the 21st Century*, 62 U. COLO. L. REV. 775 (1991); CHARLES F. WILKINSON, CROSSING THE NEXT MERIDIAN: LAND, WATER, AND THE FUTURE OF THE WEST

This need exists alongside a long-dormant tension between the historic state primacy in water law and a recently-exercised federal prerogative to develop environmental and natural resource laws that affect western water rights.[2] This set of needs and tensions is apparent in implementation of the Clean Water Act, the Wild and Scenic Rivers Act, and the ESA, as well as in assertion of federal water rights for public lands.[3]

The ESA poses a particularly difficult test for western water allocation for three reasons. First, the ESA *requires* competing interests to bend, at least to some degree, to the goal of species preservation. The criterion for ESA listing defines a strictly biological standard for invoking the law, with clear procedural guidelines for critical habitat designation and recovery plan development to be followed after listing. Second, the history of western water resource development, in tandem with obligations delineated in the ESA, places the federal government near the center of riverine and riparian species protection in the West. The federal Bureau of Reclamation played a prominent, critical role in western river development. Its responsibilities continue, in the form of managing a vast infrastructure of water projects and supplies throughout the West. Simultaneously, the ESA imposes special obligations on federal agencies for ESA compliance. And third, the absolute number of ESA-listed species reliant on western water resources appears to be large. Sixty-eight fish species are listed as endangered or threatened in the 17 western states,[4] with an additional 86 fish species officially designated as candidate species.[5] Although not addressed systematically in this analysis, many plants and animals from other taxonomic groups also rely on western rivers for critical habitat.[6] In particular, 184 individual species with

(1992).

2. *See* Lawrence MacDonnell, *Federal Interests in Western Water Resources: Conflict and Accommodation*, 29 NAT. RESOURCES J. 389 (1989); Joseph L. Sax, *The Constitution, Property Rights and the Future of Water Law*, 61 U. COLO. L. REV. 257 (1990); A. Dan Tarlock, *The Endangered Species Act and Western Water Rights*, 20 LAND & WATER L. REV. 1 (1985).

3. On the general issue of western water rights, the ESA, and endangered fish, *see also* James H. Bolin, Jr., *Of Razorbacks and Reservoirs: The Endangered Species Act's Protection of Endangered Colorado River Basin Fish*, 11 PACE ENVTL. L. REV. 35 (1993); Melissa K. Estes, *The Effect of the Federal Endangered Species Act on State Water Rights*, 22 ENVTL. L. 1027 (1992); and SCOTT W. REED, *Fish Gotta Swim: Establishing Legal Rights to Instream Flows through the Endangered Species Act and the Public Trust Doctrine*, 28 IDAHO L. REV. 645 (1991-1992).

4. Western fish species comprise over 70 percent of the ESA-listed fish; fish species, in turn, comprise roughly 25 percent of the ESA-listed animal species. *See* FISH AND WILDLIFE SERVICE, U.S. DEPARTMENT OF THE INTERIOR, ENDANGERED AND THREATENED SPECIES RECOVERY PROGRAM (1990).

5. The information presented on numbers of endangered, threatened, and candidate fish species reflects their status as of August 1993.

6. This article focuses on ESA-listed species of fish because of their obvious link to water allocation tradeoffs in western river systems. However, other endangered species also

habitat affected by federal Reclamation projects and water service areas are either listed or proposed for listing under the ESA.[7]

The role of reallocating water from agriculture to habitat restoration should be assessed when crafting plans for species protection in western river systems. For more than a century, river development and diversion provided the foundation for agricultural settlement of arid lands in the "second opening of the West."[8] Irrigated agriculture now dominates western water consumption, with agricultural use comprising 91 percent of total regional consumption of freshwater resources.[9] Surface water—water diverted from rivers and streams—provides more than 60 percent of the water for irrigated agriculture.[10] The Bureau of Reclamation delivers more than one-third of the surface water consumed by western irrigated agriculture.[11]

Riverine and riparian wildlife, not to mention the rivers themselves, were sacrificed for western river development.[12] For instance, dams or water diversions impede the migration patterns of several endangered fish species: Chinook salmon in both the Columbia River Basin and California's Sacramento River Basin, the cui-ui in Nevada's Truckee River, and the Colorado River squawfish.[13] The link to irrigated agriculture is documented in many cases, with "agricultural activities" recorded as one of the "factors in decline" for 50 of the 68 western ESA-listed fish species.[14]

---

rely heavily on riverine and riparian ecosystems in the West. For example, one study concentrated on threatened animal species in Arizona and New Mexico that rely on riparian zones for habitat (where "threatened" is used generally rather than narrowly in the context of ESA listing). AUBREY S. JOHNSON, THE THIN GREEN LINE: RIPARIAN CORRIDORS AND ENDANGERED SPECIES IN ARIZONA AND NEW MEXICO, *in* IN DEFENSE OF WILDLIFE: PRESERVING COMMUNITIES AND CORRIDORS (Gay Mackintosh ed., 1989). In addition to 49 fish species in the two states, the list includes 50 birds, 17 mammals, 15 amphibians, and 12 reptiles. Other examples of ESA species dependent on riparian habitat include the whooping crane and bald eagle. More generally, water development is identified as the cause or potential cause of endangerment of approximately one-third of all ESA-listed plant and animal species. ELIZABETH LOSOS ET AL., TAXPAYERS' DOUBLE BURDEN: FEDERAL RESOURCE SUBSIDIES AND ENDANGERED SPECIES (1993). Thus, the description and analysis of this report presents only a partial screen of the endangered species-agriculture water allocation dilemma.

7. BUREAU OF RECLAMATION, U.S. DEP'T OF THE INTERIOR, PROPOSED ACREAGE LIMITATION AND WATER CONSERVATION RULES AND REGULATIONS 3-85 (1995).

8. WALLACE E. STEGNER, THE AMERICAN WEST AS LIVING SPACE (1987).

9. WAYNE B. SOLLEY ET AL., GEOLOGICAL SURVEY, U.S. DEP'T OF THE INTERIOR, ESTIMATED USE OF WATER IN THE UNITED STATES IN 1990, 13 (1993).

10. *Id.* at 37.

11. BUREAU OF RECLAMATION, *supra* note 7.

12. STEGNER, *supra* note 8, at 50.

13. PETER MATTHIESSEN, WILDLIFE IN AMERICA 271-73 (2d ed. 1987).

14. Thirty-three *Federal Register* issues between 1973 and 1993 contained the official listings of western fish species as endangered or threatened under the Endangered Species

The potential for pervasive conflict between established water uses and endangered species for western river allocation has been recognized for over a decade. In congressional hearings on the 1982 Amendments to the ESA, western water interests raised concerns about the potential for the ESA to modify existing interstate apportionments of rivers and intrastate allocations of water rights.[15] An attempt was made to amend the ESA to make ESA-related water claims secondary to state administrative systems and their established water rights.[16] This effort proved largely unsuccessful, as Section 2 of the law was amended to make only a relatively weak policy statement: "It is further declared to be the policy of Congress that Federal agencies shall cooperate with State and local agencies to resolve water resource issues in concert with conservation of endangered species."[17] Of note, though, is that a parallel ESA policy statement was not attached concerning resolution of land resource issues. Western water interests had left their imprint on the ESA.

ESA reauthorization creates an important opportunity to reconsider the issue of endangered species recovery and western river management. Originally scheduled to occur by 1993, the 104th U.S. Congress will likely consider ESA reauthorization in the 1995-1996 legislative term. Wilkinson provides a recent perspective on the topic:

> A fast-emerging matter of federal law [concerning western water] involves the Endangered Species Act. The Endangered Species Act has only begun to play out on western rivers. It may not come to much. The last-resort statute for wildlife may, however, prove to be a sturdy hammer for dislodging long-established extractive water uses that have worked over so many western watersheds and drained them of much of their vitality.[18]

The impact of the ESA on river use continues to be one of the great uncertainties in western water resource allocation.

Act. *See, e.g.*, Endangered and Threatened Wildlife and Plants: Determination of Threatened Status for the Delta Smelt, 58 Fed. Reg. 12,854 (1993)(to be codified at 50 C.F.R. § 17).

15. Tarlock, *supra* note 2, at 19.

16. This attempt was to amend the ESA in a manner similar to § 101(g) of the Clean Water Act, which reads

> It is the policy of Congress that the authority of each State to allocate quantities of water within its jurisdiction shall not be superseded, abrogated or otherwise impaired by this chapter. It is further the policy of Congress that nothing in this chapter shall be construed to supersede or abrogate rights to quantities of water which have been established by any State.

Federal Water Pollution Control Act, 33 U.S.C. § 1251(g) (1994).

17. Endangered Species Act of 1973, 16 U.S.C. § 1531(c)(2) (1994).

18. WILKINSON, *supra* note 1, at 283.

This research addresses potential water-allocation conflicts between endangered fish species and irrigated agriculture as a systemic western issue. Section II reports baseline conditions for ESA-listed fish species and irrigated agriculture in the West. It describes several aspects of the protected fish species, including their number and geography. A parallel discussion also occurs on the extent and geography of irrigated agriculture reliant on western surface water, including water developed by the Bureau of Reclamation. Section III conducts a statistical analysis of the relationship between endangered fish species and irrigated agriculture. The analysis uses two west-wide, county-level data sets on the number of ESA-listed fish species, along with several variables on the extent of agriculture, irrigation, and Reclamation water supply in the county. Section IV considers the analytical results in light of developments in federal water policy and ESA implementation, examined through recent experience in central California and the Columbia River Basin. Section V considers two alternative roles for the Bureau of Reclamation in endangered species recovery in western river systems. Section VI summarizes the major findings and conclusions of the research.

II. DIMENSIONS OF ENDANGERED FISH-AGRICULTURE WATER ALLOCATION TRADEOFFS

A. *Physical Setting*

Freshwater fisheries require certain volumes of instream water flow to be sustained. Thirty percent of average annual flow can be considered the minimum quantity necessary to protect instream water uses.[19] River flows fail to meet this benchmark in the southern portions of California and Arizona, the headwaters of the Platte and Arkansas Rivers (in Wyoming and Colorado), the San Joaquin Valley (California), the Rio Grande (New Mexico and Texas), and in closed basins in Nevada, Utah, and California.[20] (See Figure 1 for a map of major western rivers.)

19. KEITH BAYHA, FISH AND WILDLIFE SERVICE, U.S. DEP'T OF THE INTERIOR, INSTREAM FLOW METHODOLOGIES FOR REGIONAL AND NATIONAL ASSESSMENTS 39-40 (1978). The following "rules of thumb" define three levels of habitat quality. One, 10 percent of average flow is necessary to provide short-term survival habitat for most life forms. Rivers in this category are defined as "severely depleted." Two, 30 percent of average flow will sustain good survival habitat for most life forms. Rivers at 10 to 30 percent are "under stress." Three, 60 percent of average flow will provide excellent to outstanding habitat. Rivers with flows of 30 to 60 percent are termed "degraded." DONALD TENNANT, FISH AND WILDLIFE SERVICE, U.S. DEP'T OF THE INTERIOR, INSTREAM FLOW REGIMES FOR FISH, WILDLIFE, RECREATION, AND RELATED ENVIRONMENTAL RESOURCES 19-23 (1975).

20. BAYHA, *supra* note 19, at 43.

One obvious remedy for these conditions is water reallocation from offstream, consumptive uses to instream flow.

In the narrower context of ESA-listed western fish species, factors contributing to decline of a species are reported in the *Federal Register* at the time of formal ESA listing. "Physical habitat alterations"—including

Figure 1. Major Western Rivers

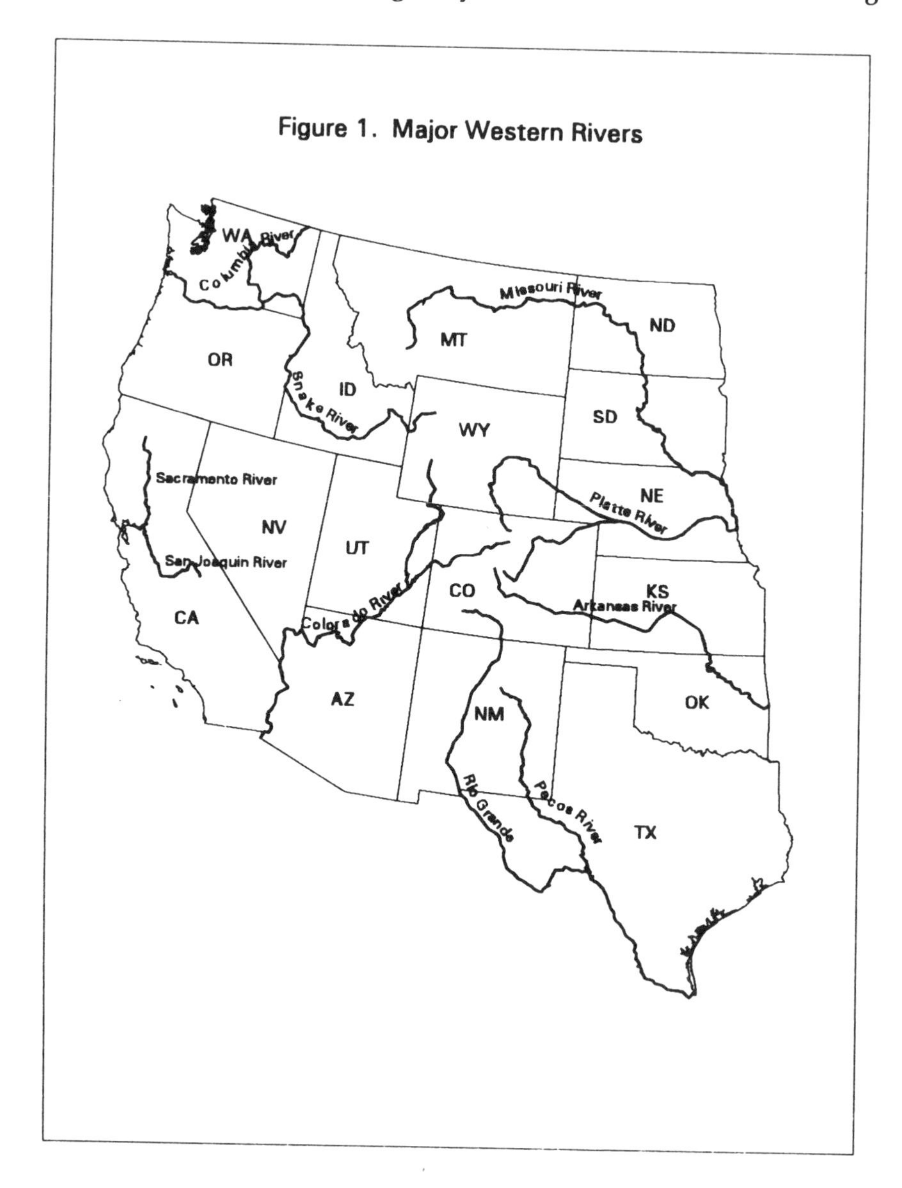

water diversions, dams, reservoirs, channeling, and watershed disturbances—are the most frequently cited factor.[21] "Agricultural activities" are a factor in the decline of almost 75 percent of these species.[22]

B. Endangered and Threatened Fish Species in the West

For purposes of analysis, this research focuses on western fish species protected under the ESA.[23] The ESA-listed fish species, though, are symptomatic of a general decline in ecosystem functioning of the West's rivers. Evidence has accumulated in the biological literature of significant decline of species richness and biodiversity in western aquatic systems.[24] Two examples illustrate the gravity of the problem. First, according to Moyle and Williams' analysis of native fish species in California:

> Of 113 native fishes, 7 are extinct, 14 are officially listed as threatened or endangered, 7 deserve immediate listing, 19 are in serious trouble in their native range and may deserve listing soon if present trends continue, 25 show declining populations but are not yet in serious trouble or naturally have very limited ranges, and 41 appear to be secure. In all, 57 percent of the existing taxa have at least some need of special management if their populations are to continue to exist indefinitely.[25]

Second, Nehlsen et al. identified "214 native naturally-spawning Pacific salmon and steelhead stocks in California, Oregon, Washington, and Idaho that appear to be facing a high or moderate risk of extinction, or are of special concern."[26] In this context, designing efforts toward

---

21. For example, *see* Federal Register, *supra* note 14, at 12,858-60.

22. *Id.* at 12,861.

23. For a somewhat related analysis that describes geographic patterns of species endangerment for the United States, *see* CURTIS H. FLATHER ET AL., FOREST SERVICE, U.S. DEP'T OF AGRICULTURE, SPECIES ENDANGERMENT PATTERNS IN THE UNITED STATES (1994).

24. For a general description of the health of aquatic ecosystems in the United States, *see* NATIONAL RESEARCH COUNCIL, RESTORATION OF AQUATIC ECOSYSTEMS: SCIENCE, TECHNOLOGY, AND PUBLIC POLICY (1992). *See also* Peter B. Moyle & Jack E. Williams, *Biodiversity Loss in the Temperate Zone: Decline of the Native Fish Fauna of California*, 4 CONSERVATION BIOLOGY 275 (1990); Willa Nehlsen et al., *Pacific Salmon at the Crossroads: Stocks at Risk from California, Oregon, Idaho, and Washington*, FISHERIES, March-April 1991, at 4; Jack E. Williams et al., *Fishes of North America Endangered, Threatened, or of Special Concern: 1989*, FISHERIES, Nov.-Dec. 1989, at 2; James D. Williams et al., *Conservation Status of Freshwater Mussels of the United States and Canada*, FISHERIES, Sept. 1993, at 6. For an analysis of biodiversity conservation in riverine ecosystems from a global perspective, *see* David J. Allan & Alexander S. Flecker, *Biodiversity Conservation in Running Water*, 34 BIOSCIENCE 32 (1993).

25. Moyle & Williams, *supra* note 24, at 278.

26. Nehlsen et al., *supra* note 24, at 4. A more complete summary of the biological

improving habitat of the ESA-listed fish with the broader view of riverine ecosystem restoration could improve the economic and biological effectiveness of those efforts.

Sixty-eight fish species are listed as endangered or threatened in the 17 western states.[27,28] The appendix lists these species, along with several of their characteristics, including: year listed, status, current habitat, state(s) in which they are found, whether agriculture was a cause of population decline, and 1990 and 1991 government expenditures on recovery. Upon listing a species, ESA procedures require designation of the species' critical habitat and development and implementation of a recovery plan.[29] Recovery plans have yet to be approved for 60 percent of the ESA-listed western fish species.

The number of ESA-listed fish species in the West has grown steadily over time (Figure 2). The number is cumulative: after formal recognition, a species remains on the list until either the recovery effort is successful (with the threat of extinction diminished markedly) or extinction occurs. Twelve western fish were listed as endangered in 1967.[30] The cumulative total then increased over 26 years to 42 endangered and 26 threatened fish species in 1993. The largest annual increases in protected western fish species occurred in 1970 (nine new listings) and 1985 (ten new listings).

---

evidence is available in Bureau of Reclamation, *supra* note 7, at 3-79-88.

27. Under the Endangered Species Act of 1973, "endangered species" is defined as any species in danger of extinction throughout all or a significant portion of its range, while "threatened species" includes any species likely to become an endangered species within the foreseeable future. 16 U.S.C. § 1532(6),(20) (1994).

28. In addition to the 68 ESA-listed species, 86 western fish species are formally listed as candidate species for possible future listing under the ESA. The numerical total of candidate species is compiled from Endangered or Threatened Wildlife and Plants; Animal Candidate Review for Listing as Endangered or Threatened Species, 56 Fed. Reg. 58,804 (1991). Similarly, 68 western fish are described as being of "special concern." Williams et al., *supra* note 24, at 3. These are species, other than those officially listed as threatened or endangered, that "may become threatened or endangered by relatively minor disturbances to their habitat, or that require additional information to determine their status."

29. Section 3 of the ESA defines critical habitat for a listed species as, "(i) the specific areas within the geographical area occupied by the species . . . on which are found those physical or biological features (I) essential to the conservation of the species and (II) which may require special management considerations or protection; and (ii) specific areas outside the geographical area occupied by the species . . . upon a determination by the Secretary that such areas are essential for the conservation of the species." 16 U.S.C. § 1532(5)(A)(i)-(ii) (1994).Under Section 4, recovery plans must include description of management actions for recovery, along with criteria that would demonstrate successful recovery. 16 U.S.C. S 1533(f) (1994).

30. The 1966 Endangered Species Preservation Act and the 1969 Endangered Species Conservation Act pre-dated the Endangered Species Act of 1973. *See* STEVEN L. YAFFEE, PROHIBITIVE POLICY: IMPLEMENTING THE FEDERAL ENDANGERED SPECIES ACT 39-42 (1982).

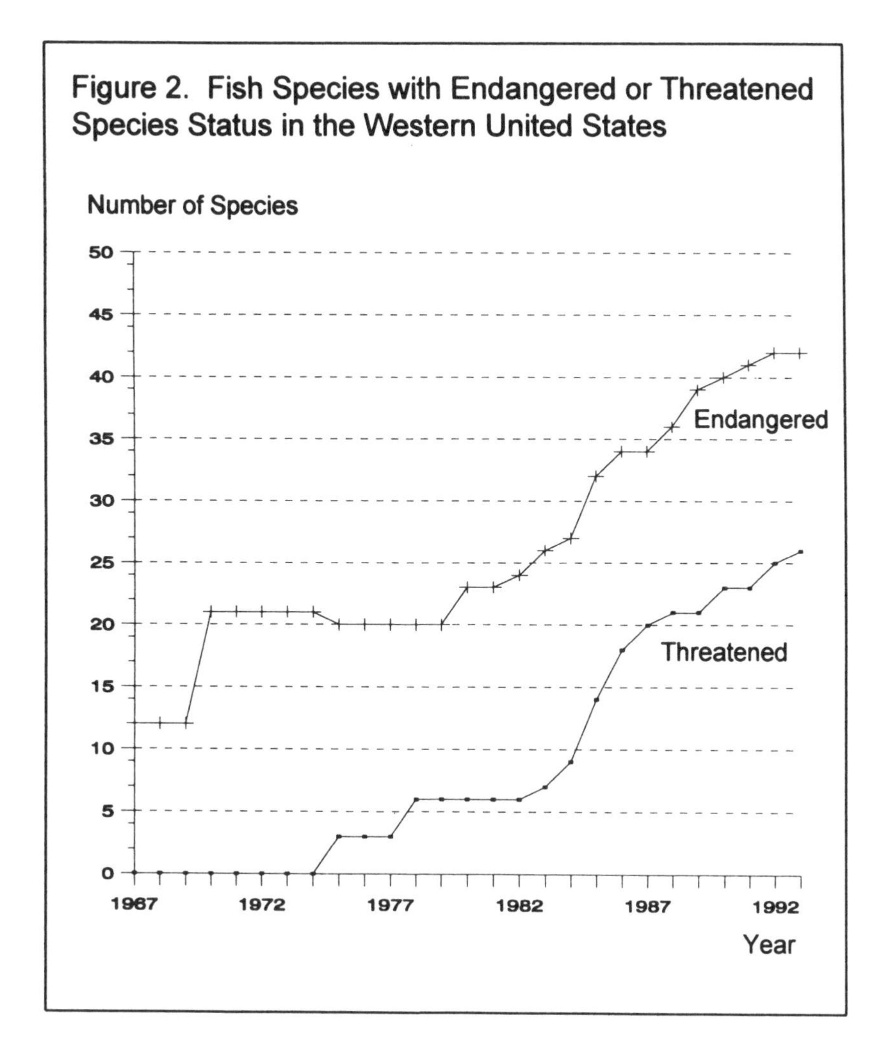

Figure 2. Fish Species with Endangered or Threatened Species Status in the Western United States

Public expenditures plus private-sector compliance costs form the total cost of ESA implementation. Federal and state governments began reporting the level of public expenditures on individual ESA-listed species in 1989.[31] For the 68 western fish species, public expenditures

31. The types of activities reported include expenditures on "fisheries, refuges, land acquisition, law enforcement, research and Regional and field operations for listing, recovery, consultation, environmental contaminant and habitat conservation activities" that could be "attributed" to an individual species. FISH AND WILDLIFE SERVICE, U.S. DEP'T OF THE

totaled $9.64 million in 1990, rising to $17.25 million in 1991.[32] Individual species receiving large expenditures in 1991 included the Sacramento River winter run Chinook salmon ($5.49 million), Colorado squawfish ($3.67 million), humpback chub ($2.77 million), and Lahontan cutthroat trout ($1.60 million).[33] The subsequent section investigates, in broad terms, costs that could arise if recovery efforts for protected western fish affected the irrigated agricultural sector.

For subsequent analysis, 50 of the 68 ESA-listed fish species are highlighted because of their direct link to agriculture; these are fish reliant on surface water for which agricultural activities is listed in the Federal Register as a "factor in decline" of the species. This link provides the basis for analysis of the tradeoffs between ESA-listed fish and agriculture.

Figure 3 maps the habitat of these 50 fish species, by county.[34] Each species indicator on the map (as depicted by a dot) simply means

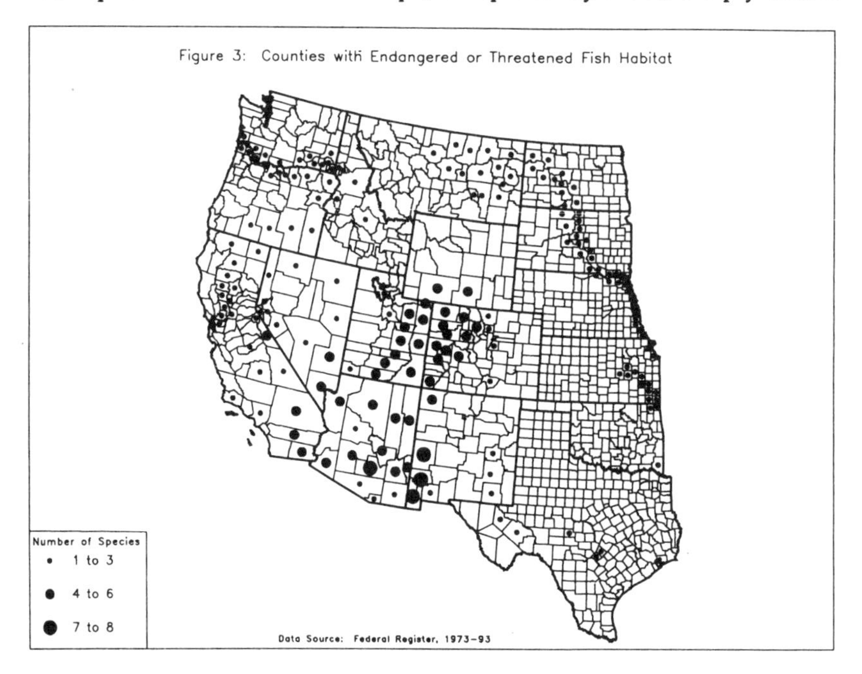

Figure 3: Counties with Endangered or Threatened Fish Habitat

---

INTERIOR, FEDERAL AND STATE ENDANGERED SPECIES EXPENDITURES: FISCAL YEAR 1991, 3 (1992).

32. FISH AND WILDLIFE SERVICE, *supra* note 31, at Table 1, 5-6. *See* the attached appendix for additional public expenditure data.

33. *Id.* at Table 1, 5-6. See the attached appendix for additional public expenditures data.

34. The geographic data displayed in Figure 3 were obtained from two sources. The primary source was miscellaneous issues of the *Federal Register* between 1973 and 1993,

that the county contains habitat for ESA-listed fish somewhere within its boundaries; the map does not portray a precise geography of habitat. Relatively larger dots represent higher numbers of species with habitat in the county. The map illustrates two interesting items. First, 198 counties—18 percent of all counties in the West—contain habitat for these 50 species. Second, the species are geographically distributed, albeit unevenly, across every state and major river basin in the West. The largest concentration is in the Colorado River Basin, where a majority of protected species reside. California provides habitat for 14 listed fish species, while Washington provides habitat for only 3 listed fish species.

C. *The Intersection Between ESA-Listed Fish Species and Irrigated Agriculture*

Several chąracteristics provide background on irrigated agriculture in the 17 western states. Nationwide, 46 million acres were irrigated in 1987.[35] The vast majority of irrigated acreage—81 percent of total irrigated acres—is located in the 17 western states.[36] Roughly half of all irrigated acreage is irrigated with surface water, with the remaining acreage irrigated with ground water.[37]

Agriculture withdraws and consumes the vast majority of western surface water resources. In western states, 76 percent of all surface water withdrawals are for agricultural purposes (Solley et al.).[38] The percentage exceeds 80 percent in eleven states and 90 percent in six states (Table 1). The six western states in which agricultural use accounts for less than 80 percent of surface water withdrawals (Kansas, Nebraska,

which publishes habitat maps in its official notice that a species is being listed. *See, e.g.*, 58 Fed. Reg. 12,854. The secondary source was species-specific recovery plans, which have been completed for some of the western fish species.

35. BUREAU OF THE CENSUS, U.S. DEP'T OF COMMERCE, 1987 CENSUS OF AGRICULTURE, electronic data file.

36. RAJINDER S. BAJWA ET AL., ECONOMIC RESEARCH SERVICE, U.S. DEP'T OF AGRICULTURE, AGRICULTURAL IRRIGATION AND WATER USE 2 (1992).

37. BUREAU OF THE CENSUS, U.S. DEP'T OF COMMERCE, 1992 CENSUS OF AGRICULTURE, VOLUME 3, RELATED SURVEYS, PART 1, 1994 FARM AND RANCH IRRIGATION SURVEY 20 (1996).

38. SOLLEY ET AL., *supra* note 9. The proportion of diverted water that is actually consumed varies by state, ranging from 22 percent in Montana to 96 percent in Kansas. The figures are calculated from *id.* at 37. (*See id.* for the distinction between water diversion and consumption). Nevertheless, irrigation accounts for 90 percent of total western water consumption (including both surface and ground water consumption). The figure is calculated from *id.* By state, this figure ranges from 58 percent in Oklahoma to nearly 100 percent in Idaho. These figures are calculated from *id.*

North Dakota, Oklahoma, South Dakota, and Texas) are in the Great Plains region, where irrigators rely heavily on ground water pumping from the Ogallala Aquifer.[39]

High productivity and, relatedly, high gross financial returns are characteristics of irrigated agriculture.[40] The 46 million irrigated acres in 1987 represented 15 percent of total U.S. harvested acreage, yet accounted for 38 percent of the $69 billion in U.S. crop sales.[41] The value of irrigated production, as measured by gross revenue per acre, varies considerably across the western states (Figure 4). Variation in climatic conditions—including length of the growing season and ability to produce high-value crops, such as fruits, nuts, and vegetables—explains much of the variation in value. California, Arizona, and Washington rank significantly higher than the other states, each with a value greater than $800 per acre. The northern Great Plains states report relatively low values.

Figure 4: Value per Acre of Total Crop Sales, by County

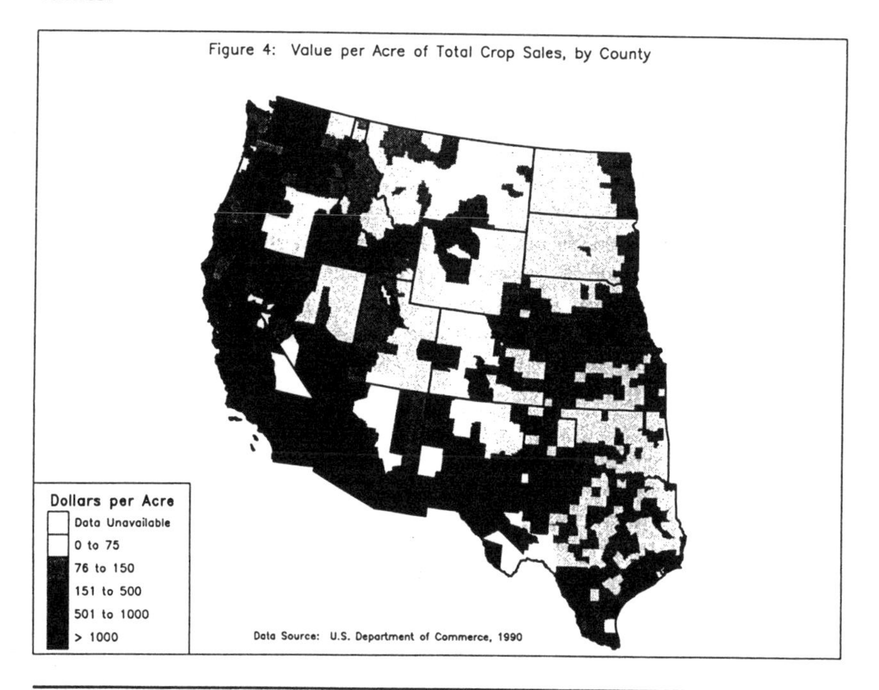

39. DONALD WORSTER, RIVERS OF EMPIRE: WATER, ARIDITY AND THE GROWTH OF THE AMERICAN WEST 313-14 (1985).

40. Higher production costs are also a characteristic of irrigated agriculture. On average, irrigated agriculture spends 2 to 3 times as much as non-irrigated agriculture on agricultural chemicals and energy inputs and 5 to 6 times as much on labor. It is more highly capitalized, with twice the value of machinery and equipment and investment in land and buildings. *See* BAJWA ET AL., *supra* note 36.

41. BAJAWA ET AL., *supra* note 36, at 3.

With this background on irrigated agriculture, we now assess the geographic relationship between areas of irrigated agriculture and ESA-listed fish species in the West. Figure 5 maps the percentage of

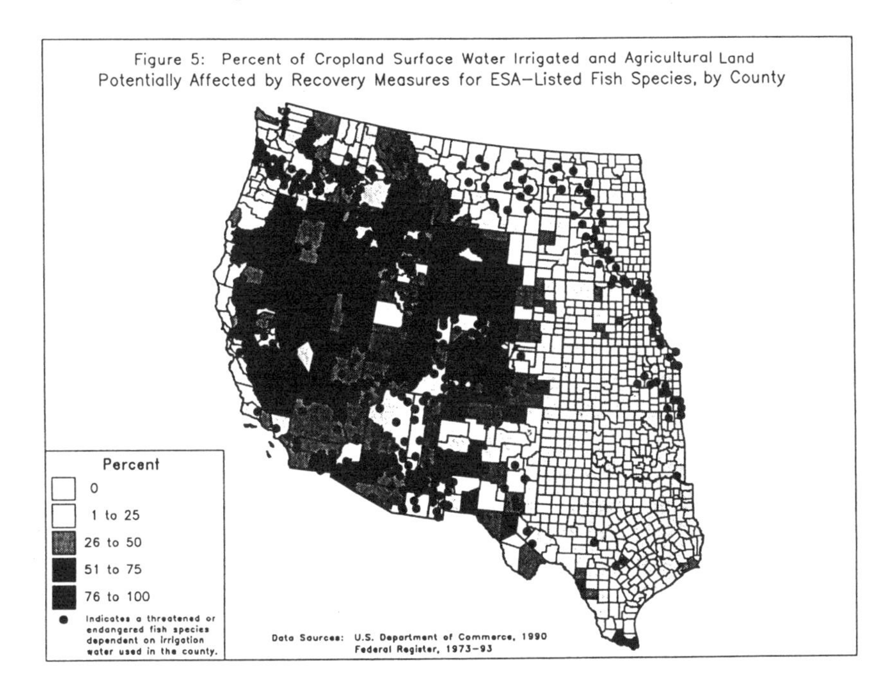

Figure 5: Percent of Cropland Surface Water Irrigated and Agricultural Land Potentially Affected by Recovery Measures for ESA–Listed Fish Species, by County

cropland that is irrigated with surface water in each county in the 17 western states. It also has an indicator for each species whose recovery effort may affect irrigated agriculture in that county. We distinguish between counties encompassing or adjacent to species habitat (as depicted in Figure 3) and counties in which irrigation water is taken from a river containing an ESA-listed species (as depicted in Figure 5). This distinction—between the proximity of irrigated areas to habitat and the reliance of irrigated acreage on diversions from rivers that provide habitat is important because the area of mitigation and recovery actions may not sufficiently describe the areas that would be affected by those actions. For example, in California, the Sacramento River winter-run Chinook salmon is endangered by changes in spawning habitat in the upper Sacramento River. Yet efforts to protect this species will likely result in reduced water supplies to farms in the San Joaquin Valley, located hundreds of miles south of the spawning habitat.

Figure 5 also illustrates the general geographic relationship between cropland irrigated with surface water and ESA-listed fish. There are 235 counties, representing 22 percent of the counties in the West, which contain agricultural production that relies on surface water from river systems with ESA-listed fish. That is, irrigated agriculture in this set of 235 counties may be affected, to some degree, by activities to recover species. These counties contain an estimated 10.35 million acres of cropland irrigated with surface water. Several comparisons provide context: 10.35 million acres exceeds one-half of all surface-water irrigated acres in the entire West; it exceeds one-third of the harvested cropland acres in the 235 counties; and it equals four-fifths of the irrigated harvested cropland acres in these same counties.

Two specific features of the geographic intersection of agriculture and species are evident from Figure 5. First, high concentrations of ESA-listed fish species correspond with areas of extensive surface water irrigation. The irrigated areas of Idaho, California, Utah, and Colorado best reflect this relationship. The concentration of ESA-listed fish species in the Colorado River Basin correlates strongly with high rates of surface-water irrigation in Colorado, Wyoming, Utah and southeastern California. This correlation is examined statistically in Section III. Second, nearly all counties reliant on surface-water irrigation draw or receive water from rivers inhabited by at least one threatened or endangered fish species. Very few counties with greater than 50 percent of cropland acreage irrigated with surface water are free from a link to an ESA-listed species.[42] These two features of Figure 5 suggest that, in general, counties with a relatively high portion of irrigated agriculture in percentage or absolute terms—will be affected to some degree by recovery measures undertaken in conjunction with ESA implementation.

The information in Figure 5 and Table 1 provides insight into the potential effects of reallocating irrigation water to endangered species habitat. A high potential for disruption of irrigated agriculture exists because of its dependence on surface water. If disruption in water supplies occurs, the potential costs are high because of the high value of irrigated agriculture. Four of the major factors that could affect the magnitude of the costs to agriculture include: the volume of water needed by ESA-listed fish; the extent to which the burden is shared

42. Exceptions to the rule typically show correlations with absolute, rather than percentage, levels of acreage. For example, counties in northern Wyoming with a high percentage of cropland surface water irrigated, but without a potential link to an ESA-listed fish species, generally have relatively low absolute acreages in crop production and relatively few irrigated acres. Conversely, counties in Texas and New Mexico (in the Pecos River Basin) show a potential link to ESA-listed fish; these counties have relatively small percentages of surface water irrigated acreage, yet the absolute acreage is as large as 20,000 acres per county. Similarly, the three counties bordering the Columbia and Snake Rivers at their confluence represent only 14, 29 and 37 percent of total cropland in the respective counties, but contain relatively large amounts of surface water irrigated acres (80,000, 131,000, and 172,000 acres, respectively).

broadly throughout the agricultural sector rather than concentrated on a relatively small number of producers; individual producer's flexibility to respond to water-supply reductions with relatively minor decreases in profitability; and the extent to which producers receive financial compensation for their water-supply reductions.

D. *The Federal Reclamation Program*

A final dimension of the issue and a recurrent focus of this research involves the role of the federal Reclamation program. The Reclamation program, a program administered by the Interior Department's Bureau of Reclamation, has been a major force in shaping river development in the West since 1902. Reclamation is the largest supplier of irrigation water in the western United States, regularly delivering more than 25 million acre feet (maf) per year to farms.[43] This water irrigates 9-10 million cropland acres, or roughly one-half of all surface-water irrigated acres in the West.[44] Reclamation-served agriculture relies on a vast network of water storage and conveyance projects. Reclamation facilities include: 355 storage reservoirs, 254 diversion dams, 16,047 miles of canals, and 37,193 miles of laterals.[45]

Reclamation controls significant percentages of river flows throughout the West. In the Colorado River Basin, for example, Reclamation delivered over 4.9 maf of water to farms in the upper and lower Colorado River Basin in 1989.[46] The river's average virgin flow equals 13.5 maf. [47] Along the upper Snake River in southern Idaho, Reclamation's Minidoka-Palisades and Boise Projects delivered almost 4.6 maf of irrigation water.[48] The Snake River's natural flow ranges between 5.2 and 12.1 maf in this region.[49] The Middle Rio Grande Project in New Mexico and the Rio Grande Project in New Mexico and Texas delivered over 0.6 maf to farms in 1989.[50] The Rio Grande flows at an average rate of almost 0.8 maf through central New Mexico. Finally, California farmers receive an average of about 7 maf of water from Reclamation's Central

43. BUREAU OF RECLAMATION, U.S. DEP'T OF THE INTERIOR, 1989 SUMMARY STATISTICS: WATER, LAND, AND RELATED DATA 66 (1990).

44. *Id.*

45. *Id.* at 1.

46. *Id.*

47. David H. Getches & Charles J. Meyers, *The River of Controversy: Persistent Issues, in* NEW COURSES FOR THE COLORADO RIVER 55 (Gary D. Weatherford & F. Lee Brown eds., 1986).

48. BUREAU OF RECLAMATION, *supra* note 43, at 62.

49. Joel R. Hamilton et al., *Interruptible Water Markets in the Pacific Northwest,* 71 AMERICAN J. AGRIC. ECONOMICS 64 (1989).

50. BUREAU OF RECLAMATION, *supra* note 43, at 62.

Valley Project.[51] Flow levels in the Sacramento and San Joaquin Rivers—the water sources for the Central Valley Project—range between 10 and 40 maf.[52]

Three features of the Reclamation program are important in the context of the ESA. One, the Bureau of Reclamation shares responsibility for ESA implementation—Section 7 of the ESA requires federal agencies to ensure that their actions are unlikely to jeopardize a listed species.[53] Potential conflicts between Reclamation operations and endangered species' recovery must be, and are, addressed in Section 7 consultation proceedings. A 1987 report found that Section 7 consultations exerted little effect on western water projects: "In terms of overall impact, between October 1977 and March 1985 only 68 consultations (out of about 3,200 consultations concerning water development projects) affected the projects with which they were associated. These consultations had varying, but normally limited, impact on the projects' timing, scope, and cost."[54] Nevertheless, this requirement potentially imposes greater restrictions on Reclamation water use, when such use may jeopardize an ESA-listed species, than would be imposed on other water development entities. Consequently, farmers relying on Reclamation water face a greater risk of reductions or disruptions in water supply.

Two, a recent government study documents the physical intersection between Reclamation areas and endangered species.[55] There were 184 federally listed or proposed species associated with Reclamation projects and water service areas in the 17 western states.[56] By state in terms of severity, the numbers are: California, 91, Oregon, 38, and Arizona, Texas, and Utah, 24 each.[57] Section III addresses, in a quantita-

---

51. Richard E. Howitt & Henry Vaux, *Competing Demands for California's Scarce Water*, *in* WATER QUANTITY/QUALITY MANAGEMENT AND CONFLICT RESOLUTION 271 (Ariel Dinar & Edna Tusak Loehman eds., 1995).

52. JOSEPH L. SAX & ROBERT H. ABRAMS, LEGAL CONTROL OF WATER RESOURCES 435 (1986).

53. Section 7(a)(2) of the ESA reads, "Each Federal agency shall, in consultation with and with the assistance of the Secretary, insure that any action authorized, funded, or carried out by such agency . . . is not likely to jeopardize the continued existence of any endangered species or threatened species or result in the destruction or adverse modification of habitat of such species which is determined by the Secretary . . . to be critical . . . ." 16 U.S.C. § 1536(a)(2) (1994).

54. U.S. GENERAL ACCOUNTING OFFICE, ENDANGERED SPECIES: LIMITED EFFECT OF CONSULTATION REQUIREMENTS ON WESTERN WATER PROJECTS 2-3 (1987).

55. *See* BUREAU OF RECLAMATION, *supra* note 7. This study co-locates endangered species and Bureau of Reclamation projects/service areas to develop the biological context for Section 7 consultations when implementing proposed western reclamation regulations under the Reclamation Reform Act of 1982.

56. *Id.* at 3-85.

57. *Id.* at 3-87.

tive analysis, the relationship between ESA-listed fish species and Reclamation-served agricultural activity.

Three, the pervasive presence of Reclamation projects throughout the West creates the potential for a general Reclamation approach to habitat restoration in western river systems. Section V pursues this idea in more detail.

III. Regression Analysis: Correlation Between ESA-Listed Fish Species and Irrigated Agriculture

This section investigates the quantitative relationship between the ESA-listed fish species and the extent of irrigated agricultural activity. A qualitative link between species endangerment and agricultural activity was already established by the Federal Register listings, which specify factors contributing to a species' decline. Figure 5, as described previously, also suggested a geographic relationship between endangered species and irrigated agriculture. Here, we examine the related quantitative issue: does the *number* of protected fish species depend on the level of irrigated agricultural activity? A positive correlation would suggest two items: (1) a higher level of irrigated agriculture may contribute to a greater number of ESA-listed fish species and (2) from a remedial perspective, the irrigated agricultural sector may be in a position to contribute significantly to recovery efforts for the more serious cases of species' endangerment (as measured by a relatively high number of protected fish species).

A. *Data, Variables, and Methods*

Two data sets are compiled for the regression analysis, a general data set on ESA-listed fish species and west-wide agricultural activity and a narrower data set on ESA-listed fish species and agricultural activity, reliant on Bureau of Reclamation water supplies. Both data sets consist of county-level data. The data on ESA-listed fish species represent the geographic dispersion of the 50 western fish species for which agriculture was a factor in species' decline. The variable formed from these data measures the number of ESA-listed fish species whose recovery could affect irrigated agriculture in that county; this variable ranges between 0 and 8.

The general data set (labeled "Census") uses county-level data on agriculture from the 1982 and 1987 *Census of Agriculture*.[58] This data set contains 1029 observations. The variables are:

58. BUREAU OF THE CENSUS, U.S. DEP'T OF COMMERCE, CENSUS OF AGRICULTURE 1982 and CENSUS OF AGRICULTURE 1987.

ESAFISH—Number of ESA-listed fish species whose recovery could affect irrigated agriculture.
AGPRD—Market value of agricultural products sold.
NIRRACR—Harvested cropland not irrigated with surface water.
PSTRACR—Pastureland and cropland used for pasture and grazing.
IRRACR—Irrigated, harvested cropland whose principal water source is surface water.
IRRACR2—*IRRACR* squared.
BOR—Indicator of whether the county contains cropland served by Bureau of Reclamation water supply.

As mentioned in the previous section, most counties in the West do not contain either ESA-listed fish habitat or agricultural land potentially affected by fish recovery efforts—803 of the 1029 observations on *ESAFISH* take on a value of 0. Table 2 contains descriptive statistics and units of variables in the Census data set.

The second data set (labeled "Reclamation") focuses on counties containing cropland irrigated with Bureau of Reclamation water supplies. It includes 1987 data describing crop production and irrigation on Reclamation-served cropland in the county.[59] This data set contains 199 observations. The definition of *ESAFISH* is identical to before, yet the variable includes only observations from Reclamation counties. *ESAFISH* takes on a value of 0 in 110 of the observations. Other variables in the Reclamation data set include:

CROPREV—market value of crops produced on Reclamation-served cropland.
IRRACR—Reclamation-served irrigated cropland.
PRJDLV—project water delivered to Reclamation-served farms.
PRJDLV2—*PRJDLV* squared.
NPRJDLV—non-project water delivered to Reclamation-served farms.
NPRJDLV2—*NPRJDLV* squared.

Table 2 also contains descriptive statistics and units of these variables.

In the regression analysis, *ESAFISH* serves as the dependent variable in the two equations estimated. *ESAFISH* is an example of "count data," i.e., data in the form of an integer quantity that measures the number of occurrences of an event. Data on recreational trips, for example, frequently come in the form of count data. To account for the integer nature of *ESAFISH*, the Poisson regression model is applied here

59. BUREAU OF RECLAMATION, U.S. DEP'T OF THE INTERIOR, 1987 SUMMARY STATISTICS: WATER, LAND, AND RELATED DATA (1988).

using GRBL econometric software.[60] The Poisson model is a statistical model that explicitly recognizes the integer nature of a dependent variable formed from count data.[61] The Poisson model is estimated using a maximum likelihood technique.

The general specification of the Census equation is

(1) *ESAFISHi = f(AGPRDi, NIRRACRi, PSTRACRi, IRRACRi, IRRACR2i, BORi, _i),*

where *i* denotes observations from the Census data set and _i is a stochastic disturbance term.

The general specification of the Reclamation equation is

(2) *ESAFISHj = g(CROPREVj, IRRACRj, PRJDLVj, PRJDLV2j, NPRJDLVj, NPRJDLV2j, _j),*

where *j* denotes observations from the Reclamation data set and _j is a stochastic disturbance term. Both (1) and (2) are estimated as linear functions of the independent variables.

B. Estimation Results

i. Census data set. Estimation of equation (1) yields a strong set of results (Table 3).
The estimated coefficient of every variable has the expected sign describing its influence on the number of ESA-listed fish species (*ESAFISH*). Only the coefficient on *AGPRD* is not significantly different from 0 at the 0.01 significance level. In this model, the negative coefficient on *NIRRACR* implies that, to the extent that much of the cropland in the county is not irrigated, *ESAFISH* tends to be smaller. The positive coefficient on *PSTRACR* indicates that *ESAFISH* increases as livestock activity increases; this may reflect the degradation of riverine habitat quality with livestock watering and grazing activity in watersheds. The

60. DANIEL M. HELLERSTEIN, U.S. DEP'T OF AGRICULTURE, GRBL: A PACKAGE OF REGRESSION PROGRAMS (1995).

61. *See generally* Daniel M. Hellerstein, *Using Count Data Models in Travel Cost Analysis with Aggregate Data*, 73 AM. J. OF AGRIC. ECON. 860 (1991). The Poisson model imposes the assumption that the mean and variance of the dependent variable are equal. The negative binomial regression model, a more general form of the Poisson, does not impose this assumption. The negative binomial model, however, did not converge to a solution when the data in this study were analyzed.

coefficients on *IRRACR* and *IRRACR2* imply that increased irrigation activity from surface water sources tends to increase *ESAFISH* (positive *IRRACR*), although at a diminishing rate as surface-water irrigated acreage increases (negative *IRRACR2*). Finally, the presence of irrigation with Reclamation water supply in a county (*BOR*) also increases *ESAFISH*.

The main finding of the estimation is simply the strong correlation between *ESAFISH* and agricultural activity. In particular, *IRRACR* serves as a proxy for several effects of irrigation on riverine and riparian habitat quality: water diversions deplete habitat quality; irrigation return flows tend to impair water quality through salinity, nutrient, and pesticide loadings; and dams and reservoirs alter natural river-flow patterns, destroy natural habitat, and can block migrations. From a restoration perspective, a decline in *IRRACR* and offstream water use, accompanied by an increase in instream flow, can contribute to recovery of ESA-listed fish species by reversing this set of harmful impacts on the natural environment.

A second finding relates to the elasticities, each one of which measures the percent change in *ESAFISH* given a one percent change in an independent variable holding all other variables constant. Every elasticity is moderately to highly inelastic. If the model is interpreted literally as showing a causal effect of agricultural activity on *ESAFISH*, the elasticities indicate that irrigation levels, as well as other agricultural activity represented in the independent variables, would have to change significantly to cause a marked decline in ESA-listed fish species.

In equation (1), the dummy variable for the presence of Reclamation activity in a county (*BOR*) has a positive coefficient with statistical significance at the 0.01 level (Table 3). That is, even controlling for surface-water irrigated acreage via *IRRACR*, the number of *ESAFISH* increases with the presence of Reclamation activity. To some degree, this relationship may be associated with a trademark of Reclamation development: construction of storage and diversion dams. Dams, and associated reservoirs, typically harm native fish habitat. Estimating equation (2) with the Reclamation data set enables closer examination of the role of the Reclamation program.

ii. Reclamation data set. The estimation of equation (2) with the Reclamation data set generally yields statistically significant relationships between *ESAFISH* and Reclamation activity. Four of the variables (*IRRACRE, PRJDLV, PRJDLV2, NPRJDLV*) are significant at the 0.01 level, and two (*NPRJDLV2* and the intercept) are significant at the 0.10 level. *CROPREV* has the only estimated coefficient that is not significantly different from 0 at the 0.10 level. The negative coefficient on *IRRACR* indicates that *ESAFISH* declines as irrigated acreage increases. This is slightly counterintuitive: after controlling for water supply, one might expect the scale of irrigated acreage to exert little effect on *ESAFISH*. The water supply variables show similar curvature properties as the irrigated acreage variables in the Census data set. Coefficients on the linear terms, *PRJDLV* and *NPRJDLV*, are positive, while coefficients on the squared terms, *PRJDLV2* and *NPRJDLV2*, are negative.

As with the first regression, the elasticities of *ESAFISH* relative to the independent variables in the Reclamation data set are moderately to highly inelastic.

The water supply variables provide the most interest. They are measures of on-farm irrigation water use on Reclamation-served cropland. The strong statistical significance of *PRJDLV* and *PRJDLV2* (significant at the 0.01 level) indicates a link to *ESAFISH*. Levels for non-project surface water supply (*NPRJDLV*) are much lower than those for project water supply, with a mean for *NPRJDLV* of only 11,706 acre-feet per Reclamation county relative to a mean for *PRJDLV* of 127,600 acre-feet. Nevertheless, the two variables for non-project supply correlate quite strongly with ESA-listed fish species.

Overall, the quantitative analysis answers the section's original question in the affirmative: the number of protected fish species correlates positively with the level of irrigated agricultural activity. This evidence suggests that irrigation water conservation—with reallocation to instream flow for habitat improvement—could be an important element of endangered fish species recovery programs in the West.

IV. The Endangered Fish Species-Irrigated Agriculture Nexus: Recent Developments in Government Policies and Programs

This section consists of two short case studies on recent developments in central California and the Columbia River Basin. The case studies complement the quantitative analysis by providing concrete illustrations of fish-agriculture water allocation tradeoffs.

A. Federal Water Policy Reform in Central California

Significant federal, state, and private investments were made in water storage and conveyance projects on the Sacramento and San Joaquin River systems in California's Central Valley. These projects transformed arid land into some of the world's most productive farmland, with agriculture consuming the vast majority of the region's developed water resources in its complete reliance on irrigation.[62]

62. CALIFORNIA DEP'T OF WATER RESOURCES, 1 CALIFORNIA WATER PLAN UPDATE 49 (1994).

However, 22 fish species are declining or have gone extinct in these river systems: three fish species (Sacramento River winter-run Chinook salmon, delta smelt, and Little Kern golden trout) are listed as threatened or endangered under the ESA;[63] 16 additional fish species either qualify for listing or are in severe decline with real potential for becoming endangered; and three fish species have already gone extinct.[64]

The Bureau of Reclamation's Central Valley Project (CVP) is integral to agriculture in central California. The CVP consists of 20 dams and more than 500 miles of major canals. Its agricultural service area encompasses 2.6 million acres, the largest of any single Reclamation project.[65] In an average year, CVP irrigation water supply equals 6.6 maf,[66] or more than 95 percent of CVP water deliveries. Revenue from the sale of crops produced with CVP water typically exceeds $3 billion per year.[67] This equals 5 percent of the value of U.S. crop production.

The decline in central California's fishery influenced passage of new federal water policy for the CVP. In 1992, the Central Valley Project Improvement Act (CVPIA) was instituted to improve fishery habitat, in addition to achieving several other water management objectives. [68] In a significant step for a Reclamation project, the law designated "fish and wildlife mitigation, protection, and restoration" as an explicit purpose of the CVP.[69] To achieve this objective, the law permanently allocates 0.8 maf of CVP water in normal water-supply years (almost 20 percent of CVP contracted irrigation water supply) for restoration of fish habitat.[70]

---

63. The official status of the Sacramento River winter-run Chinook salmon changed from threatened to endangered on January 4, 1994. 59 Fed. Reg. 440 (1994). A proposal to list the Sacramento splittail (a fish endemic to California's Central Valley) as threatened was made on January 6, 1994. 59 Fed. Reg. 862 (1994).

64. *See* Moyle & Williams, *supra* note 24, at 282-84.

65. BUREAU OF RECLAMATION, *supra* note 43, at 166.

66. The 6.6 million acre-feet of irrigation water supply splits into 4.3 maf for CVP agricultural contractors and 2.3 maf for water rights holders and exchange contractors. Water rights holders and exchange contractors are senior to CVP agricultural contractors, and thus receive CVP water supply with greater certainty. For example, CVP water supply in 1994 was below average because of low precipitation during the 1993-94 winter. Water rights holders and exchange contractors received 75 percent of full supply in 1994, while CVP agricultural contractors received only 35 percent of full supply. *See* BUREAU OF RECLAMATION & FISH AND WILDLIFE SERVICE, U.S. DEP'T OF THE INTERIOR, IMPLEMENTATION OF THE CENTRAL VALLEY PROJECT IMPROVEMENT ACT 8 (1994).

67. BUREAU OF RECLAMATION, *supra* note 43, at 166.

68. Central Valley Project Improvement Act, Pub. L. No. 102-575, § 3402, 106 Stat. 4706 (1992).

69. *Id.* at 4714.

70. The Ninth Circuit Court of Appeals recently held that the federal government's reduction of 1993 irrigation water supplies did not violate the terms of the plaintiff's CVP water service contracts decision where the reduction was made to meet the requirements of the ESA and CVPIA. *O'Neill v. United States*, 50 F.3d 677 (9th Cir. 1995).

This water is intended to address several aspects of fishery improvement. For example, the CVPIA states that the water would contribute both to addressing obligations under the ESA and, more generally, to achieving a goal of doubling natural populations of the region's anadromous fish species.[71] The key water-management requirement for Central Valley fish species involves regulating the volume and timing of flows in both the Sacramento and San Joaquin River systems until they pass through the Sacramento-San Joaquin Delta and into San Francisco Bay.[72]

Reallocation of agricultural water will likely supply most of the water for fish habitat. Although the CVPIA does not specify this explicitly, most observers believe that, in years when water-use reductions are required to meet the 0.8 maf requirement, most or all of the reductions would come from irrigation water use. This conclusion follows directly from the existing pattern of CVP water allocation, in which agricultural use dominates urban and industrial water by over a 10-to-1 ratio.[73] In addition, the CVPIA requires the Secretary of the Interior to implement by the year 2007 a least-cost plan to replace the 0.8 maf allocated to fish habitat.[74] Voluntary water transfers, voluntary agricultural land retirement, and water conservation requirements must be among the options considered for the plan.[75] Both the economic value of urban water use and the cost of expanding project capacity exceed the value of irrigation water in this region.[76] Thus, relying primarily on reduced irrigation water use appears consistent with a least-cost plan for acquiring 0.8 maf of water.

71. Central Valley Project Improvement Act § 3406. Debate is on-going regarding the intent of the fish and wildlife water allocation provision in relation to water allocations for species listed under the ESA. Interpretations of the CVPIA vary from the view that all water used to protect ESA-listed fish should be counted against the 800,000 acre-feet of water set aside in the CVPIA to the view that water quantities attributable to implementation of the ESA should be independent of the 800,000 acre-feet. The act is ambiguous on this point, containing provisions in support of both views. In practice, the answer may lie between the opposing views. In the act's first year, approximately one-half of the 800,000 acre-feet was used to protect ESA-listed fish. For example, 300,000 acre-feet were attributed to the temporary shut-down of a major CVP water pumping facility because of excessive takings of endangered winter-run Chinook salmon; the shut-down resulted in water-supply reductions to many farmers using CVP water. *Id.*

72. Anthony C. Fisher et al., *Integrating Fishery and Water Resource Management: A Biological Model of a California Salmon Fishery*, 20 J. OF ENVTL. ECON. AND MGMT. 234 (1991).

73. BUREAU OF RECLAMATION, *supra* note 43.

74. Central Valley Project Improvement Act § 3408(j).

75. *Id.*

76. Zach Willey & Thomas Graff, *Federal Water Policy in the United States—An Agenda for Economic and Environmental Reform*, 13 COLUMBIA J. OF ENVTL. L. 335 (1988).

B. Salmon Recovery in the Columbia River Basin

Salmon populations in the Columbia River Basin have declined severely as a result of fish harvesting and river development activities. Populations of salmon and steelhead have fallen to roughly 20 percent of the peak level of 10-16 million spawning adults per year.[77] Wild and naturally spawning salmon are at 2 percent of historic levels. Since 1991, three Snake River salmon runs have been listed as threatened or endangered under the ESA,[78] with an additional 10 salmon runs considered in critical condition.[79]

The primary fish-agriculture nexus in the Columbia River system occurs in the upper Snake River Basin (composed of southern Idaho and east-central Oregon). This region is one of the major areas of irrigated agriculture in the United States, with roughly three million acres irrigated with surface water and almost one million acres irrigated with ground water.[80] The majority of the irrigated acreage is in hay, wheat, and barley production, although the region is important nationally in production of Irish potatoes and sugar beets.[81] The Bureau of Reclamation operates nine water projects in the upper Snake River Basin.[82] These projects deliver water to one-half of the region's surface-water irrigated acres: about 1.5 million acres regularly receive more than 5 maf of Reclamation water.[83] The Minidoka-Palisades Project in southern Idaho contains over one million of these acres.[84]

One salmon recovery measure under consideration involves flow augmentation from the upper Snake River to improve hydrologic conditions for juvenile salmon migration through the lower Snake and

---

77. Michael C. Blumm & Andy Simrin, *The Unraveling of the Parity Promise: Hydropower, Salmon, and Endangered Species in the Columbia Basin*, 21 ENVTL. L. 657, 663 (1991).

78. The National Marine Fisheries Service formally listed the Snake River sockeye salmon as endangered on November 20, 1991, and the Snake River spring/summer-run Chinook salmon and fall-run Chinook salmon as threatened on April 22, 1992. On August 18, 1994, the two runs of Chinook salmon were converted from threatened to endangered under an emergency interim rule. NAT'L MARINE FISHERIES SERVICE & NAT'L OCEANIC AND ATMOSPHERIC ADMIN., U.S. DEP'T OF COMMERCE, PROPOSED RECOVERY PLAN FOR SNAKE RIVER SALMON I-6 (1995).

79. Nehlsen et al., *supra* note 24, at 11.

80. Joel R. Hamilton & Norman K. Whittlesey, Contingent Water Markets for Salmon Recovery (Feb. 1992) (unpublished report, University of Idaho and Washington State University).

81. BAJWA ET AL., *supra* note 36; MARCEL P. AILLERY ET AL., ECONOMIC RESEARCH SERVICE, U.S. DEP'T OF AGRICULTURE, SALMON RECOVERY IN THE PACIFIC NORTHWEST: A SUMMARY OF AGRICULTURAL AND OTHER ECONOMIC EFFECTS 4 (1994).

82. BUREAU OF RECLAMATION, *supra* note 43, at 60, 63.

83. *Id.*

84. *Id.* at 145.

Columbia Rivers. The Northwest Power Planning Council[85] recommended a minimum of 0.427 maf per year in flow augmentation from the upper Snake River, with additional augmentation of 1 maf by 1998, for a total of 1.427 maf.[86] These two levels of flow augmentation were studied in an economic evaluation of the Snake River salmon recovery plan that was proposed by the Snake River Salmon Recovery Team.[87] To achieve these two targets, 0.127 to 1.127 maf in flow augmentation would be acquired from the irrigated agriculture sector in the upper Snake River Basin.[88] Meeting the 1.127 maf target would reduce irrigation water supply by over 15 percent in the basin.

The Bureau of Reclamation has been assigned responsibility for acquiring water from irrigators in the upper Snake River Basin. The Northwest Power Planning Council originally directed Reclamation and the Bonneville Power Administration to share equally the cost of acquiring the necessary water resources to meet the flow augmentation targets.[89] However, using its authority to recover the three endangered Snake River salmon, the National Marine Fisheries Service (NMFS) designated Reclamation solely responsible for acquiring 0.427 maf per year from 1995-97, and additional water as needed after 1998, particularly in low-flow years.[90] Reclamation agreed to these terms.[91]

---

85. The Northwest Power Planning Council was created by the Pacific Northwest Electric Power Planning and Conservation Act (1980) to effect the act's mandate of treating fish and wildlife on an equal basis with hydropower and other traditional river uses.

86. NORTHWEST POWER PLANNING COUNCIL, 1994 COLUMBIA RIVER BASIN FISH AND WILDLIFE PROGRAM 5-21 (1994).

87. DANIEL D. HUPPERT & DAVID L. FLUHARTY, NAT'L MARINE FISHERIES SERVICE, U.S. DEP'T OF COMMERCE, ECONOMICS OF SNAKE RIVER SALMON RECOVERY: A REPORT TO THE NAT'L MARINE FISHERIES SERVICE (1995). *See also* Hamilton & Whittlesey, *supra* note 80 and AILLERY ET AL., *supra* note 81. The Snake River Salmon Recovery Team was a seven-member group appointed by the National Marine Fisheries Service in 1992 to develop independent recommendations for a Snake River recovery plan. Its final recommendations were made in May 1994.

88. HUPPERT & FLUHARTY, *supra* note 87, at 3-40.

89. NORTHWEST POWER PLANNING COUNCIL, *supra* note 86, at 5-32. Bonneville Power Administration markets hydroelectricity produced at the eight major mainstream dams on the lower Snake and lower Columbia Rivers. These dams are viewed as a major contributor to the decline of the Snake River salmon fishery. NAT'L MARINE FISHERIES SERVICE & NAT'L OCEANIC AND ATMOSPHERIC ADMIN., *supra* note 78, at V-2-2 to V-2-4.

90. NAT'L OCEANIC AND ATMOSPHERIC ADMIN. & NAT'L MARINE FISHERIES SERVICE, U.S. DEP'T OF COMMERCE, BIOLOGICAL OPINION: REINITIATION OF CONSULTATION ON 1994-1998 OPERATION OF THE FEDERAL COLUMBIA RIVER POWER SYSTEM AND JUVENILE TRANSPORTATION PROGRAM IN 1995 AND FUTURE YEARS 99-100 (1995) (Endangered Species Act—Section 7 Consultation); NAT'L OCEANIC AND ATMOSPHERIC ADMIN. & NAT'L MARINE FISHERIES SERVICE, *supra* note 78, at V-2-25.

91. BUREAU OF RECLAMATION, U.S. DEP'T OF THE INTERIOR, BUREAU OF RECLAMATION'S RECORD OF DECISION IMPLEMENTING ACTIONS PURSUANT TO

Voluntary water transfers appear to be the method for acquiring water from irrigators in the region. The Northwest Power Planning Council originally suggested that both incentive and regulatory programs be considered as mechanisms for obtaining water for flow augmentation. Some close observers recommended water acquisition through expansion of voluntary water markets in the region.[92] Reclamation, under guidance from NMFS, is pursuing an approach of purchasing water from willing sellers.

V. CONSIDERING RECLAMATION'S ROLE IN SPECIES RECOVERY

At this juncture, we alter course from description of current events to consideration of future approaches to recovery of endangered fish species. This section considers possible roles for the federal Reclamation program at the interface of the water-species-agriculture issue in the West. Two alternatives are characterized. One approach is reactive, in which the Bureau of Reclamation responds to its ESA obligation not to jeopardize listed species. A second approach is proactive, involving the Bureau of Reclamation as a leader in ecosystem restoration activities that would address the health of multiple species, including those not presently endangered.[93]

A certain context is pertinent to this section: namely, the Bureau of Reclamation's transition from a water development organization to a water management organization. Beginning in the late 1980s, Reclamation initiated a planning process to redefine the agency's mission in the contemporary era of western water management.[94] That effort defined the principles, objectives, and program opportunities that would create a water resource management agency. Reclamation's Strategic Plan (1992) describes a subsequent stage of the process.[95] It includes five pillars of the Reclamation mission, including one dedicated broadly to "protecting

---

BIOLOGICAL OPINIONS OF MARCH 1, 1995 11 (1995).

92. Hamilton & Whittlesey, *supra* note 80; Ray Huffaker et al., *Institutional Feasibility of Contingent Water Marketing to Increase Migratory Flows for Salmon on the Upper Snake River*, 33 NAT. RESOURCES J. 671 (1994).

93. The National Research Council recently proposed a proactive approach to aquatic ecosystem restoration in the United States. NATIONAL RESEARCH COUNCIL, *supra* note 24. The council devised a "National Aquatic Ecosystem Restoration Strategy" that would be financed by a surcharge on hydropower sales from federal facilities. *Id.* at 350-376.

94. BUREAU OF RECLAMATION, U.S. DEP'T OF THE INTERIOR, ASSESSMENT '87: A NEW DIRECTION FOR THE BUREAU OF RECLAMATION (1987).

95. BUREAU OF RECLAMATION, U.S. DEP'T OF THE INTERIOR, RECLAMATION'S STRATEGIC PLAN A LONG-TERM FRAMEWORK FOR WATER MANAGEMENT, DEVELOPMENT, AND PROTECTION (1992).

the environment."[96] Most recently, a director of the Bureau of Reclamation, Commissioner Daniel P. Beard, emphasized again the paramount importance of environmental considerations in the new, reformed Reclamation program.[97]

A. *Reclamation in a Reactive Mode: Using ESA Consultations to Influence Water Project Development and Operation*

One approach to protection of ESA-listed species in western rivers involves using the ESA Section 7 interagency consultation process to influence Bureau of Reclamation activities. Section 7 consultations currently are the primary way for Reclamation to participate in ESA implementation. They place the agency in a reactive mode of considering whether listed species might be jeopardized by construction or operation of Reclamation projects.

The consultation process appeared to function effectively in influencing the nature and scale of water project development in two river basins, the upper Colorado River and Platte River.[98] In a study of consultations related to western water projects between 1977 and 1985, GAO found that, while no consultations resulted in project termination, they modified 38 projects in the upper Colorado Basin and two major projects in the Platte Basin.[99] The consultation process also was employed to impose a water depletion fee on new projects in the upper Colorado Basin. Forty six projects paid these fees in the 1981-86 period;[100] the fees continue, at the one-time rate of $10 per acre-foot, as part of the official Recovery Implementation Program for Endangered Fish Species in the Upper Colorado River Basin.[101] Fee revenues are ear-

96. Environmental protection, under the Bureau of Reclamation's *Strategic Plan*, consists of water management for instream flows, fish and wildlife resources, recreational opportunities, and water quality. *Id.* at 10-15. In late 1992, particular elements of the strategic plan were specified in two draft implementation plans. BUREAU OF RECLAMATION, U.S. DEP'T OF THE INTERIOR, BUREAU OF RECLAMATION, AN IMPLEMENTATION PLAN FOR FISH AND WILDLIFE RESOURCES (1992); BUREAU OF RECLAMATION, U.S. DEP'T OF THE INTERIOR, AN IMPLEMENTATION PLAN FOR INSTREAM FLOWS (1992). These two implementation plans articulate commitments to fulfilling the Bureau's Section 7 obligations under the ESA, collaborating with federal and state wildlife management agencies on endangered species protection, and searching for opportunities to improve aquatic habitat through modifications in reservoir operations and other Reclamation operations.

97. DANIEL P. BEARD, BUREAU OF RECLAMATION, U.S. DEP'T OF THE INTERIOR, BLUEPRINT FOR REFORM: THE COMMISSIONER'S PLAN FOR REINVENTING RECLAMATION 1-2 (Nov. 1, 1993).

98. U.S. GENERAL ACCOUNTING OFFICE, *supra* note 54.

99. *Id.* at 17-18.

100. *Id.* at 29.

101. FISH AND WILDLIFE SERVICE, U.S. DEP'T OF THE INTERIOR, RECOVERY IMPLEMENTA-

marked for Colorado River endangered fish recovery.

The ESA consultation process has the strength of assigning clear leadership responsibility to the fisheries agencies, USFWS and NMFS. They are responsible for developing scientifically-based "reasonable and prudent alternatives" for other federal agencies to undertake to avoid placing a protected species in jeopardy of extinction.[102]

B. *Reclamation in a Proactive Mode: Elevating Fish and Wildlife Resources to a* **Project Purpose**

The proactive approach contrasts distinctly to operating solely within the context of the ESA consultation process: instead of posing the question of whether Reclamation activity might *jeopardize* a listed species, it poses the question of how Reclamation activity can contribute to *avoidance* of additional ESA-listings.

In both central California and the Columbia River Basin, as described previously, external events thrust the Bureau of Reclamation into a decidedly proactive role. Under the CVPIA, Reclamation is responsible for implementing several provisions geared toward improving fish populations, including water reallocation to instream flow for habitat improvement and water price surcharges earmarked for a habitat restoration fund.[103] In the Columbia River Basin, the NMFS directed Reclamation to acquire at least 0.427 maf of water per year from irrigators in the upper Snake River Basin, with the water earmarked to augment river flows for salmon migration.[104] In both cases, the water reallocations should improve populations of fish species whose populations are declining yet not endangered, as well as assisting the ESA-listed species.

The argument for a proactive role for the Bureau of Reclamation depends on three points. One, the biological evidence cited earlier suggests that species decline and endangerment are serious concerns in the West's riverine ecosystems. Two, a proactive role creates flexibility to restore ecosystems rather than to manage for individual species. Many observers correctly criticize the ESA's individual-species approach as piecemeal. A multi-species or ecosystem approach to species protection in western rivers appears desirable on both biological and economic grounds. Three, proactive measures may alleviate pressure for more drastic measures that frequently accompanies official ESA listing. In its *An Implementation Plan for Fish and Wildlif Resources*, Reclamation makes

---

TION PROGRAM FOR ENDANGERED FISH SPECIES IN THE UPPER COLORADO RIVER BASIN 4-6 (1987).

102. YAFFEE, *supra* note 30, at 97-98.

103. *See supra*, text accompanying notes 49-50.

104. *See supra*, text accompanying notes 61-66.

this point explicitly:

> It will be advantageous for Reclamation and its traditional constituents to cooperate in efforts to support or re-establish plant and animal habitat before species become listed as threatened or endangered. Once listing occurs and critical habitat is identified, the legal requirements for protection and recovery take effect and resulting operational restrictions may severely affect established uses. The prospect of mandated actions creates incentive for a proactive role.[105]

These general points make an intuitively appealing case for a proactive approach by Reclamation to habitat restoration in western river systems.

The CVPIA provides a feasible model for a proactive approach on a Western scale.[106] Several CVPIA elements pertaining to fish and wildlife conservation, and endangered species protection in particular, could be replicated across Reclamation projects:

- mandate fish and wildlife conservation as an explicit purpose of each Reclamation project,
- allocate water to fish and wildlife (where allowed by state water law),
- establish a Reclamation Fish and Wildlife Restoration Fund dedicated to acquiring
- water, land, and management resources for fish and wildlife conservation, and
- finance the fund through modest charges paid by recipients of Reclamation water and power supplies (for example, the CVPIA increases irrigation water prices by up to $6 per acre-foot).

As with the CVPIA, the U.S. Congress would need to authorize or direct the Bureau of Reclamation to develop a fish and wildlife recovery program of this scope.

---

105. BUREAU OF RECLAMATION, AN IMPLEMENTATION PLAN FOR INSTREAM FLOWS, *supra* note 96, at 34-35.

106. John B. Loomis, *Water Transfer and Major Environmental Provisions of the Central Valley Project Improvement Act: A Preliminary Economic Evaluation*, 30 WATER RESOURCES RES. 1865 (1994). Another example of proactive water management is in the Yakima River Basin in Washington. A major Bureau of Reclamation project provides water service to roughly 370,000 irrigated acres in the Yakima Basin. BUREAU OF RECLAMATION, *supra* note 96, at 34-35. Presently, the basin does not contain any ESA-listed fish species (although the basin's stocks of coho salmon, sockeye salmon, and the summer run of Chinook salmon are extinct. Nehlsen, *supra* note 24. In 1994, the U.S. Congress enacted the Yakima River Basin Water Enhancement Project, Pub. L. No. 103-434, § 12, 108 Stat. 4526 (1994). One purpose of the law is to restore the basin's anadromous fishery. Among other measures, the act requires that water savings of 110,000 acre-feet per year be dedicated to improving fish and wildlife habitat.

VI. CONCLUDING COMMENTS

The decline of natural aquatic ecosystems in the American West is one legacy of a century of river development for water and power supply. This article extracts two fundamental elements related to this legacy—fish species protected under the ESA and irrigated agriculture served by surface water sources—for analysis. The ESA provides the final "safety net" for wildlife when other protections afforded by wildlife policy or resource use regulations prove insufficient. More broadly, the ESA offers one tool to achieve, perhaps, ecosystem protection. Endangered fish species, through the legal imperative to allocate resources to their recovery, may create conditions to improve the natural functioning of western river ecosystems—much as the spotted owl may help to preserve the Pacific Northwest's ancient forests.

Our focus on irrigated agriculture stems from several facts. Irrigated agriculture dominates surface-water consumption in the West. Moreover, in basin after basin, empirical evidence shows that agriculture uses more water than would be the case in an economically efficient allocation of the resource across economic sectors. Finally, in many areas of the West, irrigation water conservation will be necessary to increase instream water flow for fish habitat improvement.

Three principal descriptive and analytical results emanated from the Western perspective developed here. First, the numbers that describe the potential conflict seem large when totaled across the West: 50 ESA-listed fish species are linked to agricultural activity; 235 counties, representing 22 percent of the counties in the West, contain irrigated production that relies on water from rivers with these ESA-listed species; and another 86 western fish species are officially designated as candidate species for listing. Second, the western scale permitted a statistical analysis of the agriculture-endangered fish species relationship using two county-level databases. Analysis of both databases found that the number of ESA-listed fish species in a county correlated positively with the level of irrigated agriculture reliant on surface water in the county. In particular, the number of species depended positively on water-supply levels of the Bureau of Reclamation. Finally, the western perspective—as well as individual case studies of central California and the Columbia River Basin—focused attention on the federal Reclamation program. Its pervasive presence throughout the West and specific responsibilities under the ESA will lead to Reclamation involvement in the recovery of many, if not most, ESA-listed western fish species. Perhaps more important, though, is the proactive role that could be defined for the Bureau of Reclamation in ecosystem protection and restoration of western river systems. A program designed to avoid the endangered species of tomorrow, while protecting the endangered fish of today, could minimize the cost and disruption inherent in resolution of western water allocation conflicts.

The western perspective highlights common elements of water allocation conflicts between endangered fish species and irrigated agriculture across the western rivers. At the same time, it ignores details related to individual species or river segments, information that would be important for ascertaining the exact tradeoffs present in individual river basins. The analysis thus serves as a screening device, with additional quantitative research needed on the topic.

Table 1. Surface Water Withdrawals for Irrigated Agriculture, 17 Western States

	Surface water withdrawals for irrigation, 1990	Irrigation water withdrawals as a percentage of total surface water withdrawals, 1990
	(1,000 acre-feet)	(%)
Arizona	3,640	85
California	19,300	84
Colorado	10,100	91
Idaho	13,500	99
Kansas	224	12
Montana	9,990	98
Nebraska	1,950	42
Nevada	2,190	86
New Mexico	1,840	95
North Dakota	96	3
Oklahoma	121	14
Oregon	7,060	82
South Dakota	281	74
Texas	3,250	23
Utah	3,460	90
Washington	5,920	82
Wyoming	7,760	96

Data source: Solley, et al., 1993.

Table 2. Variables in Westwide County-level Data Sets

Variable	*Units*	*Mean*	**Standard** *Deviation*
	1. Census data set[1]		
ESAFISH[2]	number of species	0.48	1.13
AGPRD	$1,000	58317	1.098E05
NIRRACR	acres	1.044E05	1.149E05
PSTRACR	acres	4.176E05	5.101E05
IRRACR	acres	18485	51065
IRRACR2	acres squared	2.947E09	2.290E10
BOR[3]		0.190	0.393
	2. Reclamation data set[4]		
ESAFISH[5]	number of species	1.11	1.56
CROPREV	dollars	3.968E07	1.199E08
IRRACR	acres	52814	87989
PRJDLV	acre-feet	1.276E05	2.592E05
PRJDLV2	acre-feet squared	8.313E10	4.704E11
NPRJDLV	acre-feet	11706	51409
NPRJDLV2	acre-feet squared	2.767E0	92.020E10

1 The Census data set contains 1029 observations.

2 803 of the observations on *ESAFISH* are 0 in the Census data set. The mean for *ESAFISH* is 2.19 when these observations are removed.

3 *BOR* is a dummy variable that takes a value of 1 when a county contains cropland served by Bureau of Reclamation water supply.

4 The Reclamation data set contains 199 observations.

5 110 of the observations on *ESAFISH* are 0 in the Reclamation data set. The mean for *ESAFISH* is 2.47 when these observations are removed from the Reclamation data set.

Table 3. Poisson Model Estimates, Dependence of ESA-Listed Fish *Species on Agricultural Activity in the West*[1]

1. Census data set

Variable	Estimated Coefficient	*t*-ratio	Elasticity[2]
AGPRD	4.184E-07	0.98	0.02
NIRRACR	-3.703E-06	-5.95	-0.39
PSTRACR	2.582E-07	4.39	0.11
IRRACR	9.238E-06	6.02	0.16[3]
IRRACR2	-1.741E-11	-3.87	
BOR	0.637	5.62	0.12
Constant	-0.989	-11.77	
LLF[4]	-986.03		

2. Reclamation data set

Variable	Estimated Coefficient	*t*-ratio	Elasticity[2]
CROPREV	1.237E-09	1.25	0.05
IRRACR	-5.496E-06	-2.99	-0.29
PRJDLV	3.413E-06	6.08	0.41[3]
PRJDLV2	-8.093E-13	-3.89	
NPRJDLV	8.430E-06	2.55	0.09[3]
NPRJDLV2	-2.166E-11	-1.74	
Constant	-0.153	-1.62	
LLF[4]	-306.82		

1 Dependent variable is *ESAFISH*, the number of ESA-listed fish species in the county whose recovery could affect irrigated agricultural activity.
2 The elasticity formula for the Poisson regression model is _X, where _ is the estimated coefficient and X is the independent variable. (Footnote 3 to the table discusses an exception to this formula for the case of an independent variable that enters both in linear and squared terms.) Each independent variable is evaluated at its mean value for the elasticities reported here.
3 The elasticity formula for the Poisson regression model when a variable enters both in linear and squared terms is (_1X + 2_2X2), where _1 is the estimated coefficient for X and _2 is the estimated coefficient on X2.
4 Value of the log-likelihood function for the regression.

APPENDIX:

WESTERN ENDANGERED AND THREATENED FISH

Several sources of information are used in compiling the appendix: official listings of 33 individual species that appear in the *Federal Register*; official Recovery Plans of 32 individual species published by U.S. Fish and Wildlife Service (USFWS) (U.S. Department of the Interior, U.S. Fish and Wildlife Service, miscellaneous recovery plans); official reports to the U.S. Congress made by USFWS (U.S. Department of the Interior, U.S. Fish and Wildlife Service, 1990) and National Marine Fisheries Service (U.S. Department of Commerce, National Marine Fisheries Service, 1991); and USFWS publications on expenditures on species recovery (U.S. Department of the Interior, U.S. Fish and Wildlife Service, 1991, 1992).

SPECIES	YEAR LISTED AND STATUS	CURRENTLY OCCUPIED HABITAT	STATE	AG LINK	FEDERAL EXPENDITURES ($1,000)	
					FY1990	FY1991
YAQUI CATFISH	1984-T	SAN BERNARDINO CREEK	AZ	YES	33.0	37.5
OZARK CAVEFISH	1984-T	SPRINGFIELD PLATEAU CAVES	OK	YES	29.4	8.1
BONYTAIL CHUB	1980-E	UPPER AND LOWER COLORADO RIVER SYSTEM	AZ CA CO NV UT WY	YES	292.3	243.9
BORAX LAKE CHUB	1982-E	BORAX LAKE	OR	YES	30.0	5.8
CHIHUAHUA CHUB	1983-T	MIMBRES RIVER	NM	YES	18.4	30.2
HUMPBACK CHUB	1967-E	UPPER AND LOWER COLORADO RIVER SYSTEM	AZ CO UT	YES	403.6	2773.0
HUTTON SPRING TUI CHUB	1985-T	HUTTON SPRING THREE EIGHTHS SPRING	OR	YES	.4	1.3
MOJAVE TUI CHUB	1970-E	SODA SPRINGS	CA	YES	24.0	35.0
OWENS TUI CHUB	1985-E	OWENS RIVER SYSTEM	CA	YES	15.4	2.0
PAHRANAGAT ROUNDTAIL CHUB	1970-E	PAHRANAGAT RIVER	NV	YES	16.7	36.8

SPECIES	YEAR LISTED AND STATUS	CURRENTLY OCCUPIED HABITAT	STATE	AG LINK	FEDERAL EXPENDITURES ($1,000)	
					FY1990	FY1991
SONORA CHUB	1986-T	SYCAMORE CREEK SYSTEM	AZ	YES	2.9	3.0
VIRGIN RIVER CHUB	1989-E	VIRGIN RIVER	AZ NV UT	YES	27.8	46.6
YAQUI CHUB	1984-E	RIO YAQUI BASIN	AZ	YES	50.0	0.5
CUI-UI	1967-E	TRUCKEE RIVER PYRAMID LAKE	NV	YES	692.6	310.3
ASH MEADOWS SPECKLED DACE	1983-E	ASH MEADOWS SPRINGS	NV	YES	45.0	10.9
CLOVER VALLEY SPECKLED DACE	1989-E	CLOVER VALLEY SPRINGS	NV	YES	0.1	0.0
DESERT DACE	1985-T	SOLDIER MEADOWS SPRINGS	NV	YES	0.6	13.6
FOSKETT SPECKLED DACE	1985-T	WARNER VALLEY SPRING	OR	YES	0.0	0.8
INDEPENDENCE VALLEY SPECKLED DACE	1989-E	INDEPENDENCE VALLEY SPRING	NV	YES	0.1	0.0
KENDALL WARM SPRINGS DACE	1970-E	KENDALL WARM SPRING	WY	YES	2.0	0.3
MOAPA DACE	1967-E	SPRINGS OFF THE MOAPA RIVER	NV	YES	54.6	32.5
FOUNTAIN DARTER	1970-E	SAN MARCOS RIVER COMAL RIVER	TX	YES	42.7	10.2
LEOPARD DARTER	1978-T	LITTLE RIVER	OK	YES	58.4	23.5
BIG BEND GAMBUSIA	1967-E	SPRINGS IN BIG BEND NATIONAL PARK	TX	NO	13.3	18.0
CLEAR CREEK GAMBUSIA	1967-E	SAN SABE RIVER CLEAR CREEK	TX	YES	2.8	0.0
PECOS GAMBUSIA	1970-E	PECOS RIVER LOST RIVER	NM TX	YES	8.1	17.2
SAN MARCOS GAMBUSIA	1980-E	SAN MARCOS RIVER	TX	NO	14.7	4.0

SPECIES	YEAR LISTED AND STATUS	CURRENTLY OCCUPIED HABITAT	STATE	AG LINK	FEDERAL EXPENDITURES ($1,000)	
					FY1990	FY1991
PAHRUMP KILLIFISH	1967-E	PAHRUMP VALLEY SPRINGS *TRANSPLANTED??	NV	YES	10.5	10.7
NEOSHO MADTOM	1990-T	NEOSHO RIVER	KS OK	YES	14.2	30.5
LOACH MINNOW	1986-T	GILA RIVER SYSTEM	AZ NM	YES	34.5	36.7
ASH MEADOWS AMARGOSA PUPFISH	1983-E	ASH MEADOWS SPRING SYSTEM	NV	YES	48.0	22.9
COMANCHE SPRINGS PUPFISH	1967-E	SPRINGS IN THE PECOS RIVER DRAINAGE	TX	YES	10.0	15.6
DESERT PUPFISH	1986-E	SALTON SEA SYSTEM AND QUITOBAQUITO SPRING	AZ CA	YES	53.3	46.3
DEVIL'S HOLE PUPFISH	1967-E	DEVIL'S HOLE SPRING	NV	YES	54.3	18.7
LEON SPRINGS PUPFISH	1980-E	DIAMOND Y SPRING AND OUTFLOW	TX	YES	9.9	18.7
OWEN'S PUPFISH	1967-E	FISH SLOUGH SPRING SYSTEM	CA	YES	15.0	0.0
WARM SPRINGS PUPFISH	1970-E	ASH MEADOWS SPRING SYSTEM	NV	YES	45.0	22.8
SACRAMENTO WINTER RUN CHINOOK SALMON	1990-E	SACRAMENTO RIVER SYSTEM	CA	YES	2306.5	5487.7
SNAKE RIVER FALL CHINOOK SALMON	1992-T	COLUMBIA RIVER SYSTEM	ID WA OR	YES	NA	NA
SNAKE RIVER SPRING/SUMMER CHINOOK SALMON	1992-T	COLUMBIA RIVER SYSTEM	ID OR WA	YES	NA	NA
SOCKEYE SALMON	1992-E	COLUMBIA RIVER SYSTEM	ID OR WA	YES	NA	NA

SPECIES	YEAR LISTED AND STATUS	CURRENTLY OCCUPIED HABITAT	STATE	AG LINK	FEDERAL EXPENDITURES ($1,000)	
					FY1990	FY1991
BEAUTIFUL SHINER	1984-T	RIO YAQUI AND GUZMAN BASINS	AZ NM	YES	27.0	22.0
PECOS BLUNTNOSE SHINER	1987-T	PECOS RIVER	NM	YES	33.8	100.4
DELTA SMELT	1993-T	SACRAMENTO RIVER	CA	YES	NA	NA
SPIKEDACE	1986-T	GILA RIVER SYSTEM	AZ NM	YES	35.4	18.7
BIG SPRING SPINEDACE	1985-T	MEADOW VALLEY WASH	NV	YES	32.5	11.6
LITTLE COLORADO SPINEDACE	1987-T	LITTLE COLORADO RIVER SYSTEM	AZ	YES	57.7	876.0
WHITE RIVER SPINEDACE	1985-E	UPPER WHITE RIVER SPRING SYSTEM	NV	YES	6.6	17.6
HIKO WHITE RIVER SPRINGFISH	1985-E	CRYSTAL SPRINGS	NV	YES	8.6	10.3
RAILROAD VALLEY SPRINGFISH	1986-T	RAILROAD VALLEY SPRINGS	NV	YES	7.3	14.6
WHITE RIVER SPRINGFISH	1985-E	ASH SPRINGS	NV	YES	6.5	8.8
COLORADO SQUAWFISH	1967-E	UPPER AND LOWER COLORADO RIVER SYSTEM	AZ CA CO NM UT WY	YES	2168.6	3669.5
UNARMORED THREESPINE STICKLEBACK	1970-E	SANTA CLARA RIVER SYSTEM SAN ANTONIO CREEK	CA	YES	23.2	14.2
PALLID STURGEON	1990-E	MISSISSIPPI MISSOURI YELLOWSTONE	KS MT NB ND SD	YES	268.0	478.2
JUNE SUCKER	1986-E	UTAH LAKE AND TRIBUTARIES	UT	YES	30.8	131.4
LOST RIVER SUCKER	1988-E	KLAMATH LAKE AND TRIBUTARIES	CA OR	YES	24.7	188.0
MODOC SUCKER	1985-E	PIT RIVER SYSTEM	CA	YES	31.4	8.8

SPECIES	YEAR LISTED AND STATUS	CURRENTLY OCCUPIED HABITAT	STATE	AG LINK	FEDERAL EXPENDITURES ($1,000)	
					FY1990	FY1991
RAZORBACK SUCKER	1991-E	UPPER AND LOWER COLORADO RIVER SYSTEMS	AZ CA CO NM NV UT	YES	NA	NA
SHORT-NOSE SUCKER	1988-E	KLAMATH LAKE AND TRIBUTARIES	CA OR	YES	22.9	188.0
WARNER SUCKER	1985-T	WARNER BASIN DRAINAGE	OR	YES	30.3	55.6
GILA TOPMINNOW	1967-E	GILA RIVER SYSTEM	AZ NM	YES	159.0	99.3
APACHE TROUT	1975-T	HEADWATERS OF THE SALT, VERDE AND LITTLE COLORADO RIVERS	AZ	NO	281.2	84.3
GILA TROUT	1967-E	GILA RIVER SYSTEM	AZ NM	YES	27.7	69.1
GREENBACK CUTTHROAT TROUT	1978-T	HEADWATERS OF SOUTH PLATTE AND ARKANSAS RIVER SYSTEMS	CO	YES	100.5	90.8
LAHONTAN CUTTHROAT TROUT	1975-T	LAHONTAN BASIN SYSTEM	CA NV	YES	1645.8	1597.7
LITTLE KERN GOLDEN TROUT	1978-T	LITTLE KERN RIVER	CA	YES	43.6	1.8
PAIUTE CUTTHROAT TROUT	1975-T	SILVER KING BASIN	CA	YES	50.0	41.8
WOUNDFIN	1970-E	VIRGIN RIVER	AZ NV UT	YES	37.0	75.6

218-29

(93)

U.S.

[15]

The Potential for Water Market Efficiency when Instream Flows Have Value

Ronald C. Griffin and Shih-Hsun Hsu

Most of the effort being expended to revise western water policy concerns the maintenance of instream waters to the exclusion of traditional diversionary interests. Absent from the economics literature is a theoretical treatment addressing the interface between diversionary and instream water uses. At issue is the potential for refining market operations to accomplish efficient allocation in the presence of both diversionary and instream uses. Optimization methods are employed to examine this issue in a highly generalized framework. If a specific structure is adopted, markets and other incentive-based policies are demonstrated to be capable of efficient water allocation.

Key words: instream flow, water allocation, water markets.

Water allocation problems are best perceived as an evolving set of interdependencies illuminated by growing scarcities. These scarcities pertain to each of the many different dimensions of water from which we derive value. Growing scarcity accentuates individual interdependencies (externalities) in a progressive fashion thereby motivating the search for institutions of ever-increasing scope. Transaction costs impede progress towards more comprehensive institutions, but evolving scarcity raises internalization benefits relative to transaction costs. Simultaneously, transaction costs can be lowered by technological advance and institutional investment. These ongoing processes motivate constant change as the adoption of untried institutions becomes justified.

It is in this context that the economics literature has gradually acknowledged the problems associated with property rights structures which focus on diversion quantities. The intricate water-related interdependencies imposed by physics and chemistry imply that institutions dealing solely through diversion quantities are inadequate. Fundamental, quantitative and qualitative aspects of water use cause diverse interrelationships among users. Some of these relations are not accommodated when property rights are measured as diversions. As a result, our literature and our institutions have both evolved to embody more complicated visions of water interdependencies. Notable among the conditions of water use needing special attention are return flows, water quality, and instream flow maintenance.

The so-called "return flow externality" has been long acknowledged as a potential cause of water misallocation in a system of property rights based on allowed diversions (Milliman; Howe, Alexander, and Moses). Reallocations of diversion rights often involve third party impacts for other diverters. In response, the literature argues for a consideration of the consumptive use of each diversion although this is difficult information to obtain. Institutions have also responded to this problem. Water right transfers occurring in western states must obtain approval in established administrative or quasi-judicial processes designed to ferret out negative return flow externalities (Johnson and DuMars).

Though treated narrowly in this paper, water quality considerations have generated considerable debate in research and policy. This attention usually addresses quality issues in isolation from the allocation of water quantities. Quality issues have achieved a life of their own in that both literature and institutions have become quite specialized and do not integrate qualitative and quantitative externalities. Issues

The authors are associate professor, Department of Agricultural Economics, Texas A&M University, and associate professor, Department of Agricultural Economics, National Taiwan University. Senior authorship is shared.

Texas Agricultural Experiment Station Technical Article No. 30154.

Review coordinated by Richard Adams.

Amer. J. Agr. Econ. 75 (May 1993): 292–303

of quantitative and qualitative water scarcity are interrelated, however and, as such, are in need of coordinated policy (Colby). As an obvious example, the assimilative capacity of every watercourse is linked to the quantity of its flows.

Instream flow protection has emerged recently as the major demand for refining research and allocative institutions. There is growing social concern about the maintenance of river, stream, and lake levels for habitat, environmental, and recreational purposes (MacDonnell, Rice, and Shupe). Even when return flow externalities are expunged and quality issues are ignored or are somehow irrelevant, market activity may not lead to economic efficiency when instream flows have value. Part of the problem is that particular nonrival and nonexclusive instream uses are inadequately represented. The other part, interfacing traditional diversionary water uses with new nondiversionary interests, presents a perplexing issue because the opposing parties care about different, though linked, dimensions of the resource. Some of the literature has argued for allowing market participation by instream users.[1] We will demonstrate that water marketing cannot efficiently allocate both diverted and instream water unless the market is administered in a particular fashion.

Alternative nonmarket policies for confronting the instream flow issue are being adopted and revised (Livingston and Miller). Most evident among these is the public trust doctrine which is being given new life through an extension to instream demands other than navigation. The public trust doctrine provides constitutional grounds for subjugating appropriative water rights to demands for increased stream flows. This represents more than the establishment of new initial endowments from which water marketing may proceed. Application of the public trust doctrine involves costly judicial actions as "vested" water right holders seek to protect their permits from uncompensated transferal to the public sector. Furthermore, uncertain tenure created by the threat of additional annexations by instream interests weighs heavily upon the value of water rights and the ability of the market to respond to future reallocative needs among diverters.

Our purpose here is to provide a more complete model of water use and human interrelationships involving water than has previously been presented. The focus is on the development of a property rights system that can serve as an interface among water diverters and instream flow users. The intent is to more generally model the efficient allocation of water in a world *sans* transaction costs. Such a model is not argued to produce immediately commendable social actions. Rather, the model is construed as offering insights for the direction in which we can head.

Review of Literature

Early water resource economists recognized the intricate interdependencies of water users. Hirschleifer, De Haven, and Milliman provide an enlightening discussion in an appendix. They refer to the ordinary hydrologic depiction of "consumptive use" as a "crude approximation" (p. 66). "Water may contain a number of qualities of economic significance; among them are location in time and space, temperature, and purity in the chemical or bacteriological sense" (p. 68). Hartman and Seastone state that "physical interdependencies of water users preclude simple property-right systems such as exist for most productive assets." They initially emphasize third party impacts and nonmarket values as reasons for the inadequacy of ordinary property rights and proceed to examine third party relationships in a much simplified modeling framework. The Hartman and Seastone model identifies the importance of consumptive use but also highlights the importance of return flow impacts especially relating to the significance of the relative locations of users. Both of these works note the potential for a two-tiered system of water pricing involving charges for input (diverted) water and credits for output (return flow) water. Both are also consistent in indicating the dependence of allocative efficiency on temporal and spatial characteristics of uses even though their modeling abstractions prohibited detailed investigations. Finally, both note the importance of instream demands but provide no real inspection of the impact of this observation.

It is revealing to contrast this early literature with the approach of most contemporary research. Some recent literature focuses on stochastic characteristics of water flows and the relationship between this variability and institutional needs— particularly relating to the appropriations doctrine (Burness and Quirk). A second body of literature emphasizes institutional prop-

[1] Water users who do not divert water have been treated as lesser interests, because they are typically not allowed to purchase or file for water rights (Colby). Even when ownership of water rights is permitted for nondiversionary uses, it is not on equal footing with diverters.

erties and needs in a more or less deterministic setting. Rather than neglecting it, this latter approach takes the importance of water flow stochastics as a separable issue. Here, typical writers maintain that the return flow issue can be readily accommodated within a simple market system if rights are defined by "consumptive use" rather than permitted diversion quantities (Anderson). This represents a subtle yet critical departure from the previous literature of Hirschleifer, Haven, and Milliman, and Hartman and Seastone, which attends to at least two quantitative dimensions of water use: diversion quantities (D) and return flow (R). The recent approach is to collapse these two metrics into a single dimension, consumptive use ($C = D - R$), and argue for C rights.

The two approaches are only superficially equivalent. Some aspects of the earlier work seem lost on recent research. For example, the older work observed the dependency of return flow credits upon location along a river; this spatial dependency of optimal price is absent from most current work. The simplification employed by the more contemporary research may be at least partially responsible for its theoretical support of marketing. The simplification is appropriate if the utility/profit of all users is dependent solely upon C and if interdependencies among users occur only in the form of C impacts. It is intuitively clear, however, that this is not the case.[2]

The legal profession is making a concerted effort to shore up western water law with concepts such as the public trust doctrine, which may supplant recent advances in water market development (Blumm; Johnson and DuMars). Although economists possess tools with which to address such nonrival and "nonconsumptive" use issues, we have failed to integrate this knowledge within our conceptual models of competing water uses. Accomplishing this requires more than transferal of public good theory to a new setting. The joint interrelationships of diversion, consumption, and instream water levels must be simultaneously modeled in order to obtain a workable paradigm with promise for prescription. Thus far, the extent of progress largely rests in claims that individuals demanding instream flow could adequately provide their needs through the purchase and retirement of consumptive rights (Anderson; Anderson and Leal; Huffman; Tregarthen). The difficulties with these claims are not due solely to the public good character of instream flows. Account must be taken of the third party effects of such retirements, just as for exchanges of water rights among diverters (Livingston and Miller). By not addressing these effects explicitly, the promarketing literature may have cavalierly treated instream water levels as a factor influencing the value of water. A more rigorous examination is needed.

The Model

To deal adequately with the instream flow issue requires a new economic depiction of water interrelationships. A model of these interrelationships should recognize that the withdrawal of water resources from a stream impacts those people who derive value from instream flow even if the entire diversion is returned downstream. The issues of return flow interdependencies and instream flow demands are inseparable.

Elements of the needed model are as follows. Water is a multidimensioned good in that people derive value from multiple properties of water; each water use has substitution opportunities implying that diversion, consumption, and return flow quantities can be controlled; the withdrawal of water for use and the subsequent return flow alters the availability of water at downstream locations; each water use can have a unique character with respect to when and where its return flow will reenter the stream; and the flow characteristics of the stream, especially natural inflows/outflows and speed, can be important.

These fundamentals suggest an optimization problem which should incorporate spatial details. The relative locations of water users and return flows is relevant to any theory attempting to integrate diversionary and instream water demands.

The "bare bones" framework employed here holds that people care about the amount of water they divert, d; the amount of water they consume, c, from this diversion; and the amount of water residing in the stream/river, w. Different users weigh these metrics differently, but all three are potentially important. Furthermore, they likely have a nonseparable influence in many, if not most, uses. Other water properties such as quality or velocity are not considered in the present model. The decision variables are d and c; w is influenced by the decision variables. We begin with a general river basin containing many water

---

[2] In the same vein, some models assume that across all users consumptive use (or return flow) is an immutable percentage of the quantity of water diverted. If this were true, then consumptive use could serve as an adequate basis for defining property rights.

Table 1. Variable Definitions

Variable	Definition
a_j^i	proportion of diverter s_i's return flow returning at s_j
$c(s_i)$, $d(s_i)$	amounts of water consumption and diversion by diverter s_i
$\hat{c}_{si}$, $\hat{d}_{si}$.	diverter s_i's endowment of consumption and diversion rights
c_{ij}, d_{ij}	amounts of rights transferred from diverter s_i to diverter s_j
$F_{s_i}(\ )$	diverter s_i's utility function
$G_{s_i}(\ )$	aggregate instream use utility function for the segment (s_i, s_{i+1})
p_i, q_i	prices for consumption and diversion rights
p^{I_i}, q^{I_i}	instream water district i's unit payments for c and d transfers
s_i	a location along the river beginning at s_0 and ending at s_N
$w(s_i)$	amount of instream water at location s_i after s_i's diversion
$\bar{w}(s_i)$	natural amount of instream water at location s_i (no human use)
$\bar{w}(s_i)$	amount of instream water at location s_i after existing use patterns and prior to market participation by instream users
λ_i	Lagrange multiplier indicating the marginal value of instream water across the segment $(s_i, s_{i+1}]$

users with various desires. All notation defined below is also summarized in table 1.

Let there be two classes of water users to consider, diverters and instream users. Diverters (e.g., farms, families, factories) are located at a finite number of points along a river basin that extends from the most upstream location, $s = s_0$, down to the end of the river, $s = s_N$ (figure 1). Diverters at places $s_1, s_2, \ldots, s_N$ have utility functions

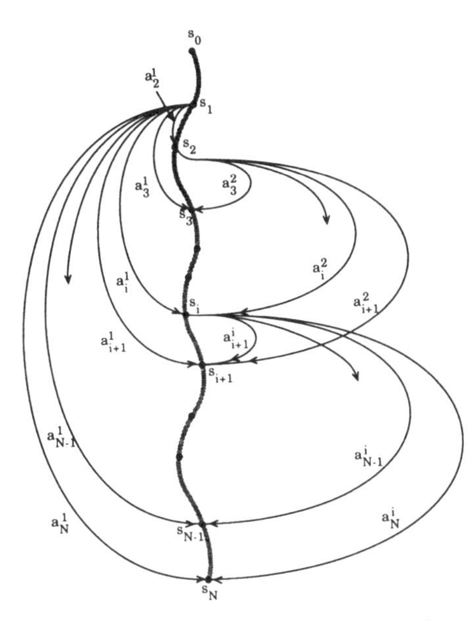

Figure 1. Model structure and return flow coefficients

$$(1)\quad U_{s_i} = F_{s_i}(w(s_{i-1}), d(s_i), c(s_i)),\ i = 1, \ldots, N,$$

where $w(s_{i-1})$ is the amount of instream water before diverter s_i removes any water from the stream, $d(s_i)$ is the diversion by user s_i from the river, and $c(s_i)$ is the amount of consumptively used water from the diversion. If more than one diverter is present at s_i, then F_{s_i} represents aggregate welfare. At the river's end, $s = s_N$, the user or user group need not be a true diverter. F_{s_N} is similar to salvage value in a temporal context. If s_N is the terminal point of a tributary or stream segment, then F_{s_N} is the value of water downstream. If s_N is the place where a river system empties into the sea, then F_{s_N} could include bay and estuary values.

Instream users are continuously located on the intervals separating neighboring diverters. At each point within these intervals, instream water value is given by a location-specific function of water quantity, $g_s(w(s))$. Instream values can be aggregated across each segment of the basin to form segment-specific welfare functions for the nonrival uses undertaken by instream users. This aggregation produces N utility functions specified by

$$(2)\quad V_{s_i} = G_{s_i}(w(s_i)) = \int_{S_i}^{S_{i+1}} g_s(w(s_i))ds, \qquad i = 0, \ldots, N-1,$$

where $w(s_i)$ is again the amount of instream water after diversions at s_i.

The model is structured so that diversion, consumption, and return flow occur only at locations $s_1, s_2, \ldots, s_N$. Nothing prevents F_{s_i} from being zero at some locations, so it is possible to model return flow locations or special stream segments not demarcated by diverters. All variables are parameterized by locations. Conventional regularity conditions, e.g., the concavity of the utility functions G_{s_i} and F_{s_i} with respect to all arguments, are assumed.

The inclusion of several water-related variables in the preceding utility functions permits a more accurate depiction of actual water uses. For example, irrigators generally benefit from larger stream flow (w) because pumping lifts are reduced, inlet pipes are submerged, and diversion gates are functional. Diversion quantities (d) are useful to the irrigator because more uniform application rates can be achieved with greater diversions and soil salinity can be better controlled. Greater consumptive use (c) permits more irrigated acreage, higher yields, and/or more valuable crops. In urban uses, larger w typically assists water withdrawal by municipal utilities. Urban waste treatment authorities value w for its ability to receive pollutants. Many household uses are primarily dependent upon d (most aspects of sanitation) although a portion of such diversions are inevitably consumed during conveyance or in waste treatment processes. Some household uses are more dependent upon c—lawn irrigation, for example. These examples show that a single water use can simultaneously derive its value from multiple water parameters.

Typical diversions of water for any purpose will have positive derivatives of F with respect to all the identified water variables in the ranges of interest. For instream uses, however, we generally expect

$$\frac{\partial G}{\partial w} \neq 0, \frac{\partial G}{\partial d} = 0, \text{ and } \frac{\partial G}{\partial c} \cong 0$$

so d and c are omitted as arguments of G.

The amount of water flow at the initial point, s_i, of each segment is determined by the condition

$$(3) \quad w(s_i) = \bar{w}(s_i) + \sum_{k=2}^{i} \sum_{j=1}^{k-1} a_k^j (d(s_j) - c(s_j)) - \sum_{j=1}^{i} d(s_j)$$

where $\bar{w}(s_i)$ is the stream flow without any diversion by or return flow from previous users.[3]

[3] Naturally occurring water inflows and losses along the watercourse can be handled within this formation as long as they are fully exogenous. Tributary inflows can normally be treated in this way. Water movement to and from a hydrologically connected groundwater body is often not exogenous, however, as these exchanges are commonly dependent upon $w(s)$.

The convention adopted here is that $w(s_i)$ excludes water withdrawn by diverter s_i, so that $w(s_i)$ is available throughout the segment $(s_i, s_{i+1}]$. The coefficient as a_i^j characterizes the proportion of return flow of user s_j, i.e., that amount of $(d(s_j) - c(s_j))$ that comes back into the river at location i where user s_i is located (see figure 1). (Water flows down from the superscript to the subscript). Return flow can only occur downstream of the diversion point. Return flow coefficients are viewed as fixed constants representing the distribution of each user's return flow across downstream locations. We clearly must have

$$\sum_{i=j+1}^{N} a_i^j = 1, \quad \text{and } 0 \leq a_i^j \leq 1.$$

Equation (3) describes the influence of diversions and subsequent return flows upon instream flow for $s \in (s_i, s_{i+1})$. Recognition that a diverter's return flow can be distributed across multiple locations is a desirable model feature. This contributes to analyses concerning instream flows, for the location of return flow can matter greatly when instream flows are valuable.

The optimal consumption of water in the presence of return flows can be elucidated through the maximization of a regional welfare function.[4] The regional objective is assumed to be to choose $c(s_i)$ and $d(s_i)$, $i = 1, \ldots, N$, so as to maximize

$$(4) \quad J = \sum_{i=0}^{N-1} \{G_{s_i}(w(s_i)) + F_{s_{i+1}}(w(s_i), c(s_{i+1}), d(s_{i+1}))\}$$

subject to the N equations (3). Two obvious constraints, $d \leq w$ and $c \leq d$, at every diversion point are not included in the present model. Analytical incorporation of these constraints was not found to contribute substantially to the nature of the results, so they are omitted from the model presented here. The concept of economic efficiency embodied in this particular model is clearly one of potential Pareto optimality.[5]

[4] We express our appreciation to an anonymous *AJAE* reviewer for simplifying our original optimal control model over spatial coordinates into the calculus model developed here.

[5] This efficiency criterion is commonly employed in models investigating optimal resource use, yet we should be mindful that it is a more restrictive criterion than Pareto optimality. The selected allocation will be Pareto optimal, but there are an infinite number of Pareto optima. Like the numerous other economic models of this type, our model will select the allocation where the net benefits of reallocation are the highest without regard to the distribution of losses or gains.

Combining the system equality constraints with the objective function, we obtain the Lagrangian expression

$$(5)\quad L = \sum_{i=0}^{N-1} G_{s_i}(w(s_i)) + \sum_{i=0}^{N-1} F_{s_{i+1}}(w(s_i), c(s_{i+1}), d(s_{i+1})) - \sum_{i=0}^{N-1} \lambda_i \Big\{ w(s_i) - \bar{w}(s_i) - \sum_{k=2}^{i}\sum_{j=1}^{k-1} a_k^i \,(d(s_j) - c(s_j)) + \sum_{j=1}^{i} d(s_j) \Big\}$$

Grouping terms differently to facilitate the derivation of necessary conditions, equation (5) becomes

$$L = \sum_{i=0}^{N-1} G_{s_i}(w(s_i)) + \sum_{i=0}^{N-1} F_{s_{i+1}}(w(s_i), c(s_{i+1}), d(s_{i+1})) - \sum_{i=0}^{N-1} \lambda_i\{w(s_i) - \bar{w}(s_i)\} - \sum_{i=1}^{N-1} d(s_i)\Big\{\sum_{k=1}^{N-1} \lambda_k - \sum_{j=i+1}^{N-1} a_j^i\Big(\sum_{k=j}^{N-1}\lambda_k\Big)\Big\} - \sum_{i=1}^{N-1} (c(s_i)\Big\{\sum_{j=i+1}^{N-1} a_j^i\Big(\sum_{k=j}^{N-1}\lambda_k\Big)\Big\}.$$

Taking derivatives with respect to w, d, and c, the following set of necessary conditions are obtained:

$$(6a)\quad \lambda_i = \frac{\partial G_{s_i}}{\partial w} + \frac{\partial F_{s_{i+1}}}{\partial w}, \quad i = 0, \ldots, N-1;$$

$$(6b)\quad \frac{\partial F_{s_i}}{\partial d} = \sum_{k=i}^{N-1} \lambda_k - \sum_{j=i+1}^{N-1}\Big(a_j^i \sum_{k=j}^{N-1} \lambda_k\Big), \quad i = 1, \ldots, N-1; \text{ and}$$

$$(6c)\quad \frac{\partial F_{s_i}}{\partial c} = \sum_{j=i+1}^{N-1}\Big(a_j^i \sum_{k=j}^{N-1} \lambda_k\Big), \quad i = 1, \ldots, N-1.$$

Interpretations

The Lagrangian multiplier λ_i is the opportunity cost or shadow price of small changes in the variable $w(s)$ on the stream segment $(s_i, s_{i+1}]$. Thus, if diverter s_i were to forego a small unit of diversion and thereby enhance w across the entire segment $(s_i, s_{i+1}]$, the social value of this change across the immediately following stream segment is given by λ_i.

With this understanding of the economic meaning of λ_i, equation (6c) can be readily interpreted. With maximized regional welfare, user s_i's marginal benefit from consuming water that has been diverted (left hand side of (6c)) must be equal to the foregone total benefits from location s_{i+1} down to the end of the river. That is, user s_is decision not to consume a unit of diverted water bestows benefits upon all downstream users through increased return flows.

To further inspect the effect of return flow externalities upon efficient water allocation, consider any two diverters at different locations, e.g., $s_i < s_{i+h}$. A reformulation of equation (6c) implies

$$(7)\quad \left.\frac{\partial F_s}{\partial c}\right|_{s=s_i} = \sum_{j=i+1}^{i+h} a_j^i \Big\{\sum_{k=j}^{N-1} \lambda_k\Big\} + \sum_{j=i+h+1}^{N-1} (a_j^i - a_j^{i+h}) \Big\{\sum_{k=j}^{N-1} \lambda_k\Big\} + \left.\frac{\partial F_s}{\partial c}\right|_{s=s_{i+h}}$$

According to (7), the marginal benefit from consuming water at location s_i must equal the sum of the foregone instream benefits from s_{i+1} down to s_{i+h+1} plus the difference between user s_i and s_{i+h} in foregone return flow benefits from location s_{i+h+1} down to the end of the river plus the foregone marginal consumptive benefit accruing to user s_{i+h}. Clearly, the return flow externalities are affected by both types of marginal instream benefits, $\partial G/\partial w$ and $\partial F/\partial w$, embodied in the λ_k's and the return flow coefficients a_j^i. In general, unless public good values such as instream flow values are insignificant, the marginal benefit from consuming water should not be identical along the river basin. Optimally, $\partial F/\partial c$ should decline from s_0 to s_N. Upstream users would otherwise tend to consume more water than their socially optimal level, and insufficient water would be allocated to instream and downstream users. If diverters' utility functions are equivalent along the basin, the concavity of F with respect to c implies that the water manager should allocate progressively more consumptive use to each diverter as we proceed downstream.

Furthermore, the pattern of return flow is important. If most of s_i's return flow would come back to the river at locations above s_{i+h+1}, then the weighted sum of the foregone instream benefits could be considerably large. Unusual hydrological circumstances can lessen or reverse some of the conclusions just drawn. For instance, if none of s_i's return flow comes back to the river above location s_{i+h+1} and $a_j^i = a_j^{i+h}$ for $j \geq i + h + 1$, then the sum of the forgone instream benefits would be zero and the efficient allocation among consumptive uses would require equal marginal consumptive benefits of water for users s_i and s_{i+h}. In more extreme settings, s_i's return flow might primarily occur after the return flow of the downstream diverter s_{i+h}. In this unusual case, economic efficiency dictates lower marginal benefits of consumption at the upstream location.

Analyzing optimal diversions in a similar way, the first sum in the right hand side of (6b) is the direct opportunity cost of user s_i's decision to remove one unit of water from the stream. The second sum captures return flow benefits to all downstream users. Analogous to equation (7), equation (6b) can be rewritten as follows:

$$(8) \quad \left.\frac{\partial F_s}{\partial d}\right|_{s=s_i} = \sum_{k=i}^{i+h-1} \lambda_k - \sum_{j=i+1}^{i+h} a_j^i \left[\sum_{k=j}^{N-1} \lambda_k\right] - \sum_{j=i+h+1}^{N-1} (a_j^i - a_j^{i+h}) \left[\sum_{k=j}^{N-1} \lambda_k\right] + \left.\frac{\partial F_s}{\partial d}\right|_{s=s_{i+h}},$$

where $s_i < s_{i+1}$. Thus, diversions should be allocated so that the marginal opportunity cost of diverting water at the upstream location s_i is equal to a rather complex set of instream water values weighted by return flow coefficients plus the marginal opportunity cost of diverting water at the downstream location s_{i+h}. A more intuitive interpretation of (8) is difficult to obtain immediately, but further clarification is forthcoming.

Water Markets with Instream Demand

Equations (6) set forth a desired norm for all allocative institutions addressing water scarcities. In this section we investigate the ability of an idealized water market to obtain these first order equations simultaneously. Intuitively, the dimensionality of policy instruments (price and quantity guides) must at least equal the dimensionality of water attributes (d, c, w) if social optimality is to be achieved. Therefore, we presume that by establishing three distinct, but coordinated, water right structures economic efficiency can be obtained. But because instream water levels can always be enhanced by reallocating diversion and consumption to points farther downstream, it may be possible to achieve optimality with only two instruments. This approach is also consistent with Johnson, Gisser, and Werner's observations concerning marketable water rights in situations where upstream transfers might harm intervening diverters.

Current legal doctrines are undergoing rapid change, varying from state to state, and these doctrines do not clearly rely on d, c, and/or w rights. Diversion and consumption rights do appear to have some emphasis, however, so it is interesting to investigate whether d and c rights systems can be managed to achieve efficient results. The objective here is to assess the feasibility of such a notion as well as to examine salient features of efficient water markets when instream flows possess value.

A Simplified Scenario

The exchange of consumption or diversion rights between any two water users produces complex effects upon others. These effects are importantly related to return flow patterns. To render the analysis more understandable and presentable, we introduce the following temporary assumption: all return flows from a diversion at s_i reenter the stream at s_{i+1}. That is, $a_{i+1}^i = 1$ and $a_j^i = 0$ for all $j > i + 1$. Equations (6) now become (9).

$$(9a) \quad \lambda_i = \frac{\partial G_{s_i}}{\partial w} + \frac{\partial F_{s_{i+1}}}{\partial w}, \quad i = 0, \ldots, N-1;$$

$$(9b) \quad \frac{\partial F_{s_i}}{\partial d} = \lambda_i, \quad i = 1, \ldots, N-1; \text{ and}$$

$$(9c) \quad \frac{\partial F_{s_i}}{\partial c} = \sum_{k=i+1}^{N-1} \lambda_k, \quad i = 1, \ldots, N-1.$$

Interestingly, (7) and (8) now become

$$(10) \quad \left.\frac{\partial F_s}{\partial c}\right|_{s=s_i} = \sum_{j=i+1}^{i+h} \lambda_j + \left.\frac{\partial F_s}{\partial c}\right|_{s=s_{i+h}} \text{ and}$$

$$(11) \quad \left.\frac{\partial F_s}{\partial d}\right|_{s=s_i} = \lambda_i - \lambda_{i+h} + \left.\frac{\partial F_s}{\partial d}\right|_{s=s_{i+h}}.$$

These latter equations are intuitively lucid. In (10) we can clearly see that water consumption reallocated from s_i to a downstream diverter at s_{i+h} also produces an aggregate instream value summed across all stream segments intermediate to return flow locations s_{i+1} and s_{i+h+1}. In (11), we likewise see that the downstream reallocation of diversion water from s_i to s_{i+h} contributes instream value along $(s_i, s_{i+1}]$ while losing instream value along $(s_{i+h}, s_{i+h+1}]$.

Idealized d and c Markets

A market system using d and c rights (two distinct instruments) and capable of achieving (10) and (11) is sought. Instream users who value w can conceivably participate in the c market so as to purchase and "retire" c rights—thereby increasing w. Retiring some of s_i's consumptive use does not actually increase stream flow until the point where s_i's return flow occurs (at s_{i+1} under the current simplified scenario). Stream flow is then increased for the remaining length of the river. Diversion rights can be likewise retired, but in this case stream flow is immediately but temporarily enhanced. For example, retiring one acre-foot of s_i's diversions increases w by one acre-foot across the segment (s_i, s_{i+1}) but has no impact on stream flow from s_{i+1} to s_N under the simplified scenario because the diverted water is returned at s_{i+1} (unless consumption rights are also retired). Actually, there is no need to retire either c or d rights, for it suffices to reallocate consumption and/or diversion to the farthest diverter, s_N. Such a reallocation increases w as much as retirement, and it is also available for full use by s_N. Aside from the limiting case of reallocation to s_N, there are more moderate, and more interesting, alternatives to reallocate diversion and consumption to downstream points above s_N.

Suppose water rights are initially granted only to diverters as is the prevailing custom in the west.[6] Although water availability may limit upstream transfers in true settings, let each diverter purchase rights from any other diverter. Each diverter is granted two types of rights: consumption rights, $\hat{c}_{s_i}$, and diversion rights, $\hat{d}_{s_i}$. Economic efficiency requires that prices be allowed to vary spatially, as will be demonstrated shortly. The market-clearing price for consumption rights at s_i is p_i, and q_i is the price of diversion rights at s_i.

Consider all possible transfers of consumption and/or diversion rights from diverter s_i to and from all other diverters. Let c_{ij} and d_{ij} designate transfers from diverter s_i to diverter s_j. Negative values of c_{ij} or d_{ij} represent transfers from s_j to s_i. Therefore, $c_{ij} = -c_{ij}$ and $d_{ij} = -d_{ij}$. Essential details of diverter s_i's participation in these two water markets are captured within the following individual optimization problem:

$$\max F_{s_i}(w(s_{i-1}), c(s_i), d(s_i)) + \sum_{\substack{j=1 \\ j\neq i}}^{N-1} c_{ij} p_i + \sum_{\substack{j=1 \\ j\neq i}}^{N-1} d_{ij} q_i$$

$$\text{subject to } c(s_i) = \hat{c}_{s_i} - \sum_{\substack{j=1 \\ j\neq i}}^{N-1} c_{ij}$$

$$\text{and } d(s_i) = \hat{d}_{s_i} - \sum_{\substack{j=1 \\ j\neq i}}^{N-1} d_{ij}.$$

Buying and/or selling is allowed within this structure. Absent from this formulation is any recognition of the impact of the diverter's market activities upon the stream flows experienced by the diverter.[7] For analytical convenience, it is presumed that these indirect impacts are small relative to the direct effects of c and d transfers. In any case, the stream flow effects upon diverters are soon addressed in combination with the more important stream flow concerns of nondiverters. Necessary conditions generated for this problem are given by (12) and (13).

$$(12) \quad \frac{\partial F_{s_i}}{\partial c}\frac{\partial c}{\partial c_{ij}} + p_i = 0 \Rightarrow \frac{\partial F_{s_i}}{\partial c} = p_i \quad \text{for all } s_i;$$

$$(13) \quad \frac{\partial F_{s_i}}{\partial d}\frac{\partial d}{\partial d_{ij}} + q_i = 0 \Rightarrow \frac{\partial F_{s_i}}{\partial d} = q_i \quad \text{for all } s_i.$$

Thus, if $\partial F_{si}/\partial c < p_i$, diverter s_i should sell consumption rights. If prices were constant along the basin, then these two sets of necessary conditions would imply

$$\frac{\partial F_{s_i}}{\partial c} = p = \frac{\partial F_{s_j}}{\partial c} \quad \text{for all } i, j, \text{ and}$$

$$\frac{\partial F_{s_i}}{\partial d} = q = \frac{\partial F_{s_j}}{\partial d} \quad \text{for all } i, j$$

[6] By this assumption we intend no sanction of this custom nor does this assumption bias the results of the investigation in any meaningful way. The allocation of diversion and consumption rights among diverters has the side effect of declaring a status quo arrangement regarding sanctioned instream flow levels. It is from these initial endowments that the analysis of this section proceeds. Alternative endowments are equally addressed by this analysis in that the same first order conditions apply to all starting points.

[7] Under the simplified scenario, if diverter s_i exchanges c rights with a diverter downstream or d rights with any other diverter, then $w(s_i)$ is unaltered. If, however, the exchange involves c rights with an upstream diverter, then $w(s_i)$ is changed, and the diverter would wish to account for this benefit or harm in considering the prospective exchange.

which are inconsistent with efficiency conditions (10) and (11). Therefore, market design requires spatial pricing of consumption and diversion rights. Moreover, instream water users must participate in this market in a particular manner if the efficiency conditions are to be achieved.

The modeling of water markets requires careful attention to the externalities associated with these transactions. Within the simplified scenario, transfers of diversion quantities have no third party impacts on other diverters and will not usually affect nondiverters. Only the stream segment immediately beneath each transacting diverter will be affected by a transfer of d. Transfers of d from s_i to s_j will increase w along the stream segment (s_i, s_{i+1}) and reduce w along (s_j, s_{j+1}). This is true regardless of whether it is upstream or downstream transfer.

Transfers of consumption rights influence all intermediate diverters and nondiverters. The implications of these observations for the design of efficient markets are as follows. First, all intermediate water users should participate in transfers of consumption rights. Second, two stream segments should be involved in transfers of diversion rights.

To further develop this market setting into one capable of supporting maximum regional water value, let us introduce N "Instream Water Districts," each representing collective desires relating to instream flows along portions of the river. Instream Water District i, hereafter I_i, is an organization of all water users along $(s_i, s_{i+1}]$. Organization I_i is constructed so as to include diverter s_{i+1}'s instream interests, although these interests will generally be small in relation to the nondiverting membership of the district.

Transfers of consumption rights influence all intermediate diverters and nondiverters. If $j < i < k$, then downstream transfers of c_{jk} (>0) are beneficial to I_i. As discussed above, only stream segments (s_j, s_{j+1}) and (s_k, s_{k+1}) are affected by a transfer of d_{jk}. In order to support the efficient provision of stream flows, suppose that I_i participates in water marketing by subsidizing consumption transfers benefiting the district and accepting compensation for diversion transfers harming the district. The unit price paid by the district for c_{jk} is p^{I_i}, and the unit price paid for d_{jk} is q^{I_i}. The district's optimization problem is then to maximize

$$G_{s_i}(w(s_i)) + F_{s_{i+1}}(w(s_i), c(s_{i+1}), d(s_{i+1})) - \sum_{j=1}^{i} \sum_{k=i+1}^{N} p^{I_i} \cdot c_{jk} - \sum_{j=i+1}^{N} q^{I_i} \cdot d_{ij}$$

with respect to the selection of $w(s_i)$, c_{jk}, and d_{jk} subject to

$$w(s_i) = \bar{w}(s_i) + \sum_{j=1}^{i} \sum_{k=i+1}^{N} c_{jk} + \sum_{j=i+1}^{N} d_{ij}$$

where $\bar{w}(s_i)$ identifies preexchange water levels. Forming the appropriate Lagrangian and using the same multipliers as before for convenience,

$$L_{I_i} = G_{s_i}(w(s_i)) + F_{s_{i+1}}(w(s_i), c(s_{i+1}), d(s_{i+1})) - \sum_{j=1}^{i} \sum_{k=i+1}^{N} p^{I_i} \cdot c_{jk} - \sum_{j=i+1}^{N} q^{I_i} \cdot d_{ij} - \lambda_i \left\{ w(s_i) - \bar{w}(s_i) - \sum_{j=1}^{i} \sum_{k=i+1}^{N} c_{jk} - \sum_{j=i+1}^{N} d_{ij} \right\}$$

District responses to this price system are embodied in the resulting first order conditions which presume $j \le i < k$:

$$\text{(14a)} \quad \lambda_i = \frac{\partial G_{s_i}}{\partial w} + \frac{\partial F_{s_{i+1}}}{\partial w}, \quad i = 0, \ldots, N-1$$

$$\text{(14b)} \quad p^{I_i} = \lambda_i, \quad i = 1, \ldots, N-1; \quad 1 \le j \le i < k \le N;$$

$$\text{(14c)} \quad q^{I_i} = \lambda_i, \quad i = 1, \ldots, N-1; \quad 1 \le j \le i < k \le N.$$

These conditions are valid for all c_{jk} and d_{ij} regardless of sign, that is, both upstream and downstream transfers. By (14b) and (14c), $p^{I_i} = q^{I_i}$. One of these price variables is superfluous, because the district is indifferent to the type of transfer impacting it—a unit of water of increase/decrease in stream flow possesses the same value regardless of whether it results from a c or d exchange. Dropping q^{I_i} and comparing economic efficiency conditions (10) and (11) with market equilibrium conditions (12), (13), (14b), and (14c), optimal market pricing is feasible if

$$\text{(15)} \quad p_i = \sum_{j=i+1}^{i+h} p^{I_j} + p_{i+h} \quad \text{and}$$

$$\text{(16)} \quad q_i = p^{I_i} - p^{I_{i+h}} + q_{i+h}.$$

For either downstream or upstream reallocations, an idealized market appears to require two special features. First, a spatially dependent price is needed. Second, collaborative market participation by instream users is needed. Each of N distinct "Instream Water Districts" representing the aggregate interests of each stream segment $(s_i, s_{i+1}]$ should receive compensation in the amount of λ_i for unit transfers of c or d rights harming the district and pay a like amount for beneficial transfers. As w is a nonrival and nonexclusive good, there are obvious problems associated with the functionality of Instream Water Districts, but the need is clear and the device is illustrative for another incentive-based policy to be identified later.

Spatial price differentiation and participation by Instream Water Districts are interrelated, and district participation takes on different forms for these two markets. Overall, an efficient price system for consumption and diversion rights is captured by (15) and (16). Within the simplified scenario, all instream water districts lying between the return flow locations of transacting diverters should participate in transfers of c rights. Other districts are irrelevant to this type of exchange. When d rights are exchanged, only two districts need be involved. A transfer of d_{ij} (>0) should be accompanied by compensation for district I_j and copayment by district I_i. In equilibrium, it is optimal for a district to employ the same marginal valuation of c and d rights in both markets for all transfers affecting the district.[8] These conclusions are specific to the simplified scenario.

Spatial differentiation of optimal prices is only required when instream flows possess value at the margin. Where preferences do not extend to instream flow quantities or when instream flows are at high levels where $\partial G/\partial w$ and $\partial F/\partial w$ are everywhere zero (implying zero λ_i's), there is no need for differentiated prices or Instream Water Districts. In general though, uniform prices across a basin are incompatible with economic efficiency. Price differences between locations should reflect instream values of affected third parties. When prices are appropriately established, economic efficiency can be achieved with two instruments, c and d property rights, and a third instrument pertaining to instream flow rights would be redundant. If prices are spatially invariant, however, the market system will promote an allocation pattern where water is consumed farther upstream than is optimal. In general, it cannot always be said that diversion prices should increase systematically as one proceeds upstream (equation (16)). Similarly, a uniform diversion price, while inconsistent with efficiency, does not necessarily cause water to be diverted farther upstream than is optimal.[9]

Marketing beyond the Simplified Scenario

Whether the general or simplified scenario is best suited to a given basin depends upon the distributions of both diverters and return flows which serve to partition the river into segments. In the simplified case where $a^i_{i+1} = 1$ and $a^i_j = 0$ for all $j > i + 1$, the original social optimality conditions become greatly clarified, and the potential for water marketing is lucid. This optimal marketing contrasts with the current conduct of water marketing in the west. Current marketing procedures typically provide for the administrative or quasi-judicial hearing of objections by potentially harmed diverters, and if a claim of potential harm is substantiated the proposed transfer is denied. Positively affected diverters have no apparent roles, however, and nondiverters have little standing regardless of whether they are positively or negatively impacted (Colby).

Where return flow is distributed across multiple downstream locations, the required institutions are somewhat more complex. In the generalized situation, water markets or other institutions must pursue equations (7) and (8) rather than the simplified (10) and (11). In the more general case, it remains clear that consumption and diversion prices require spatial definition. Furthermore, under an efficient market system c prices will typically decrease in a particular manner as one moves downstream. Again, use of a single diversion price and a single consumption price implies that consumption of water will occur farther upstream than is optimal (i.e. instream levels will be too small). These conclusions are substantively unaltered from the simplified case.

The potential for constructive participation by Instream Water Districts is clouded in the most general case however. Required district participation is now complex, for the return flow coefficients now serve as weights upon the marginal benefits/costs of all districts receiving direct or secondary return flows from either con-

[8] Admitting transaction costs would alter this conclusion and justify compensation prices that are larger than copayment prices (Foley).

[9] If all diverters have equivalent utility functions and all instream users do too, and if $w(s_i)$ increases as $i \mapsto N$ as is commonly true, the concavity of utility functions with respect to w can be employed to argue that λ_i decreases as $i \mapsto N$(equation (14a)). Under these conditions p^{I_i} also decreases as $i \mapsto N$, and equation (16) implies that diversion prices should be lower as $i \mapsto N$. Concavity of diverters' utility functions with respect to d then implies that uniform diversion prices would allocate water diversions farther upstream than is optimal.

tracting party. District participation is no longer constrained to those districts lying between the buyer and seller. Every district downstream of the uppermost transactor is potentially relevant to the transaction. This conclusion becomes more apparent when (7) and (8) are revised so as to collect terms for each district's λ_i. First, define a new variable,

$$A_k^i = \sum_{j=i+1}^{k} a_j^i$$

representing s_i's accumulated return flow coefficients down to $s_k \cdot A_k^i$ is the proportion of s_i's return flow having returned to the stream anywhere from s_{i+1} to s_k, inclusive. After careful manipulation, equations (7) and (8) can be rewritten as

$$(17) \quad \left. \frac{\partial F_s}{\partial c} \right|_{s=s_i} = \sum_{k=i+1}^{i+h} A_k^i \lambda_k + \sum_{k=i+h+1}^{N-1} \{A_k^i - A_k^{i+h}\} \lambda_k + \left. \frac{\partial F_s}{\partial c} \right|_{s=s_{i+h}}, \text{ and}$$

$$(18) \quad \left. \frac{\partial F_s}{\partial d} \right|_{s=s_i} = \lambda_i + \sum_{k=i+1}^{i+h-1} \{1 - A_k^i\} \lambda_k - A_{i+h}^i \lambda_{i+h} - \sum_{k=i+h+1}^{N-1} \{A_k^i - A_k^{i+h}\} \lambda_k + \left. \frac{\partial F_s}{\partial d} \right|_{s=s_{i+h}}$$

The λ_k are ordered from λ_i to λ_{N-1} in both of these equations to facilitate examination. Here, it is observed that all instream districts downstream of the uppermost transactor are potential parties to a c or d transaction and that optimal participation by districts requires prices which are weighted λ_i's. Weights applied to the λ_i's are potentially different in the two markets, particularly for districts lying between the transactors. Also, districts downstream from both transactors do not become irrelevant until that point on the river where all of both transactors' return flow has reentered the river ($A_k^i = A_k^{i+h} = 1$). Depending on the hydrological features of the basin, market operations capable of producing efficient results can be complex.

Administratively Established Incentives

The formation of Instream Water Districts is problematic in light of the information costs for elliciting accurate and funded valuation statements from member individuals. It may also be worrisome that each district possesses the market power to prevent upstream transfers by establishing a suitably high asking price.[10] Fortunately, while the preceding analysis has been presented in the context of direct water market participation by instream users, clear implications for the design of related institutions are also present.

In lieu of Instream Water Districts, a state or regional agency can erect and manage a system of economic incentives based upon hydrological information regarding return flow coefficients and economic research concerning marginal instream values for each river segment. Agency-promulgated incentives, the λ_i, would likely vary along the river. Diverters who are considering market-based exchanges with other diverters would have to carry out their bargains with the knowledge that agency incentives represent subsidies for downstream transfers and charges for upstream transfers in accordance with equations (17) and (18). This type of policy mechanism is advantageous in that it still relies upon the decentralized market for managing water allocation responsive to consumption and diversion values. That is, the agency need not concern itself with any valuations other than for instream flows. Because it sidesteps the demand relevation problem to be experienced by Instream Water Districts, the economic incentives approach may be a superior policy.

Conclusions

The model developed here employs a highly generalized framework capturing essential details of hydrologic interdependencies among water users. The results are intuitively supported and offer an insightful perspective concerning policy opportunities for achieving economic efficiency when some user groups benefit from instream flow levels. To summarize the most policy-relevant results, water marketing is capable of promoting potential Pareto optimality if the fol-

[10] It is noteworthy, however, that these districts are envisioned as participants in upstream and downstream transfers of rights. Such two-way activity creates a more practical policy. Market involvement by districts is predicated on the districts' statements of λ_i—an amount indicating willingness to pay for downstream transfers and willingness to accept for upstream transfers. Misstatement of λ_i is potentially harmful to the district: overstatement may serve districts' interests by limiting upstream transfers, but to do so would constrain district income—income employed to subsidize downstream exchanges.

lowing elements are included in market design

(*i*) Transferable diversion andconsumption rights must be established. These rights can be exchanged independently or together. (Because of inseparabilities in most uses, market transactions would likely involve both *d* and *c* rights simultaneously.)

(*ii*) Return flow coefficients must be established to identify where each diverter's return flow reenters the water body. This information is required for every diverter engaged in water marketing.

(*iii*) An institutional mechanism such as Instream Water Districts or administratively established economic incentives is needed to establish market presence for those individuals with preferences concerning instream flows. In the case of either districts or incentives, equations (17) and (18) dictate how unrestricted market exchanges between diverters need to be corrected in *c* and *d* markets, respectively.

These are the three fundamental components of an efficient system of water marketing. When instream flows do not have value, a system of *c* and *d* rights with trade only involving diverters is capable of managing return flow externalities—much like the early literature's reference to two-tiered pricing of diversion rights and return flow quantities.

The optimality results identified here demonstrate the complexity of achieving economic efficiency when instream flow is valuable. To simply allow instream users or user groups to purchase water rights is not a complete solution for the issue of allocating water to instream flows. The problem is substantially more complicated. Even when the implications of nonrivalness and nonexclusiveness are set aside, optimal market participation by instream users is a complex affair (recall equations (17) and (18)). Current trends to permit water right ownership by instream users can serve to improve resource allocation, but the consequences appear inadequate. The primary reason is that market transfers among diverters will continue to neglect instream water values in the absence of the third fundamental component above. It is noteworthy that both the public good character of instream water use and the absence of a proper interface between instream users and diverters result in the underallocation of water for instream purposes.

[Received January 1991. Final revision received July 1992.]

References

Anderson, T. L. *Water Crisis: Ending the Policy Drought.* Baltimore: The Johns Hopkins University Press, 1983.

Anderson, T. L., and D. R. Leal. "Going with the Flow: Marketing Instream Flows and Groundwater." *Columbia J. Environ. Law* 13 (1988):317–24.

Blumm, M. C. "Public Property and the Democratization of Western Water Law: A Modern View of the Public Trust Doctrine." *Environ. Law* 19 (Spring 1989):573–604.

Burness, H. S., and J. P. Quirk. "Appropriative Water Rights and the Efficient Allocation of Resources." *Amer. Econ. Rev.* 69 (March 1979):25–37.

——. "Economic Aspects of Appropriative Water Rights." *J. Environ. Econ. and Manag.* 7 (December 1980):372–88.

Colby, B. G. "Enhancing Instream Flow Benefits in an Era of Water Marketing." *Water Resources Research* 26 (June 1990): 1113–20.

Foley, D. K. "Economic Equilibrium with Costly Marketing." *J. Econ. Theory* 3 (September 1970): 276–91.

Hartman, L. M., and D. Seastone. *Water Transfers: Economic Efficiency and Alternative Institutions.* Baltimore: The Johns Hopkins University Press, 1970.

Hirshleifer, J., J. C. De Haven, and J. W. Milliman. *Water Supply: Economics, Technology, and Policy.* Chicago: The University of Chicago Press, 1960.

Howe, C. W., P. K. Alexander, and R. J. Moses. "The Performance of Appropriative Water Rights Systems in the Western United States During Drought." *Nat. Res. J.* 22 (April 1982): 379–89.

Huffman, J. "Instream Water Use: Public and Private Alternatives." In *Water Rights: Scarce Resource Allocation, Bureaucracy, and the Environment*, pp. 249–82, edited by T. L. Anderson. Cambridge: Ballinger Publishing Co., 1983.

Johnson, N. K., and C. T. DuMars. "A Survey of the Evolution of Western Water Law in Response to Changing Economic and Public Interest Demands." *Nat. Res. J.* 29 (Spring 1989): 347–87.

Johnson, R. N., M. Gisser, and M. Werner. "The Definition of a Surface Water Right and Transferability." *J. Law and Econ.* 24 (October 1981): 273–88.

Livingston, M. L., and T. A. Miller. "A Framework for Analyzing the Impact of Western Instream Water Rights on Choice Domains: Transferability, Externalities, and Consumptive Use." *Land Econ.* 62 (August 1986): 269–77.

Milliman, J. W. "Water Law and Private Decision-Making: A Critique." *J. Law and Econ.* 2 (October 1959): 41–63.

MacDonnell, L. J., T. A. Rice, and S. J. Shupe, eds. *Instream Flow Protection in the West.* Boulder: Natural Resources Law Center, University of Colorado School of Law, 1989.

Tregarthen, T. D. "Water in Colorado: Fear and Loathing of the Marketplace." In *Water Rights: Scarce Resource Allocation, Bureaucracy, and the Environment*, pp. 119–36, edited by Terry L. Anderson. Cambridge: Ballinger Publishing Co., 1983.

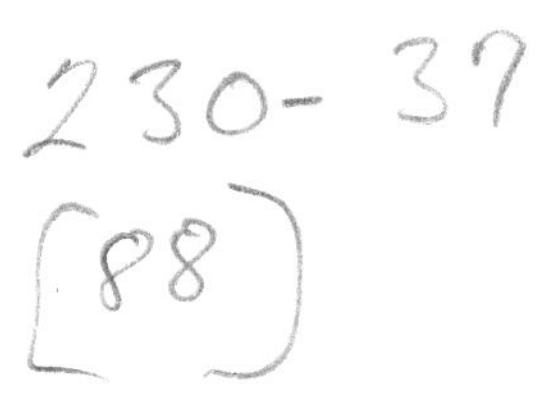

[16]

WATER RESOURCES RESEARCH, VOL. 24, NO. 11, PAGES 1839–1846, NOVEMBER 1988

Benefits of Increased Streamflow: The Case of the John Day River Steelhead Fishery

NEAL S. JOHNSON AND RICHARD M. ADAMS

Department of Agricultural and Resource Economics, Oregon State University, Corvallis

Conflicts between instream water uses such as fish production and traditional out-of-stream uses are an important water resource issue. One criterion for evaluating the merits of alternative water allocations is economic efficiency. This study uses an integrated approach to measure the recreational steelhead fishery benefits of incremental streamflow changes in the John Day River in Oregon. The analysis combines a steelhead fishery production model with a contingent valuation assessment of changes in fishing quality to obtain estimates of the marginal value of water in producing fishing quality. The results suggest that increased summer flows to enhance fishing have a marginal value of about $2.40 acre-foot. When expressed in terms of water actually consumed, the value may be up to 10 times higher. These values are sensitive to the location of flow alterations in the river, potential for downstream uses and number of anglers in the fishery.

INTRODUCTION

Most contemporary water resource issues involve allocations of water among competing uses. In the Pacific Northwest, one such use is minimum streamflow for anadromous fish production. Increases in streamflow are viewed as one means of meeting judicially and legislatively mandated improvements in fish production as compensation for losses suffered due to hydroelectric projects. Any increase in allocations to instream use, however, imply conflicts with current or future off-stream uses.

One criterion for evaluating the merits of these alternative water allocations is economic efficiency [*U.S. Water Resources Council*, 1979]. By comparing net economic benefits, policy makers can assess the social gains and losses of current and future water allocations. Unfortunately, information to implement the economic efficiency criterion is often lacking when dealing with water allocations involving nonmarketed public goods. The value of instream water is a case in point. Until recently, few studies explicitly sought to estimate a streamflow-economic benefits relationship [*Gibbons*, 1986]. As a consequence, instream flow reservations for maintenance or enhancement of fishery stocks are based largely on biologic and hydrologic, rather than economic, criteria [*Ward*, 1987]. In comparison, the value of water in traditional off stream uses, such as agriculture, is well-documented [see *Young and Gray*, 1972; *Gibbons*, 1986]. This lack of a common measure of economic value across all water uses contributes to potential misallocations of water.

OBJECTIVES

The overall objective of this analysis is to evaluate the recreational fishing benefits of incremental streamflow changes using biologic and economic assessment methods. Specifically, the study provides estimates of the marginal value of water with respect to the production of an important game fish, the steelhead trout (*Salmo gairdneri*). The empirical focus is on realistic portrayal of the relationship between streamflow, fishery production, and sport fishing quality during the critical summer flow period for the John Day River of north central Oregon. The interdisciplinary methodology employed here consists of two tasks: (1) quantification of the relationship between streamflow and fishery productivity (and hence the quality of the recreational fishery) and (2) valuation of incremental changes in the quality of the steelhead recreational fishery using nonmarket valuation techniques. Combining the results of these two tasks provides an estimate of the marginal value of summer streamflow in "producing" recreational steelhead angling. These values can then be compared with the value of that water in off-stream uses to assess the economic efficiency of such allocations. While the methodology is generalizable, the empirical results have specific policy implications for Pacific Northwest water allocations that affect anadromous fish production, particularly within the Columbia River system.

BACKGROUND

Economists have developed several methodologies to circumvent the lack of market data for public goods. These methodologies either indirectly impute a price to the good in question, as illustrated by the travel cost method [*Bockstael et al.*, 1985], or directly query consumers as to their willingness to pay for stated levels of a nonmarketed good, as in the contingent valuation method (CV) [see *Cummings et al.*, 1986; *Mitchell and Carson*, 1987]. The theoretical validity, strengths and limitations of each procedure are well-documented. Applications of these methodologies include studies of the economic benefits attached to water-based recreation, such as the valuation of salmon sport fishing [*Sorhus et al.*, 1981], waterfowl hunting [*Hammack and Brown*, 1974; *Bishop and Heberlein*, 1979], and recreational boating [*Sellar et al.*, 1986].

The total value of a recreational experience elicited in such studies, however, is rarely attributable to one input, such as fish catch or streamflow. As a result, the majority of extant studies are not directly applicable for valuing instream flow, though a subset of this literature attempts to link streamflow levels to recreation benefits (see *Gibbons* [1986] and *Loomis* [1987] for a survey and summary of the existing streamflow-economic benefits literature). Of these studies, only a few have explicitly considered the relationship between streamflow and the resultant benefits to anglers: *Daubert and Young* [1981], *Ward* [1987], and *Walsh et al.* [1980] are notable examples.

Paper number 88WR03268.
0043-1397/88/88WR-03268$05.00

Acknowledged weaknesses of these latter studies involve specification of the relationship between streamflow, fishery productivity, and the resultant quality of the sport fishery. For example, the biological model employed by *Daubert and Young* [1981] only provides information on potential, as opposed to actual, catch rates under alternative flow levels. As a consequence, anglers questioned in the CV portion of the study were asked to reveal their willingness to pay for potential catch rates. In general, however, anglers' experiences are limited to actual catch rates.

A related difficulty of existing streamflow-benefit studies pertains to the timing of benefits. *Milhous* [1983] notes that the flow of benefits calculated by *Daubert and Young* [1981] was assumed to occur concurrently with streamflow. This ignores the dynamic nature of the streamflow-fishery productivity relationship; i.e., by altering current flows, future benefits may also be affected. This is particularly true for anadromous fish, such as steelhead, where recreational benefits typically accrue at locations temporally and/or spatially removed from the site of primary production. *Ward* [1987], in acknowledging this same weakness in his study, concludes that "important work linking the time path of both streamflows and the resultant catchable fish density needs to be conducted . . ." [*Ward*, 1987, p. 383].

VALUATION FRAMEWORK

Assessing the economic value of incremental changes in streamflow involves a multistep procedure, starting with an understanding of how changes in physical and hydrological conditions of the stream affect fish production. Changes in fishery productivity, in turn, need to be linked to items valued by anglers, such as changes in fishing quality. The first task is thus to quantify relationships between streamflow changes and steelhead production. The next task is to assign values to improvements in fishing quality arising from an increase in fish populations. Combining the outputs from each task provides a measure of the value of incremental changes in streamflow in producing recreational fishing success.

This multistage bioeconomic framework is common to much of resource economics (see, for example, *Freeman* [1979]). Economic valuation is based on the premise that the individual's autonomous preferences, as revealed by his behavior, reflects the value attached to particular items or activities. Consumers select and consume certain mixes and levels of recreational activity because those combinations provide greatest enjoyment given time and budget constraints. They are thus "willing to pay" a proportion of individual time and income endowments for each activity. This willingness to pay can be measured from observed behavior (market data). For public goods, the lack of orderly markets requires use of methods based either on travel cost as a proxy for the price of the experience or on willingness to pay provided by direct questioning of consumers.

The theoretically correct measures of willingness to pay (WTP) for increments in fishing quality or willingness to accept (WTA) compensation for decrements in fishing quality are the Hicksian compensating measures [*Hicks*, 1943]. Of the four measures proposed by *Hicks* [1943], the two of interest here are compensating variation and compensating surplus. Each represents income adjustments that maintain the individual at some initial utility level. In this study, value measures in the form of Hicksian compensating variation, designated as WTPc, will be used to assign economic value to increments in fishery productivity. Detailed discussions of these and other welfare issues are provided elsewhere (see, for example, *Brookshire et al.* [1980]).

Once obtained, WTPc estimates can then be used to assess the marginal value of the resource or environmental change in question. For water or any other input allocation decision, interest is on the value of that input in each use at the margin. Economic efficiency is greatest when the marginal value of water is equated across all uses, i.e., no greater level of socially valued outputs could be achieved by reallocating water to other uses. To test whether efficiency gains are possible with alternative allocations requires knowledge on the current values of water across uses. Once obtained, these estimates can be compared with other uses of water (e.g., agricultural irrigation) to determine first, the possibility of gains from reallocations and second, the magnitude of those gains.

APPLICATION TO THE JOHN DAY STEELHEAD FISHERY

The empirical setting for this study is the John Day River of north central Oregon, a major tributary of the Columbia River. The John Day River basin encompasses 8010 square miles (mi^2; 1 mi^2 = 2.590 Km^2) in north central Oregon, ranging in elevation from 150 feet (1 foot = 30.48 cm) above sea level at the mouth of the John Day River to 9038 feet on Strawberry Mountain. The basin supports the largest runs of wild spring chinook salmon (*Oncorhynchus tshawytscha*) and summer steelhead in eastern Oregon (Oregon Department of Fish and Wildlife, unpublished report, 1985*a*). Decreases in summer flows due to riparian damage, coupled with the basin's semiarid climate, exacerbate potential conflicts between instream and out-of-stream water users during the critical summer flow period. The basin's economy, largely centered around crop and livestock production, heightens the need for reliable data on the value of instream water. This combination of hydrologic, liminologic, and economic conditions led to the basin's recent selection as the first river basin in Oregon to undergo a comprehensive water management plan.

A Fishery Production Model

The quality of a salmon or steelhead fishery depends, in part, on the number of fish returning to spawn. These populations, in turn, are influenced by numerous environmental conditions throughout their life cycles. The influence of ocean conditions and streamflow on the survival of salmon (*Oncorhynchus* sp.) have been extensively studied [see *Anderson and Wilen*, 1985; *Nickelson*, 1986; *Peterman*, 1981]. Our interest is limited to assessing the effect of increased or decreased streamflow on the quality of the John Day steelhead sport fishery. The approach selected utilized time series data to estimate a fishery production model, following the procedure found in the work by *Anderson and Wilen* [1985].

For steelhead, the number of adult fish entering the John Day River in year t, N_t, can be represented as a function of parental stock size, P_{t-n}, where $t-n$ indicates the year the parental stock spawned, environmental conditions affecting survival, E, and fishing pressure, FP:

$$N_t = f(P_{t-n}, E, FP) \qquad (1)$$

A stock-recruitment model is required to express this relationship in a format amenable to statistical estimation. Stock-recruitment models express recruitment as a function of parental stocks and generally include explanatory variables measuring density-dependent and density-independent mortality.

Two commonly employed models are the *Ricker* [1975] and the *Beverton and Holt* [1957]. The Ricker model is employed here.

The Ricker stock-recruitment model expresses the relationship between the number of recruits (progeny) and the size of the parental stock (the spawners) as

$$R_t = \alpha P_t e_t^{-\beta P} \quad (2)$$

where

R number of recruits to the fishery;
P size of parental stock;
α a dimensionless parameter;
β a parameter with dimensions of $1/P$ that relates stock density to mortality.

Since the interest here is on the number of adults returning to spawn N_t, as opposed to the original number of recruits R, it is necessary to expand expression (2) to account for the mortality occurring between the spawning of the parental stock and the return of their offspring in 5 years. (Steelhead in the John Day typically spend 2 years in freshwater as juveniles, 2 years in the marine environment, and 1 year in migration between each environment.) Accounting for mortality, N_t can be related to R_{t-5} as

$$N_t = R_{t-5} \prod_{i=1}^{n} (1 - m_i) \qquad 0 \leq m_i \leq 1 \quad (3)$$

where m_i is the conditional mortality rate associated with the ith environmental factor and $1 - m_i$ is the corresponding conditional survival rate. Each conditional survival rate is assumed to take the functional form exp $[\beta_i \times E_i]$. Substituting this into (3) and combining with (2) yields

$$N_t = \alpha P_{t-5} e^{-\beta P_{t-5}} \exp\left[\sum_{i=1}^{n} (\beta_i \times E_i)\right] v_t \quad (4)$$

where v_t is a random error term. Specifying the error term as a multiplicative lognormal distribution is in keeping with the assumed multiplicative nature of mortality (see *Peterman* [1981] for further explanation and empirical support). By taking the natural logs of both sides this expression can be converted to a linear form and the parameters estimated via linear regression methods:

$$\ln N_t = \ln \alpha + \alpha' \ln P_{t-5} - \beta_0 P_{t-5} + \sum_{i=1}^{n} (\beta_i \times E_i) + \ln v_t \quad (5)$$

where the β_i are a measure of the mortality attributable to the ith environmental factor. Note that a coefficient on ln P_{t-5} has been included to increase model flexibility.

Data

Ideally, annual data on adult escapement for various reaches of the John Day should be used to estimate (5). Lack of a time series of such observations requires an alternate approach. As noted by *Ricker* [1975], fishing success is often related to stock level as

$$C/E = qN \quad (6)$$

where C is catch, E is effort, and q is known as the "catchability" coefficient. Catch statistics are available from creel surveys conducted annually in the John Day Basin (Errol Claire, unpublished data, Oregon Department of Fish and Wildlife, 1987) and offer an index of stock level. The index for returning stock N_t was defined as steelhead caught per 100 hours of fishing in year t. Use of this index provides the critical link with the economic valuation results to be discussed subsequently, which are based on fishing quality as opposed to stock levels.

A second index of stock levels used in the analysis is redd counts. (A redd is the spawning nest dug out by the female steelhead.) Redd counts are available starting in 1959 [*Johnson*, 1988]. The number of redds per mile, lagged 5 years, was included in the final model as a measure of parental stock, P_{t-5}.

Use of separate indices to measure stock levels in (5) (for N_t and P_{t-5}) offers several advantages. First, use of separate indices to measure parental and returning stocks minimizes correlated errors between periods $t - 5$ and t. Second, using the creel survey index as a measure of parental stock would necessitate accounting for fishing pressure during the sport fishing season. Use of the redd count index bypasses this difficulty.

A hypothesis of this analysis is that increases in summer flow lagged 5 years (SU_{t-5}) increase fish production. Increased flow will reduce stream temperatures and increase habitat area. Currently, summer stream temperatures in the upper 70s and lower 80s (°F) have been measured on some stream reaches within the basin [*U.S. Bureau of Reclamation*, 1985], well above the optimal range for steelhead of 45° to 58° F. However, other factors also influence steelhead survival. A priori, their effects are summarized as follows.

1. Spring streamflow, lagged 5 years, (SP_{t-5}) is assumed to be a negative factor in that high flows may scour spawning beds and destroy newly laid eggs [*Shephard and Withler*, 1958].

2. Winter flow, lagged 4 years, (W_{t-4}) is positive in that higher streamflows reduce the probability of ice-ups and might also be indicative of warmer temperatures.

3. Spring flow, lagged 4 years, (SP_{t-4}) is negative, as in SP_{t-5}.

4. Marine productivity, $U_{t-1} + U_{t-2}$ is positive as improved marine productivity increases fish survival and hence the number of returning spawners. Ocean upwelling as defined by *Nickelson* [1986] is employed as a proxy for measuring marine productivity. It is the sum of the monthly upwelling volumes (in cubic meters per second per 100/m^2) for March through September. The sum of the upwelling indices 1 and 2 years prior to return of the parental stock was employed. This variable exhibited a strong correlation with marine survival indices derived from other steelhead stocks [see *Johnson*, 1988].

5. Migration route influences (i.e., dams on the Columbia River) increase both upstream and downstream mortality. For this latter effect, a dummy variable was employed to account for the construction of the John Daly Dam in 1968. To accommodate the hypothesized life cycle, a 3-year lag is employed in specifying this variable.

All flow data are from the United States Geological Survey (various years) which maintains several stream gauges within the John Day Basin. Since 40% of John Day steelhead are produced in the North Fork and 25% in the Middle Fork, streamflow measurements at Monument, on the North Fork below the confluence of the Middle Fork, were used to construct all flow variables. Spring flow (April–June) was defined as the average flow, expressed in cubic feet per second (1 ft^3 = 0.028 m^3). Summer (July–September) and winter

(January–March) flow variables were similarly constructed. All data are reproduced in the work by *Johnson* [1988].

Results

The model estimated was specified as

$$\ln N_t = \ln \alpha + \alpha' \ln (P_{t-5}) + \beta_0 P_{t-5} + \beta_1 SP_{t-5} + \beta_2 SU_{t-5} + \beta_3 W_{t-4} + \beta_4 SP_{t-4} + \beta_5 (U_{t-1} + U_{t-2}) + \beta_6 D3 \quad (7)$$

where each variable corresponds to definitions above. Equation (7) was estimated with data from 1964 to 1983 using ordinary least squares (OLS) via the SHAZAM statistical package [*White*, 1978].

OLS estimation results are presented in Table 1. The results of this estimation are statistically robust, in that all mortality coefficients had signs in agreement with expectations and are statistically significant ($\alpha = 0.05$). These coefficients are used to construct streamflow-angler success elasticities, which provide a means of combining the fishery model results with the CV results to be presented subsequently. These elasticities are defined as

$$\varepsilon_i = \frac{\%\Delta \text{HRSFISH}}{\%\Delta \text{FLOW}_i} = \beta_i \times \text{FLOW}_i \quad (8)$$

where β_i is the coefficient on the *i*th streamflow variable, FLOW_i, and HRSFISH is the catch rate in hours per steelhead. Table 1 reports these values for each of the streamflow periods, as calculated at mean values. The interpretation is straightforward. Spring streamflow, for example, if increased 1% will lead to a 1.15% reduction in angler success four years later and a 1.58% reduction in angler success the fifth year. Summer flow, in contrast, if increased 1%, will increase angler success 5 years later by 0.88%.

Two caveats must be attached to the above streamflow-angler success elasticities. As used in the model, the flow variables measure the average flow over 3-month periods. This ignores critical periods during these months when an increment in flow is more productive (or destructive, for spring flows) when compared to increments in other periods. Whether this makes a difference from a management standpoint depends on the ability to accurately identify critical periods and to what degree managers can "target" additional flows to occur in these periods. The more accurate the identification and targeting, the more the magnitudes of the above elasticities should be increased. Along the same lines, the fishery production model presented in (7) does not allow for threshold effects. Spring flows, for example, may have no adverse effect on fishery productivity until a critical flow level is reached. The elasticities are thus most accurate for average flow levels. Estimated elasticities for higher or lower flows should be used with caution. Identification of critical flow levels and periods is an important future research issue.

TABLE 1. Estimated Ricker Model of John Day Summer Steelhead, 1964–1983, Dependent Variable ln (N_t)

Variable	Estimated Coefficients	Standardized Coefficient	Elasticity at Means
Constant	0.737 (1.46)		
P_{t-5}	−0.166 (0.125)	−0.804	
ln (P_{t-5})	1.292 (1.12)	0.751	
SP_{t-5}	−0.000586* (0.000149)	−1.066	−1.581
SU_{t-5}	0.00432† (0.00168)	0.694	0.882
W_{t-4}	0.000327† (0.000142)	0.504	0.515
SP_{t-4}	−0.000411* (0.000135)	−0.786	−1.15
D_{t-3}	−0.504* (0.168)	−0.460	
$U_{t-1}+U_{t-2}$	0.00102* (0.00037)	0.429	1.22

Observations = 20 (1964–1983); adjusted R squared = 0.72. Elasticities at other flow levels can be calculated as the product of the coefficient and the flow level in question. Standard errors are presented in parenthesis.

*Significance at 0.02 level.

†Significance at 0.05 level.

Economic Valuation Procedure

The contingent valuation procedure was used in this study for two reasons. First, our focus is on estimating benefits for improvements above the current angler success level. This is motivated by the Northwest Power Planning Council's stated objective of doubling the Columbia River's fish runs and Oregon Department of Fish and Wildlife goal of increasing average John Day steelhead production from the current escapement level of 15,000 adults to 23,000 (Oregon Department of Fish and Wildlife unpublished report, 1985). The hypothetical structure of the CV questionnaire allows these unobserved quality levels to be valued. Second, the John Day steelhead fishery is not composed of one distinct angling area. This effectively ruled out use of the travel cost method. A third methodology, the household production function approach, was not used due to potential empirical estimation difficulties [see *Bockstael and McConnell*, 1981].

Survey Design

The survey used here involved personal interviews with John Day anglers to collect information on current and past visitation rates to the John Day as well as socioeconomic data such as age, education, and income. The main body of the survey was devoted to measuring the angler's compensating variation (WTPc) values for stated increments in fishing quality. To elicit WTPc, the angler was first given information on the average success rate on the John Day River in each of the previous 5 years. These data came from creel surveys conducted annually by the Oregon Department of Fish and Wildlife (ODFW) and the Oregon State Police [*Johnson*, 1987]. Given this information, the respondent was then asked to state his own catch rate on the John Day in an average year. This gave a base level of fishing quality at which to construct the contingent market. The respondent was then told that there were three postulated increases in the number of steelhead in the river: 33, 67, and 100% above the average level. Under each of these improvement levels, identified as improvement A, B and C, respectively, the respondent was asked to state his new expected catch rate. This format allowed the angler to define the contingent market to reflect his own skill and experience level, reducing the amount of hypothetical bias present.

Once the contingent market was defined, the following type of question was posed to elicit estimates of WTPc: "What is the maximum (in the form of a steelhead stamp) fee you would be willing to pay for improvement A in the John Day steelhead fishery? $ ______ ." Subsequent WTPc questions focused

TABLE 2. WTP and Mean Expected Catch Rate Summary

Stock Level	Mean Expected Catch Rate, hours per steelhead	Mean Bid 1986 dollars
Current conditions	9.3	
A	7.1	WTPc(A) = $8.58
B	5.0	WTPc(B) = $11.11
C	2.9	WTPc(C) = $13.59

Mean bids were calculated as the average of the 62 usable surveys.

on payments for additional improvements, i.e., B and C. After the CV questionnaire was pretested on the Alsea River, Oregon salmon fishing in August and September 1985, it was administered in person to 67 steelhead anglers during the 1986/1987 steelhead fishing season. Five anglers declined to be interviewed, resulting in an acceptance rate of 93%. Of the 67 surveys, five were deemed unusable due to key questions which remained unanswered, leaving 62 usable for analysis.

Analysis of Survey Data

The individual CV responses were used to estimate an aggregate bid function for potential improvements in steelhead production. Such a function is based on a concept presented in the work by *Bradford* [1970] and further developed in the work by *Brookshire et al.* [1980]. The curve is obtained by first calculating mean bids and quality levels to arrive at a mean individual curve. Mean bids and catch rates are presented in Table 2 and graphically shown in Figure 1. The mean bid curve was fitted to a quadratic functional form:

$$\underset{}{\text{WTPc}} = \underset{(1.22)}{0.27} + \underset{(0.91)}{4.12}\Delta\text{HRSFISH} - \underset{(0.14)}{0.32}(\Delta\text{HRSFISH})^2 \qquad (9)$$

(base = 9.3 hours per steelhead), where ΔHRSFISH is the improvement in the success rate from the base level of 9.3 hours/steelhead. An improvement in fishing quality is indicated by reductions in HRSFISH. Values in parentheses are standard errors. The current success level was included as an observation (i.e., WTPc = 0, ΔHRS = 0). The adjusted R^2 is approximately 0.95.

The total number of anglers in a given year is needed to convert the mean individual bid function into an aggregate bid function. This number is approximated from several sources. By dividing the average total annual angler hours (derived from creel surveys and ODFW catch estimates) by the average hours per angler, an estimate of total anglers per season can be derived. From the survey responses average hours per angler per day was 6.8 hours. Average days of 5.7 per angler per season was estimated via a weighting procedure to account for the likelihood of being sampled. On the basis of these values, the estimated number of anglers each year is 888 anglers with a standard deviation of 300. An increase in success rates may be accompanied by an increase in angling effort, resulting in higher aggregate benefits than would otherwise be calculated by assuming a fixed number of anglers. Alternatively, this increase in effort may also lead to congestion, lowering individual benefits [*Anderson*, 1980]. Using standard statistical tests, the null hypothesis that annual catch is a linearly homogeneous function of the catch rate could not be rejected. Thus fishing pressure on the John Day does not appear to increase with increases in success rates, given the catch rate levels used in the analysis.

The aggregate bid function for fishing improvements can now be represented as

$$\text{AGG. WTPc} = 240 + 3{,}660\Delta\text{HRSFISH} - 280(\Delta\text{HRSFISH})^2 \qquad (10)$$

(base = 9.3 hours per steelhead), which is (9) multiplied by the annual user rate, 888 anglers. This aggregate bid function represents the benefits accruing only to current users of the resource. Any existence, option, or bequest value held by nonusers is not represented in these values.

To integrate the CV results with the fishery production model requires that the ΔHRSFISH variable employed in both the economic and biologic analyses to be of comparable units. This is accomplished via an adjustment factor, calculated as the average expected catch rate from the CV survey (9.3 hours/steelhead) divided by the average observed catch rate from the creel surveys (17 hours/steelhead). Equation (10) is thus changed to

$$\text{AGG. WTPc} = 240 + 2{,}002\Delta\text{HRSFISH} - 84(\Delta\text{HRSFISH})^2 \qquad (11)$$

(base = 9.3 hours per steelhead). This equation is then combined with the fishery production model estimated in the previous section to derive the marginal value of instream water.

The WTPc estimates for fishing quality captured in (9) can be used to measure the value of an additional steelhead caught. From the survey the average angler catches 4.2 steelhead in a season. An additional steelhead implies a success rate equal to 7.5 hours/steelhead compared with 9.3 hours currently. From (9) the willingness to pay for this increased fishing quality is $6.65. Hence the typical value of an additional sport-caught steelhead is $6.65 under current catch conditions. This WTP_c estimate for an additional steelhead, while close to some recent estimates [e.g., *Samples and Bishop*, 1985; *Cameron and James*, 1987] is in contrast to the much higher values currently used for policy analysis (see, for example, *Scott et al.* [1987]). These latter values, however, are typically average, rather than marginal values.

Value of Water in the Production of Fishing Quality

A value function for incremental changes in streamflow can be obtained by combining the streamflow-angler success elas-

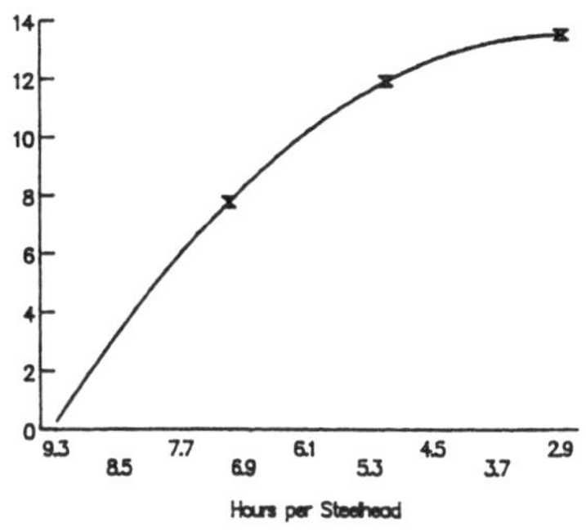

Fig. 1. WTP versus success rate (hours per steelhead).

ticities (equation (8)) with the results from the CV analysis (equation (11)). By multiplying (8) through by $\%\Delta FLOW_i$ and rearranging, one obtains

$$\Delta HRSFISH = \beta_i \times \Delta FLOW_i \times HRSFISH \qquad (12)$$

Combining with (11), adjusting for the percentage of steelhead produced in the North Fork of the John Day and letting HRSFISH equal the current average angler success rate (of 17 hours/steelhead) results in

$$AGG.\ WTPc = 156 + 22{,}122(\beta_i \times \Delta FLOW_i) - 15{,}780\beta_i^2 \times \Delta FLOW_i^2 \qquad (13)$$

(base = 17 hours per steelhead). The marginal benefits function is derived by taking the first derivative of (13) with respect to $\Delta FLOW_i$:

$$MARG.\ VAL.\ WATER = 22{,}122 \times \beta_i - 31{,}560\beta_i^2 \times \Delta FLOW_i \qquad (14)$$

(base = 17 hours per steelhead). Equation (14) is used to calculate the marginal value of instream water for steelhead enhancement. These marginal values of water for summer, spring, and winter streamflow in the production of steelhead fishing 5 years after the flow's occurrence, are presented in Table 3, first column. As is evident from the table, the values for increased summer flow are positive and considerably greater than other periods. In absolute terms, however, the value of an incremental change in summer flow of \$0.53/acre-foot (1 acre-foot = 1233 m^3) appears quite modest. These flow values are directly affected by the relatively small numbers of anglers fishing the John Day River.

As noted earlier, recreational benefits may accrue at locations far removed from the primary point of production. Many John Day-reared steelhead are caught outside the basin in the Columbia River sport and Indian gillnet fisheries. The *U.S. Bureau of Reclamation* [1985] assumes that 1.5 John Day steelhead are caught in these other fisheries per escaping John Day steelhead. Using this estimate and assuming the marginal value per additional sport caught steelhead of \$6.65 can be transferred to these other fisheries, a more complete accounting of the marginal value of instream water can be obtained:

$$\begin{aligned} &MARG\ VAL.\ WATER \\ &= \Delta FISH\ CATCH \times MARG.\ VAL.\ FISH \\ &= \Delta ESCAPEMENT \times 115 \times \$6.65 \\ &= [\beta_i \times \Delta FLOW_i \times ESCAPEMENT \times 1.5 \times 0.65] \times \$6.65 \\ &= \beta_i \times \Delta FLOW_i \times \$97{,}256 \qquad (15) \end{aligned}$$

A reasonable assumption is that the streamflow-angler success

TABLE 3. Marginal Value of Instream Water in Production of John Day River Steelhead

Period	β_i	Mean Flow Level, cfs	Dollars per acre-foot* A	Dollars per acre-foot* B
Spring	−0.000586	2700	− \$0.073	− \$0.32
Summer	0.00432	204	\$0.53	\$2.36
Winter	0.000327	1573	\$0.041	\$0.18

*Calculated by assuming a 1 cfs change over a 3-month period. Converted to value per acre-foot by dividing by 178.

elasticities from (8) can also be interpreted as streamflow-escapement elasticities. Equation (15) can then be rewritten as

$$MARG.\ VAL.\ WATER = \beta_i \times \Delta FLOW_i \times ESCAPEMENT \times 1.5 \times 0.65 \times \$6.65 = \beta_i \times \Delta FLOW_i \times \$97{,}256 \qquad (16)$$

where ESCAPEMENT has been assumed to be 15,000 (Oregon Department of Fish and Wildlife, unpublished report, 1988*a*). Results are presented in Table 3, second column. The resultant summer flow value increases approximately fourfold, to \$2.36/acre-foot. The second column calculation demonstrates that the value of instream water is sensitive not only to the estimated number of anglers but also to what benefits are included in the measurement. Excluding out-of-basin benefits leads to an undervaluation of John Day River streamflow.

IMPLICATIONS AND CONCLUSIONS

The above analysis provides an estimate of use values for an increment of water in the production of recreational steelhead fishing within the John Day River. Including some out-of-basin benefits, the value of an additional acre-foot of water in the production of recreational steelhead fishing is \$2.36 in 1987 dollars. A complete analysis of the efficiency of water use, however, requires measures of the value of water in both instream and off-stream uses. For example, other fish species in the river, such as resident trout, chinook salmon, and warm water species, would benefit from an improvement in streamflow patterns. While this study focuses only on the value in one instream use, recreational steelhead fishing, some evidence is available on the most important out-of-stream use within the Basin, agriculture.

The John Day Basin has approximately 59,000 irrigated acres (1980 acreage). As is the case with fish production, the value of additional irrigation supplies will vary by location within the Basin. As a result of frequent water shortages in the summer period, current cropping patterns include a high percentage of grain and forage crops. Where climatic and water supply conditions are favorable, more profitable crops such as mint, orchards, potatoes, and sunflowers can be grown. The *U.S. Bureau of Reclamation* [1985] estimated irrigation benefits from increased water supplies in the Basin to vary from \$10 to \$24 per acre-foot, depending on location and crop alternatives. These values are consistent with estimates from other locations with similar crops and environmental conditions. The lower estimate, based on farm budgets for a representative 320-acre family grain and forage farm, appears most appropriate as a measure of the value of an additional acre-foot of water in agricultural production under current cropping patterns.

Judged against the \$10.00/acre-foot estimate for agricultural use, the value of an additional acre-foot of water in the production of recreational fishing does not appear to support any reallocation across uses. These values, however, may be misleading. *Ward* [1987] notes that water used for recreational purposes may be used later by agriculture at a lower point on the river. In addition, due to losses in transport and application (some of which may return to the river), agriculture does not consume 100% of the water diverted. To increase instream flow by 1 cubic feet per second (cfs) it may be necessary to decrease agriculture diversions by 2 cfs or more. The correct values should thus be for acre-feet of water consumed. Unfortunately, complete hydrologic data for all reaches of the John Day River are unavailable. Furthermore, water values within the Basin will vary greatly by location. It is instructive, however, to present some reasonable estimation of use values in consumption. If irrigation consumes half of the total water diverted, agricultural water would be valued at $20/acre-foot consumed. Additions to instream flow would consume far less, say, 10% of the increased flow, raising instream values to over $23. This higher value assumes that the remaining proportion of instream flow is subsequently withdrawn or otherwise put to beneficial use after passing through the juvenile rearing area. If it is instead left instream, then the correct value for instream water reverts to $2.36, reflecting the true opportunity cost of making that water unavailable for alternate uses. Whether transfers of water from agriculture to fishery production are justified on efficiency grounds is thus highly sensitive to the assumed spatial use pattern in each competing use. Further research is needed to define the most productive spawning and rearing reaches of the John Day River Basin, as well as the hydrology of the system.

Finally, it should be noted that the conclusions drawn from this analysis reflect a recent phenomenon within the basin of increasing anadromous fish runs. While the causes of these increases are not fully understood, improved passage conditions at Columbia River dams and increased marine survival are likely causes. Current flow levels appear to be adequate to sustain a viable fishery when riparian habitat and downstream conditions are maintained. The results of this analysis, however, cannot be viewed as supporting less water for fishery production nor do the results imply that no investments should be made in enhancing the steelhead fishery. Some increase in summer streamflow in the upper reaches of the John Day, coupled with riparian habitat management and instream habitat alternatives, may be viable investments.

It should be stressed that benefits reported in this study are due to changes in streamflows, ignoring any benefits arising from habitat improvements. In reality, streamflow, water quality, adjacent riparian cover, the dynamics of the stream, and other ecosystem attributes all combine to "produce" fish. Habitat degradation caused by mining, forestry, agricultural, and range activities in the Basin have led to significant reductions in anadromous fish populations (Oregon Department of Fish and Wildlife, unpublished report, 1985*b*). The above values assigned to instream water have been estimated under the assumption that the relationship between streamflow and these other inputs remains constant. Given current and future habitat improvement projects planned for the John Day Basin, this assumption is questionable. Further research is needed to quantify the relationship between streamflow, habitat quality, and steelhead production. Such research will be of particular importance in developing least cost strategies of meeting judicially mandated increases in fish production.

Acknowledgments. This study was funded in part by a grant from the Oregon State University Agricultural Research Foundation. We are grateful to Hiram Li for assistance in obtaining and interpreting biological data, to J. David Glyer for assistance in estimation of the fishery production models and to two anonymous reviewers for helpful comments on an earlier version of this paper. Technical paper 8562 of the Oregon State Agricultural Experiment Station.

REFERENCES

Anderson, L. G., Estimating the benefits of recreation under conditions of congestion: Comment and extension, *J. Environ. Econ. Manage.*, *7*, 401–406, 1980.

Anderson, J. L., and J. E. Wilen, Estimating the population dynamics of coho salmon (Oncorhynchus kisutch) using pooled time-series and cross-sectional data, *Can. J. Fish. Aquat. Sci.*, *42*, 459–467, 1985.

Beverton, R. J. H., and S. J. Holt, On the dynamics of exploited fish populations, *Minist. Agric. Fish. Food London Fish. Invest. Ser. 2*(19), 533 pp., 1957.

Bishop, R. C., and T. A. Heberlein, Measuring values of extra-market goods: Are indirect measures biased?, *Am. J. of Agric. Econ.*, *61*, 926–930, 1979.

Bockstael, N. E., and K. E. McConnell, Theory and estimation of the household production function for wildlife recreation, *J. Environ. Econ. Manage. 8*, 199–214, 1981.

Bockstael, N. E., W. M. Hanemann, and I. Strand, Measuring the benefits of water quality improvements using recreational demand models, technical report, Econ. Anal. Div., U. S. Environ. Protect. Agency, Washington, D. C., 1985.

Bradford, D. F., Benefit-cost analysis and demand curves for public goods, *Kyklos*, *23*, 775–791, 1970.

Brookshire, D. S., A. Randall, and J. R. Stoll, Valuing increments and decrements in natural resource service flows, *Am. J. Agric. Econ.*, *62*, 478–488, 1980.

Cameron, T. A., and M. D. James, Efficient estimation methods for "closed-ended" contingent valuation surveys, *Rev. Econ. Stat.*, *69*, 269–276, 1987.

Cummings, R. G., D. S. Brookshire, and W. D. Schulze (Eds.), *Valuing Environmental Goods: A State of the Arts Assessment of the Contingent Valuation Method*, Rowman and Allanheld, Totowa, N. J., 1986.

Daubert, J. T., and R. A. Young, Recreational demand for maintaining instream flows: A contingent valuation approach, *Am. J. Agric. Econ.*, *63*, 665–675, 1981.

Freeman, A. M. III, *The Benefits of Environmental Improvement: Theory and Practice*, John Hopkins Press, Baltimore, Md., 1979.

Hammack, J., and G. M. Brown, Jr., *Waterfowl and Wetlands: Towards Bioeconomic Analysis*, John Hopkins Press for Resources for the Future, Baltimore, Md., 1974.

Hicks, J., The four consumers' surpluses, *Rev. Econ. Stud.*, *11*, 31–41, 1943.

Gibbons, D. C., *The Economic Value of Water*, Resources for the Future, Washington, D. C., 1986.

Johnson, N., A bioeconomic analysis of altering instream flows: Anadromous fish production and competing demands for water in the John Day River Basin, Oregon, unpublished thesis, Oreg. State Univ., 1988.

Loomis, J., The economic value of instream flow: Methodology and benefit estimates for optimum flows, *J. Environ. Manage.*, *24*, 169–179, 1987.

Milhous, R. T., Instream flow values as a factor in water management, *Proceedings: AWRA Symposium on Regional and State Water Resources Planning and Management*, American Water Resources Association, Washington, D. C., 1983.

Mitchell, R. C., and R. T. Carson, *Using Surveys to Value Public Goods: The Contingent Valuation Method*, Resources for the Future, Washington, D. C., 1987.

Nickelson, T. E., Influences of upwelling, ocean temperature, and smolt abundance on marine survival of coho salmon (Oncorhynchus kisutch) in the Oregon Production Area, *Can. J. Fish. Aquat. Sci.*, *43*, 527–535, 1986.

Peterman, R. M., Form of random variation in salmon smolt-to-adult relations and its influence on production estimates, *Can. J. Fish. Aquat. Sci.*, *38*, 1113–1119, 1981.

Ricker, W. E., Computation and interpretation of biological statistics of fish populations, *Bull. Fish. Res. Board Can.*, *191*, 382 pp., 1975.

Samples, K. C., and R. Bishop, Estimating the value of variations in anglers' success rates: An application of the multiple-site travel cost method, *Mar. Res. Econ.*, *2*, 55–74, 1985.

Scott, M. J., R. J. More, M. R. LePlane, and J. W. Currie, Columbia River Salmon: Benefit-cost analysis and mitigation, *Northwest Environ. J.*, *3*, 121–151, 1987.

Sellar, C., J.-P. Chavas, and J. R. Stoll. Specifications of the logit model: The case of valuation of nonmarket goods, *J. Environ. Econ. Manage.*, *13*, 382–390, 1986.

Shephard, M. P., and F. C. Withler, Spawning stocks size and resultant production for Skeena sockeye, *J. Fish. Res. Board Can.*, *15*, 1007–1025, 1958.

Sorhus, C. N., W. G. Brown, and K. C. Gibbs, Estimated expenditures by salmon and steelhead sport anglers for specified fisheries in the Pacific Northwest, *Oreg. Agric. Exp. Stat. Spec. Reprint 631*, Oreg. Agric. Exp. Stat., Corvallis, Oreg., 1981.

U.S. Bureau of Reclamation, *Planning Report Concluding the Study for the Upper John Day Project, Oregon*, U.S. Bureau of Reclamation, Boise, Id., 1985.

United States Water Resources Council, Procedures for evaluation of national economic development benefits and costs in water resources planning, *Fed. Reg.*, *44*, 72,892–72,976, 1979.

Walsh, R. G., R. K. Ericson, D. J. Arosteguy, and M. P. Hansen, An empirical application of a model for estimating the recreation value of instream flow, *Colo. Water Resour. Res. Inst. Complet. Rep. 101*, Colo. State Univ., Fort Collins, 1980.

Ward, F. A., Economics of water allocation to instream uses in a fully appropriated river basin: Evidence from a New Mexico wild river, *Water Resour. Res.*, *23*, 381–392, 1987.

White, K. J., A general computer program for econometric methods—SHAZAM, *Econometrica*, 239–240, 1978.

Young, R. A., and S. L. Gray, Economic value of water: Concepts and empirical estimates, *Tech. Rep. NWC-SBS-72-047*, Natl. Water Comm., Springfield, Va., 1972.

R. M. Adams and N. S. Johnson, Department of Agricultural and Resource Economics, Oregon State University, Corvallis, OR 97331.

(Received December 23, 1987;
revised July 8, 1988;
accepted July 18, 1988.)

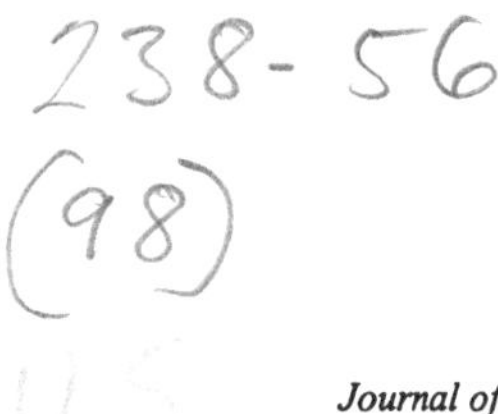

[17]

Journal of Agricultural and Resource Economics 23(1):225–243

The Effects of Water Rights and Irrigation Technology on Streamflow Augmentation Cost in the Snake River Basin

David B. Willis, Jose Caldas, Marshall Frasier, Norman K. Whittlesey, and Joel R. Hamilton

Three species of salmon in the Snake River Basin have been listed as endangered. Recovery efforts for these fish include attempts to obtain increased quantities of water during smolt migration periods to improve habitat in the lower basin. Agriculture is the dominant user of surface flows in this region. This study investigates farmer cost of a contingent water contract requiring the agricultural release of stored irrigation supplies in low flow years during critical flow periods. Results show that contingent contracts can provide substantial quantities of water at a relatively modest cost without significantly affecting the agricultural base of the area.

Key words: contingent water contracts, irrigation technology, streamflow augmentation, water rights

Introduction

Many salmon stocks, once abundant in the Columbia and Snake Rivers, are now extinct (Peterson, Hamilton, and Whittlesey). In 1993, the National Marine Fisheries Service (NMFS) listed Snake River sockeye and chinook salmon as endangered under the Endangered Species Act (ESA). This reduction in the overall stock level and specie variety has coincided with the development of a variety of multipurpose water projects within the Snake River Basin over the last century. The earliest projects were designed primarily to facilitate irrigated agriculture, but later projects paid more attention to flood control and hydropower objectives (Clairbon). These projects have severely altered the quantity and timing of Snake River flows, contributing to salmon population declines (Sims and Ossiender 1991).

Hydroelectric dams have lowered streamflow velocities so that smolt migration from Idaho to the Pacific Ocean that once took 7–14 days now takes as long as 40 days (Wernstedt, Hyman, and Paulsen). The slower travel exposes smolts to dangers of disorientation, predation, and diseases, in addition to the physical dangers of passing through each of eight large hydropower dams (Hamilton and Whittlesey 1992). Sims and Ossiender (1992a, b) found that increased stream velocity during smolt migration (April through June) could increase smolt survival and the number of returning adults.

Willis is post doctoral research associate, and Caldas is former graduate research assistant, both with the Department of Agricultural Economics, Washington State University; Frasier is assistant professor, Department of Agricultural Economics, Colorado State University; Whittlesey is professor, Department of Agricultural Economics, Washington State University; and Hamilton is professor, Department of Agricultural Economics and Rural Sociology, University of Idaho.
Review coordinated and publication decision made by B. Wade Brorsen.

Despite the current uncertainty about how much fisheries benefit from increasing streamflows, both the NMFS (through its recovery plan for Snake River salmon) and the Northwest Power Planning Council (NPPC 1994, 1995) recommend that minimum flow targets be established and maintained for smolt migration periods. Currently, up to 1.19 million acre-feet (MAF) of water from nonagricultural sources is "budgeted" for Snake River releases between 15 April and 15 June to aid salmon migration. However, these additional supplies have been insufficient to generate desired flows in the lower river during critical fish migration periods (Ewbank).

Alternative water sources must be found before a successful recovery program can be implemented. The U.S. Army Corp of Engineers is now investigating the possibility of additional flow augmentation polices for the lower Snake River. A promising water supply source is the substantial amount of agricultural water stored upstream in the Snake River Basin. About five MAF of Snake River flow is stored in reservoirs each spring and subsequently released for irrigation use later in the summer, mainly as a supplemental supply to stream diversions. Most of the upstream irrigation storage reservoirs were built when irrigation technology was relatively unsophisticated and maximum attainable irrigation efficiency was quite low, often below 25%. Generally, sufficient storage was built to serve project lands at these low efficiencies. With improved irrigation technology, per acre diversions have declined and storage capacity often exceeds the diversion needs of acreage with storage rights. Subsequently, the upper Snake River water bank was created to encourage leasing of unused storage water to other agricultural users or for instream uses. The main value of the unused storage water to current right holders is insurance against future drought.

Today, the water bank, along with normal irrigation storage, is being considered as a source of water to supplement river flows for salmon migration during years of drought. However, Idaho law contains a major impediment to using bank water to augment streamflows:

> Storage space . . . that is evacuated to supply water for nonconsumptive uses . . . shall be the last space to fill in the reservoir from which the space was originally assigned . . . in the ensuing year (Sims and Ossiender 1991, Chap. 5, p. 10).

This provision is intended to assure that water sellers (rather than nonparticipant third parties) bear the risk of future water shortage when reservoirs fail to refill due to nonagricultural water sales (Peterson, Hamilton, and Whittlesey). This risk could be an important determinant of farmer willingness to enter into a contingent water contract designed to augment streamflow levels in low flow years.

Previous Research

Several researchers have proposed that water markets be used to improve efficiency of resource use or meet instream flow requirements in the Columbia/Snake River Basin. Gardner, representing the Idaho governor's office, states:

> Water markets for rights or perpetual permits, or even for annual rentals where exchanges can be freely made, provide a solution to our allocation problems. . . . Both economic efficiency and distributional equity would be well served by allocating free transfers of . . . consumptive use (p. 25).

Whittlesey, Hamilton, and Halverson; and Hamilton, Whittlesey, and Halverson first proposed a contingent water market to move Snake River water from irrigation to hydropower. Halverson showed that such a market might also be applicable to the Columbia Basin Project of Washington State. Hamilton, Reading, and Whittlesey extended previous work, focusing on the potential of contingent water markets to benefit lower Snake River fish passage. Additional research by Peterson, Hamilton, and Whittlesey; Sommers; and Huppert, Fluharty, and Kenney has suggested that salmon recovery in the Snake River Basin could be enhanced by limiting irrigation diversions in low streamflow years to improve fishery habitat while leaving the long-term agricultural production base intact. Today, contingent water contracts are being seriously considered as a salmon recovery tool, but additional information on potential risks to farming, management issues, and political acceptance of such contracts is needed (Middaugh).

Study Purpose and Area

This study extends previous research by focusing on the influence of water right seniority and irrigation technology on the minimum compensation required to induce a risk-neutral farmer to contingently contract for release of stored irrigation water to augment streamflows for fisheries habitat. The water broker is assumed to be the designated representative of public interests for protection of salmon habitat. It is anticipated that the broker would work with and for the NPPC, NMFS, and the Bonneville Power Administration (BPA).

Two contingent water-contracting scenarios are analyzed: the first based on farmers selling portions of *excess stored water* modeled after the existing water bank structure, and the second based on selling portions of *total stored water*. Excess stored water is defined as that portion of total stored water surplus to expected irrigation requirements for the current growing season, and total stored water is the quantity of water available to the right holder for all uses within a growing season. Under each contract, a farmer commits to release a specific percentage of stored water when downstream flow levels drop below the critical threshold level in each low flow year of the contract period. The cost of releasing water under an excess stored water contract is incurred in subsequent years, whereas the cost of releasing water under a total stored water agreement may be felt in both the current year and future years.

The percentage of stored water committed to the contract, along with rainfall in subsequent years, determines whether irrigation storage refills in the following or subsequent seasons. Releasing stored water for salmon habitat in low flow years increases the probability of incurring an on-farm irrigation water shortage in the current and/or future years. This is different from the stream diversion contingent market contracts analyzed by Hamilton, Whittlesey, and Halverson that would reduce current-year crop production. They did not consider effects of releasing stored water on future years' income.

The Snake River Basin provides the empirical setting for this analysis. The Snake River is the largest tributary of the Columbia River, draining 108,500 square miles, 42% of the Snake/Columbia Basin, and contributes 20% of Columbia River flows. Average annual upper Snake River agricultural diversions exceed 16 MAF, of which up to 5 MAF are stored water diversions, and are used to irrigate about 4 million acres, with 8–10 MAF eventually becoming return flow to the river. Approximately 56% of the irrigated acres in the upper basin use gravity application systems. About two-thirds (68%) of the sprinkler systems are side-roll, followed by center pivot at 29%.

Snow melt and precipitation occur primarily in March to early June. Based on seasonal water supply projections in March, reservoirs are managed to be as full as possible in early to mid-June. It is assumed that by March, farmers have sufficient information to estimate June storage levels and available inflows over the following irrigation season. Given this information and expected crop irrigation requirements, farmers can project how much water can be released to augment April/June streamflows under an *excess* water contract without jeopardizing current season irrigation needs. Future water shortages may occur if subsequent water years are below normal and vacated storage capacity does not refill—particularly likely for junior right holders. Under the alternative contract, a specified percentage of *total* storage available in early June is released to augment flows in low flow years. The total storage contract can create on-farm irrigation water shortages in both the current and subsequent years. The entire contracted percentage is released in each low flow year. The amount of stored water committed under either contract (up to 100%), in combination with stochastic streamflows, determines the probability and severity of a water shortage in the release year and subsequent years.

Contingent contracts of this type do not guarantee a specific quantity of stored reserves will be released for instream flow augmentation in each low flow year; instead, they specify what percentage of defined reserves will be released. Quantities cannot be guaranteed since they are dependent on stochastic reservoir inflows. Refill priority right and irrigation technology will affect stored releases. Storage always refills in the order of priority right. Hence, senior right holders generally will incur less risk of future shortage than junior right holders for similar contracts. Because farms with senior refill priorities have their vacated storage refilled before farms with lower refill priorities, they generally will have higher stored reserves available for contract release in low flow years. Moreover, farms using more efficient irrigation technologies generally will have higher storage levels than farms using less efficient technologies because smaller quantities of stored supplies are required for irrigation diversion per irrigated acre. Historical changes in irrigation technology have not affected individual farm refill priority or quantity of storage rights.

Modeling Procedure

A simulation model was constructed to estimate expected farm-level water supply shortages and net income losses due to contract participation for the three dominant irrigation technologies in the basin. Representative farms were constructed for each irrigation technology using average cropping patterns and yields to represent existing agriculture in the upper Snake River Basin. These three irrigation technologies, which comprise more than 98% of all irrigated basin acreage, are (*a*) rill, (*b*) side-roll, and (*c*) center pivot. Three levels of appropriative rights for stored water (A, B, and C) were defined for each representative farm, with farm A holding the most senior right and farm C the most junior. The farm cost of contracting to release 25%, 50%, 75%, and

Table 1. Baseline Values of Net Farm Income, Crop Mix, and Water Use for the Representative Farms with Unrestricted Water Supplies

		Irrigation System		
Item	Unit	Rill	Side-Roll	Center Pivot
Net Income	$/acre	171	182	212
Irrigated Crop Acreage:				
Pasture	%	14	12	0
Sugarbeets	%	5	2	4
Dry beans	%	3	2	0
Corn	%	3	0	10
Winter wheat	%	32	45	47
Alfalfa	%	32	27	25
Potatoes	%	11	12	14
Water Use	inches/acre	76.38	38.57	30.20
Net Irrigation Requirement (NIR)	inches/acre	24.97	24.91	25.27
Irrigation Efficiency	%	32.69	64.58	83.67

100% of defined stored water reserves in low streamflow years under both the excess and total stored water contractual arrangements is estimated for each contracting level. A 10-year contract period is assumed. Per acre changes in net present value of farm returns over the life of the contract are used to measure expected contract cost.

The baseline data for each representative farm are presented in table 1. Under full water supply, rill irrigated farms annually average $171 per acre net income above variable cost. Gross margins are $182 and $212 per acre, respectively, for side-roll and center pivot farms. The higher gross margins associated with the sprinkler systems are mainly due to a higher value crop mix. Because sprinkler systems require more capital investment, the long-run net income advantage of sprinkler systems is less than indicated by gross margin values. Despite differences in irrigation efficiency and irrigated crop mix, the net irrigation requirement for each representative farm is nearly equal.

As shown in figure 1, the analytic structure consists of three linked models: (*a*) a probability model, (*b*) a hydrology model, and (*c*) an economic model. For simplification, the flowchart is drawn for a single year and a specific contract level.

Probability Model

The probability model simulates upper Snake River monthly flow levels in year *t*, and determines if late spring flows in the lower Snake River are below the target level. The entire contract commitment is released when downstream flows are below the specified target level.

A contingent water contract motivated by fish habitat needs will require contract deliveries when flows in the lower river fall below target levels for the 15 April–15 June smolt migration period. Thus, contract release conditions must be clearly established so the probability of contract-required deliveries and expected cost can be determined

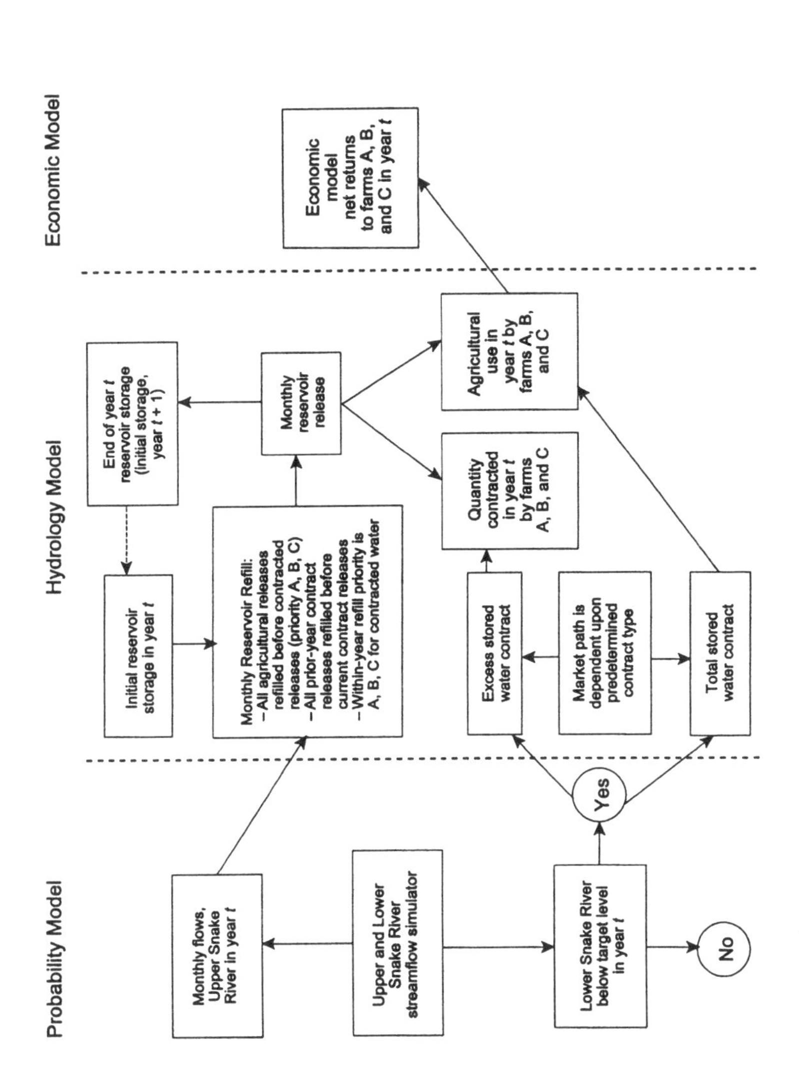

Figure 1. Flowchart of water contract simulation procedure for a given contract rate in year *t*

Table 2. Historic Relationship Between Upper Snake River Flow Level and Probability that Flow Level at Lower Granite Dam Is < 85 Kcfs Target Level

Upper Snake River Average Flow Level, April–June, 1929–88 (Kcfs)	Probability that Lower Snake River Flow < 85 Kcfs
00.00–11.24	0.8
11.24–12.44	0.7
12.44–14.47	0.3
14.47–16.04	0.1
16.04–18.59	0.0
18.59–25.00	0.0

by the owners of storage capacity. In this study, contracted water is released in years when the spring flow level falls below the NPPC minimum monthly average flow target of 85 thousand cubic feet per second (Kcfs) at Lower Granite Dam. Based on historical data for the period 1929–88, the average flow level was below the 85 Kcfs target in 32% of the years—meaning that, on average, contracting farmers would be expected to deliver stored water about three years out of 10.

Despite the apparent existence of cyclical weather patterns, a variety of ARIMA-based statistical tests on the annual streamflow data failed to detect the presence of a statistically significant serial correlation pattern in upper river flows. Hence, each 10-year sequence for upper Snake River streamflow levels, which determines the quantity of water available for reservoir refill each year, was produced by randomly drawing flows (with replacement) from the historic record.

The correlation between upper Snake River flows where the irrigation is located and the lower river flow levels where the salmon migration habitat is located is positive, but not perfect, as shown in table 2. Here the empirical cumulative density function for the three-month average flow level for April through June in the upper Snake River for the years 1929–88 is divided into six intervals. Each interval contains one-sixth (10) of the historic outcomes. The probability of observing a downstream spring flow level below the 85 Kcfs flow target level when the upper flow level is less than 11.24 Kcfs is 0.8. As seen in table 2, the probability that lower river flows are below the target level falls sharply as the spring flow level in the upper river increases. A contract release year is determined when the drawn upstream flow level is associated with a low flow year in the lower Snake River. A drawn 10-year upper Snake River flow sequence, in combination with the information on which years are low flow years on the lower Snake River, provides the hydrology model with the necessary information to simulate the effect of contract participation on on-farm water supplies for one 10-year contract. Given the stochastic nature of streamflow supplies, each 10-year upstream flow sequence with associated contract deliveries was randomly replicated 250 times to derive the expected cost of contract participation.

Hydrology Model

The hydrology simulation model developed by Frasier, Whittlesey, and Hamilton (FWH) uses the information on upper Snake River monthly flows and contract year status to simulate monthly reservoir refill by storage right priority, the quantity of contract water released, the monthly quantity of water diverted for irrigation by each farm, and the reservoir storage level at the end of the year. The FWH model allows the user to specify a typical farming region and establish water right priorities for farms within the region. Additionally, the model allows the user to (*a*) specify irrigation efficiencies, (*b*) control the portion of applied water lost to evaporation or phreatophytes, (*c*) control the fraction of applied water constituting return flow, and (*d*) control the share of return flows via surface drain or deep percolation. This model assumes three equal-sized representative farms, denoted as A, B, and C, with similar irrigation technology and having storage water rights with seniority in the order ABC. The model uses the water rights structure, along with monthly information on storage inflows and outflows and crop water demand, to estimate the quantity of irrigation water diverted by each farm and return flow quantities.

The hydrology model assumes that in any given month, each farm will divert the full net irrigation requirement (NIR) to satisfy baseline crop demand to the limit of available stored water supplies. When stored supplies are less than irrigation requirements, an on-farm water shortage occurs. Consistent with legal statutes designed to avoid third-party effects, the simulation model completely refills all agricultural release in accordance with the specified refill priority before any contingently contracted flow releases are refilled. Moreover, contract water released in a prior year is refilled before contract water released in the current or subsequent years in accordance with the seniority structure existing at the time of the release. End-of-year reservoir storage is initial reservoir storage in the subsequent simulation year.

Baseline irrigation storage capacity is assumed to be 6.38 acre-feet per acre (AF/A) for acreage with storage rights. This per acre storage quantity equals the per acre seasonal quantity of water applied by a low-efficiency (rill) irrigated farm, and is consistent with original (and current) per acre storage capacity of farms in the region. Storage refill potential is such that under rill irrigation and in absence of a contingent water contract, priority farms A and B never experience water shortages, while farm C incurs a minor water shortage about one year in 10. Farms using sprinkler technology always have sufficient irrigation water supplies under baseline conditions.

Economic Model

The economic model is a mathematical programming model that allows farmers to change crop mix and irrigation strategies in response to contract-caused water shortages.

Farmer Response to Water Shortage. Annual choices available to irrigation management include deficit irrigation, reducing irrigated acreage, and/or changing the crop mix. The production cost and yield adjustments for "water deficit" irrigation used methods developed by Willis. Yields are linearly interpolated between expected yield at full NIR and expected yield at the maximum allowed deficit level for irrigation levels falling between the two irrigation levels. Yield-dependent production costs are proportionately reduced when yields are decreased due to deficit irrigation. Crop budgets and price/income data used in the analysis reflect the conditions of 1993. Annual net farm income was maximized subject to restrictions on land, crop rotation, and monthly water supply. Perennial crop budgets were developed assuming sufficient water is available to preserve the crop stand through its normal life. Individual farms cannot shift irrigation technology because the research objective is to measure the agricultural cost of market participation for farms with specific technologies and priority water rights under a 10-year market contract. It is acknowledged, however, that such contingent water contracts could eventually stimulate additional shifts in irrigation technology.

Under baseline conditions, rill irrigated farm C has a water shortage 8% of the time, receiving on average 98.2% of full water supplies over 250 replications of a 10-year period. These water shortages reduce expected average annual net income $0.62 below the $171 per acre return of farms A and B. Side-roll and center pivot farms A, B, and C never experience water shortages in the baseline.

Net Income Effects of Contract Participation. Contract participation leads to farm income losses when stored water releases impose irrigation water shortages in either the current or future years. The economic model calculates the participating farmer's decrease in net income each year. As the simulation proceeds through time, annual accounts are maintained on the quantity of contract releases, severity of on-farm water shortage, and net income loss. Farmer cost depends upon the percentage of stored water committed to the contract, the frequency of contract releases, and streamflow conditions in subsequent years. The yearly contract-caused net income losses over the 10-year contract period are then chronologically arranged and discounted, using a 4% real discount rate, into a per acre net present value (NPV) estimate of contract cost. This process is replicated 250 times for each contract scenario, and the 250 NPV estimates are subsequently averaged to derive expected contract cost per irrigated acre for each farm. The expected average annual cost per irrigated acre to a participating farmer is derived by converting the average NPV estimates into annualized equivalent values. An annualized equivalent cost value incurred in each contract year and discounted by the appropriate rate of interest (4% real rate) will exactly equal the NPV contract cost estimate.

Results and Analysis

Excess Stored Water Contracts

We first consider an excess water contract where a farmer agrees to release some pre-specified percentage of stored water not needed for irrigation in the current year. If the storage does not adequately refill, on-farm water shortages may occur in future years. The excess water contract assumes the full contract commitment is released in each low flow year.

Rill Irrigation. Water supply shortages from contract participation are reported in table 3 for rill irrigated farms. Senior priority right farm A is excluded from table 3 since it does not incur a water shortage at any contract participation level. The seniority of farm A's refill right assures that all water released under contract will be refilled before it is needed on-farm. The fact that farm A incurs no farm income losses could change if storage refill rules were modified, or if farm A were to release water while

Table 3. Average Percentage of Baseline Water Supplies Received by Rill Irrigated Farms B and C Under Four Excess Stored Water Contract Specifications

Percentage of Excess Stored Water Contingently Contracted	Farm B			Farm C[a]		
	Avg.[b]	Min.[c]	SD[d]	Avg.[b]	Min.[c]	SD[d]
25%	100.0	100.0	0.0	99.8	76.4	7.8
50%	99.8	98.4	0.5	99.4	72.2	8.4
75%	99.6	96.5	1.6	99.1	68.1	8.9
100%	99.1	94.1	2.8	98.7	63.1	9.5

Note: Farm A experiences no water supply shortage at any excess stored water contract level, and so is not represented here.
[a] Average supplies for farm C are reported relative to average baseline condition and not full water supply. Under baseline condition, farm C receives 98.2% of full water supply in an average year.
[b] Average water supply over 250 replications of a 10-year contract period.
[c] Minimum values are single-year minimum supplies expressed as a percentage of full water supply for each 10-year contract period, averaged over 250 replications.
[d] Standard deviation of farm water supply over 250 simulations of a 10-year contract period.

farms B and/or C choose not to participate. In the latter case, farm A would lose its priority right for refill of the portion of storage sold for nonagricultural use, while farms B and/or C would not be so affected. This alternative is not evaluated here.

With 25% of excess stored water committed to the contingent contract, farm B has no future water shortages and farm C has only minute shortages. Farm C averages 99.8% of baseline water supplies (98% of full water supply) under a 25% contract, with the worst year average in each 10-year contract averaging 77.8% of baseline water supply (76.4% of full water supply). However, when compared with the no-contract baseline situation where shortages are routine, single minimum year supplies average 95% of the baseline minimum average value over a 10-year period. The contract creates water shortages for farm B only when at least 50% of excess stored water is contracted for release in low flow years. Under the maximum 100% contract, farm B averages 99.1% of full water supply over the contract and incurs minor water deficits in only 6% of the years. Under the same contract, farm C averages 98.7% of baseline water supply (96.9% of full water supply) over the contract period.

Per acre NPV losses for farms B and C are presented in table 4. At the 100% contract level, NPV losses for farm B average \$2.10 per acre, and a maximum NPV loss of \$12.31 is incurred in one contract simulation. Average NPV losses for farm C range from \$0.93 per acre for a 25% contract to \$3.96 per acre for a 100% contract. At the 100% contract level, farm C has a maximum NPV loss of \$23.45 per acre in one simulation. Perhaps the most useful information for a farmer considering contract participation is that annualized equivalent values of NPV losses average less than \$1 per irrigated acre for all rill farms under the 100% contract. The small cost is a consequence of the contract design which assures water releases never exceed current-year stored surpluses. Hence, contract cost is from subsequent-year income losses due to incomplete reservoir refill.

The average quantity of water released in a delivery year is shown in table 5. Farms with senior water rights release more water than those with lower priority water rights

Table 4. Net Present Value and Annualized Equivalent Cost of Irrigation Water Reallocated to Instream Use for Rill Irrigated Farms B and C Under an Excess Stored Water Contract ($/acre)

Contract Level		Farm B				Farm C			
		Avg.	Max.[a]	Min.[a]	SD[b]	Avg.	Max.[a]	Min.[a]	SD[b]
25%:	Net Present Value[c]	0.00	0.00	0.00	0.00	0.93	4.63	0.17	2.54
	Annualized Cost[d]	0.00	0.00	0.00		0.11	0.55	0.02	
50%:	Net Present Value[c]	0.17	1.51	0.00	0.46	1.86	9.28	0.42	2.70
	Annualized Cost[d]	0.02	0.18	0.00		0.22	1.10	0.05	
75%:	Net Present Value[c]	0.67	4.13	0.00	1.06	2.70	15.60	0.76	3.88
	Annualized Cost[d]	0.08	0.49	0.00		0.32	1.85	0.09	
100%:	Net Present Value[c]	2.10	12.31	0.00	2.00	3.96	23.45	0.84	4.70
	Annualized Cost[d]	0.25	1.46	0.00		0.47	2.78	0.10	

Note: Farm A experiences no income loss at any excess stored water contract level, and so is not represented here.
[a] Maximum (minimum) present value cost of one 10-year contract simulation in 250 contract replications.
[b] Standard deviation of average present value net income loss for a 10-year contract period.
[c] Net present value for 10-year contract cost averaged over 250 replications using a 4% discount rate.
[d] Annualized equivalent cost is calculated using a 4% discount rate.

Table 5. Excess Stored Water Released in Low Flow Years for Alternative Participation Levels: Rill Irrigated Farms (acre-feet/acre)

Contract Level	Farm A				Farm B				Farm C			
	Avg.	Max.[a]	Min.[a]	SD[b]	Avg.	Max.[a]	Min.[a]	SD[b]	Avg.	Max.[a]	Min.[a]	SD[b]
25%	1.03	1.10	0.87	0.13	0.67	1.01	0.45	0.22	0.35	0.70	0.13	0.25
50%	1.89	2.20	0.92	0.51	1.21	2.01	0.38	0.54	0.65	1.34	0.07	0.50
75%	2.61	3.30	0.78	1.08	1.65	2.97	0.26	0.94	0.91	2.01	0.05	0.76
100%	3.22	4.41	0.45	1.70	2.05	3.93	0.18	1.36	1.13	2.67	0.03	1.05

Note: Water volumes correspond to average stored supplies released in a contract-triggered low flow year (32% of all years are low flow years).
[a] Maximum (minimum) values are single-year maximums (minimums) for each 10-year contract period, averaged over 250 contract replications.
[b] Standard deviation of average release in a low flow year over 250 simulations of a 10-year contract period.

Table 6. Percentage of Baseline Water Supplies Received by Side-Roll Irrigated Farms B and C Under 75% and 100% Excess Stored Water Contracts

	Farm B			Farm C		
Contract Level	Avg.[a]	Min.[b]	SD[c]	Avg.[a]	Min.[b]	SD[c]
75%	100.0	100.0	0.00	99.9	99.8	0.02
100%	99.9	99.5	0.14	99.5	92.1	2.50

Note: Farm A experiences no water supply shortage at any excess stored water contract level, and so is not represented here. Farms B and C experience no water supply shortage at the 25% and 50% contract levels.
[a] Average water supply, expressed as a percentage of full water supply, over 250 replications of a 10-year contract period.
[b] Minimum values are single-year minimum supplies expressed as a percentage of full water supply for each 10-year contract period, averaged over 250 replications.
[c] Standard deviation of farm water supply over 250 simulations of a 10-year contract period.

at each participation level. Average releases range from 3.22 AF/A for farm A to 1.13 AF/A for farm C under a 100% contract. There is considerable variation in the quantity released by each farm. For example, at the 100% contract level, the maximum single-year quantity released by farm A in a 10-year contract period averages 4.41 AF/A, but the minimum quantity released averages only 0.45 AF/A. Average single-year maximums and minimums for the quantity released by farm C are less, averaging 2.67 AF/A and 0.03 AF/A, respectively, and are attributable to the lower refill priority. That is, priority of refill affects the amount of surplus water available over time.

The standard deviation of the quantity released increases with the contract participation level. At high percentage rates of participation, the standard deviation is larger for farm A than for farm C, primarily because the senior right holder releases about three times as much water. However, at the lowest participation level, the standard deviation of the released quantity is less for farm A than for farm C because seniority of the refill right nearly guarantees that farm A will be able to release its maximum contract commitment in all low flow years. Regardless of the contract level, the relative variation in the quantity released, as measured by the coefficient of variation, is smaller for farm A than for farm C.

Side-Roll Irrigation. Water supply deficits imposed on side-roll irrigated farms are shown in table 6 for farms B and C. (Farm A incurs no on-farm water deficit at any contract level under side-roll irrigation technology.) The more efficient technology confines farm B contract-related water shortages to the 100% contract. Farm C incurs water supply shortages under both the 75% and 100% contracts. But these shortages are minimal, and farm C averages over 99% of full water supplies at both contract levels, incurring a contract-caused water shortage less than one year in 10.

The annual average quantities of water released by each side-roll farm in a low flow year are shown in table 7. For a 100% contract, average quantities released range from 5.14 AF/A for farm A to 3.84 AF/A for farm C. Generally, three to four times more stored water is delivered under a 100% contract than a 25% contract. Moreover, average releases per acre of irrigated land are greater than for rill irrigated farms due to the more efficient technology and resulting greater excess storage capacity.

Center Pivot Irrigation. With center pivot technology, farms A and B receive full water supplies at all contract levels, and farm C shortages are limited to the 100% contract. Farm C averages 99.7% of full water supply in each contract period and sustains small water deficits in 2% of the years. Annual releases in low flow years are greatest under center pivot technology because it has the highest irrigation efficiency and thus smaller diversion requirements, leading to increased excess stored water supplies relative to the less efficient technologies. Farm A releases an average of 5.41 AF/A in a low flow year compared with 4.23 AF/A for farm C under a 100% contract.

Review of Excess Storage Market. The average acre-foot cost of water released is computed by dividing the annualized equivalent value of NPV by the expected quantity of water released in each contract year (average quantity released in a low flow year multiplied by the probability of a low flow year). Regardless of irrigation technology, no water deficit or net income loss is incurred by farm A at any contract level. Rill irrigated farm B cost is $0.38 per acre-foot of released water under a 100% contract. Rill irrigated farm C encountered the greatest costs, with annualized costs ranging from $0.98 per acre-foot at the 25% level to $1.30 per acre-foot at the 100% level. Farm C contract cost is significantly less with a more efficient irrigation technology. For example, at the 100% contract level, annualized costs are $0.14 and $0.11 per acre-foot for the side-roll and center pivot technologies, respectively. Hence, priority rights and irrigation technologies affect the ability of farms to enter into a contingent water contract. Policy makers should target high priority rights and high efficiency technologies for such contingent markets to obtain the greatest return for water purchases to enhance fish habitat. However, in the end, it may be necessary to deal with all farm types in order to achieve streamflow targets.

Total Stored Water Contract

In this section we examine the effects of a contingent contract wherein farmers would commit 25%, 50%, 75%, or 100% of *total* stored water for flow augmentation when needed in the lower Snake River. A 75% total stored water contract requires the farmer to release 75% of all stored water supplies when triggered by the downstream flow condition. The actual release would range from 0% to 75% of the total storage right, depending upon how full storage reservoirs are when water delivery is mandated. In contrast to an excess stored water contract, a total stored water contract can impose water shortages in both the current and subsequent years. As before, when lower Snake River flows are projected to be below the 85 Kcfs target level, all contracted water is released in an effort to support the target flow level.

Rill Irrigation. Table 8 indicates that contract-caused water shortages are minor at the 25% contract level for rill irrigated farms, but significantly increase at higher contract levels. Water shortages are common under a 100% contract, and farms A, B, and C can expect some water supply deficit 32%, 38%, and 46% of the time, respectively. At this contract level, farms A and B average 86.7% and 83% of baseline water supplies, respectively, compared with only 74% for farm C. Expected minimum single-year farm water supplies in each 10-year contract period are significantly lower, respectively averaging 54.5%, 40.6%, and 18.6% of baseline requirements for farms A, B, and C.

Contract costs are either zero or minimal for all three rill irrigated farms until contract obligations exceed 50% of storage. Under a 100% contract, per acre NPV losses

Table 7. Excess Stored Water Released in Contract Years for Alternative Participation Levels: Side-Roll Irrigated Farms (acre-feet/acre)

	Farm A				Farm B				Farm C			
Contract Level	Avg.	Max.[a]	Min.[a]	SD[b]	Avg.	Max.[a]	Min.[a]	SD[b]	Avg.	Max.[a]	Min.[a]	SD[b]
25%	1.34	1.34	1.34	0.00	1.32	1.34	1.28	0.03	1.20	1.32	1.09	0.11
50%	2.67	2.68	2.66	0.01	2.62	2.68	2.48	0.12	2.32	2.61	1.60	0.32
75%	3.97	4.01	3.71	0.14	3.72	3.91	2.30	0.59	3.10	3.83	0.94	1.01
100%	5.14	5.35	3.43	0.63	4.53	5.15	1.12	1.50	3.84	5.02	0.24	1.90

Note: Statistical results are for contract release years only (approximately 32% of all years).
[a] Maximum (minimum) values are single-year maximums (minimums) for each 10-year contract period, averaged over 250 contract replications.
[b] Standard deviation of average release in a low flow year over 250 simulations of a 10-year contract period.

Table 8. Percentage of Baseline Water Supplies Received by Rill Irrigated Farms at Each Contract Participation Level Under a Total Stored Water Contract

	Farm A			Farm B			Farm C[a]		
Contract Level	Avg.[b]	Min.[c]	SD[d]	Avg.[b]	Min.[c]	SD[d]	Avg.[b]	Min.[c]	SD[d]
25%	100.0	100.0	0.0	99.9	99.1	0.3	97.1	85.0	11.1
50%	100.0	100.0	0.0	97.7	86.2	5.6	91.5	72.9	16.5
75%	98.4	95.6	2.6	91.9	71.0	13.9	83.8	56.8	23.7
100%	86.7	54.5	21.3	83.0	40.6	26.3	74.0	18.6	34.3

[a] Average supplies for farm C are reported relative to average baseline condition and not full water supply. Under baseline condition, farm C receives 98.2% of full water supply in an average year.
[b] Average water supply over 250 replications of a 10-year contract period.
[c] Minimum values are single-year minimum supplies expressed as a percentage of full water supply for each 10-year contract period, averaged over 250 replications.
[d] Standard deviation of farm water supply over 250 simulations of a 10-year contract period.

average $46 for farm A, $74 for farm B, and $174 for farm C over the 10-year contract, indicating some water required for baseline irrigation needs is released at this contract level.

Stored water releases in a low flow year are generally two to four times greater than under the excess stored water contract. Average releases by rill farm A in a low flow year range from 1.53 AF/A for the 25% contract to 6.09 AF/A for the 100% contract. Corresponding quantities are 1.33 AF/A to 4.78 AF/A for farm C. Farms with senior water rights consistently contribute more water for streamflow augmentation than those holding junior water rights, other factors equal.

Side-Roll Irrigation. Farms employing sprinkler technology are less likely to experience water shortages than farms using rill technology. Side-roll irrigated farms incur no significant water supply shortages under a 25% or 50% contract, and only small shortages at the 75% level. However, under a 100% contract, water shortages are fairly common, with farms A, B, and C averaging 90%, 86%, and 80% of baseline supplies, respectively. Minimum single-year water supplies average 66%, 32%, and 17% of baseline crop requirements for farms A, B, and C, respectively.

Income losses are directly related to these water shortages. Farms A, B, and C incur an expected per acre NPV loss of $40, $67, and $106, respectively, when contracting at the 100% level. These values are slightly less than for the rill irrigated farm because the more efficient technology reduces crop diversion requirements and buffers agricultural exposure to water supply deficits attributable to contract releases.

Water releases are slightly higher than for rill irrigated farms. Under a 25% water contract, each farm releases 1.59 AF/A in a low flow year, but releases vary by right priority at higher contract levels. At the 100% contract level, average releases are 6.31, 5.98, and 5.16 AF/A for farms A, B, and C, respectively. The variation in the quantity released by side-roll farms A and B under a 100% contract is considerably less than that of rill irrigated farms A and B, but the quantity released by farm C fluctuates widely and ranges from zero to 6.37 AF/A. The more efficient technology reduces the variance of the total quantity released by all three side-roll farms in a low flow year relative to the rill irrigated farms.

Center Pivot Irrigation. A more efficient irrigation technology creates more surplus stored water. These increased surpluses reduce the probability of contract-related water shortages. Under a 100% contract, farms A, B, and C annually average 90%, 86%, and 82% of full on-farm water requirements, respectively. Average minimum single-year supplies average only 66% of full supply requirements for farms A and B, and 45% for farm C, but are significantly higher than the corresponding side-roll minimum values. Water deficits are incurred 28%, 30%, and 42% of the time, respectively, by farms A, B, and C under a contingent contract for 100% of stored water.

Contract cost is minimal for all center pivot farms except at the 100% contract level, where per acre NPV losses average $55, $92, and $121, respectively, for farms A, B, and C. These losses are greater than for side-roll irrigated farms because center pivot farms have high-value crops.

Annual quantities of water released in low flow years are similar to those for the side-roll farm. At the 25% contract level, each farm contributes about 1.59 AF/A in a low flow year, and average quantities released under a 100% contract are 6.34, 6.07, and 5.21 AF/A for farms A, B, and C, respectively. The variability in the quantity released by farm C under a 100% contract remains high, ranging from 6.37 to 0.16 AF/A, which is slightly less than for similar farms using either rill or side-roll technologies.

Table 9. Annualized Equivalent Cost per Acre-Foot of Water Released by Upper Snake Irrigated Farms Under a Total Stored Water Contract

	Annualized Cost of Water Released ($/acre-foot)								
	Rill Farm			Side-Roll Farm			Center Pivot Farm		
Contract Level	A	B	C	A	B	C	A	B	C
25%	0.00	0.29	2.22	0.00	0.00	0.00	0.00	0.00	0.00
50%	0.05	1.38	3.91	0.00	0.00	0.03	0.00	0.00	0.00
75%	0.45	2.72	6.84	0.00	0.05	0.90	0.00	0.01	0.34
100%	3.04	5.32	14.54	2.52	4.47	8.19	3.46	6.06	9.23

Note: Annualized equivalent cost per acre-foot of water released is calculated by dividing the annualized per acre cost by the expected annual quantity released per irrigated acre.

Forgone Benefit of Water Sold. Annualized income losses per acre-foot of water released for each total stored water contract are presented in table 9. For a given irrigation technology and contract level, it is less costly for a senior right holder to enter into a contractual agreement than a junior right holder. This occurs because of the refill rules for water storage. Assuming that all storage right owners in a reservoir participate in the same contingent water contract, it is the senior right that will always refill first. Hence, the risk or cost of such participation must be greatest for the junior right holders. Of course, the senior right holders also will be able to furnish greater amounts of water over time to the contract. Per acre-foot cost is higher for center pivot farms A and B than the comparable rill farmer under a 100% contract because of the higher valued crop mix. This is not the case for farm C because the more efficient technology produces fewer and/or less severe water supply shortages than the rill technology, more than offsetting the effect of the higher valued crop mix.

Aggregate Effects

While this investigation was not designed to specifically measure the aggregate streamflow effects of a contingent water contract for irrigation storage in the Snake River Basin, some general conclusions can be gleaned from the analysis. A study by Hamilton and Whittlesey (1996) found that 1–2 million acre-feet of water would be needed in the 13% of wettest years to meet all potential late spring and summer monthly flow targets—which extends beyond the critical two-month salmon migration period of this analysis—without other changes in river operations. In the driest 25% of years, the requirements would increase to 6–10 million acre-feet to meet flow targets in all months of habitat need, which exceeds the basin's reservoir storage capacity. Hence, using stored water supplies to obtain additional water for flow augmentation is only a first step in meeting fish habitat needs.

There are approximately 5 million acre-feet of irrigation storage in the upper Snake River Basin. Of this, approximately 2–3 million acre-feet could be obtained in the manner of this analysis and distributed to improve fish habitat in a timely manner (Hamilton and Whittlesey 1996). Thus, while the contingent water contracts for stored water considered here could make a significant contribution to streamflows, it is not possible to describe a specific measure of benefit that might be obtained in this manner. The net benefit to target flow rates for fish will depend upon the type and extent of other changes in river operations that are undertaken. Environmental impact studies are now underway to evaluate partial and total drawdown of all four lower Snake River reservoirs. With such changes in the lower river, the amount of water needed to meet flow targets is greatly reduced. In fact, permanent removal of the four lower river dams would completely eliminate the need for additional water in that portion of the river.

Conclusions, Limitations, and Future Research

The farm-level costs of two alternative contingent water contracts designed to augment lower Snake River streamflows in low flow years to enhance salmon migration were examined. The source of the contracted water supplies is the upper Snake River reservoir storage system originally built for irrigation. The first water contract considered was largely modeled after the existing upper Snake River water bank, where farmers agree to sell a percentage of stored water that is excess to current-year irrigation needs. The second contract specification calls for farmers to sell a percentage of total stored water. The economic cost of releasing stored water under an excess stored water contract is limited to future years when marketed storage is not refilled before being needed on-farm, whereas releases under a total stored water contract can have an economic cost in both the current year and future years. Both contracts are different from a market that takes only surface supplies from farmers in low flow years and leaves them unaffected by the market release in subsequent years.

Contingent contract cost was estimated by assuming farmers sign 10-year contracts to release a specific portion of their stored water supplies to improve fishery habitat in designated low flow years. Each contract simulation further assumed that all storage owners sell the same percentage, not quantity, of available water in a low flow year while maintaining the same irrigation technology over the 10-year contract period.

Contract cost per acre-foot of water obtained is less under an excess stored water contract than a total stored water contract, but the excess stored water contract provides much less water for fish habitat. The total stored water contract generally contributes two to four times more water for streamflow augmentation at each contract level. Acre-foot water cost increases as the percentage of stored water contracted is increased, particularly for junior appropriators. Farms using more efficient irrigation technologies are less likely to incur water shortages or experience net income losses.

Widespread adoption of stored water contracts might encourage some farmers to adopt a more efficient irrigation technology or alternative crops. Such implementation could induce widespread changes on the basin hydrology. However, relatively few of the rill irrigated farms, currently comprising more than 55% of land irrigated in the region, can easily and quickly adopt a more efficient technology with the associated cropping patterns used in this analysis. Soils, slopes, farm financial condition, field size, and climatic factors individually and collectively constrain the choice of irrigation technology and crop rotation. For example, small field size or irregular-shaped fields can prohibit the efficient adoption of center pivot irrigation. While it is expected that the evolution of irrigation technologies will continue and that future market conditions will influence the mix of crops available to farms, it was beyond the scope of this study to consider the specific effects of long-term contracts for stored water on either technology adoption or crop selection.

Huffaker and Whittlesey note that increased irrigation efficiency can lead to water spreading if water "conserved" through irrigation efficiency increases can be applied to acreage currently not under irrigation. Water spreading, if allowed, will generally increase the marginal value of water to agriculture and the opportunity cost of contract participation. The contract cost estimates for the more sophisticated irrigation technologies do not include this potential opportunity cost because water spreading is prohibited within the study region. Stored water can be applied only to land having stored water rights.

The effects of risk preference on contract cost were not evaluated. In this regard, the expected cost estimates are only baseline values from which market exchange values for water would be negotiated. Other forms of risk are also ignored. For example, changes in absolute and relative crop prices over the duration of the contingent water contract could affect the relative costs of each contract option. Problems of this type could be solved by market contract terms sensitive to changes in crop markets.

Transaction cost was also ignored. However, in this setting, this cost would be minimal and limited to the cost of negotiating the broad-based contracts with the regional agriculture. Most likely, the designated buyer would negotiate with entities such as irrigation districts or ditch service areas rather than individual farmers.

In summary, this study evaluates the effects of irrigation technology and water right priority on contingent contracts for irrigation storage to supplement streamflow for salmon recovery. It is shown that substantial quantities of water could be obtained at relatively modest cost, with a major advantage of being able to use existing storage capacity to shape downstream flows during critical periods and, importantly, without causing the long-term retirement of some exiting irrigated acreage. Knowledge of how irrigation technology and water priority right affect the agricultural cost of releasing stored water supplies provides instream interest groups with a tool for cost-effectively achieving streamflow management goals in an irrigated river basin. While many questions about contract implementation remain unanswered, this study describes a means to increase the flexibility of water allocations in an overappropriated river basin to better meet the needs of environmental concerns.

[*Received September 1995; final revision received January 1998.*]

References

Clairbon, A. B. "Predicting Attainable Irrigation Efficiencies in the Upper Snake River Region." Unpub. master's thesis, Idaho Water Resources Research Institute, University of Idaho, Moscow, 1993.

Ewbank, M. "Development of a Simplified Model for Analysis of Lower Snake River Salmon Flow Proposals." Unpub. master's thesis, Dept. of Environ. Engr., University of Washington, Seattle, 1992.

Frasier, M., N. Whittlesey, and J. Hamilton. "The Role of Irrigation in Salmon Population Recovery in the Pacific Northwest." Work. pap., Dept. of Agr. Econ., Washington State University, Pullman, December 1994.

Gardner, D. "The California and Other Nonmarket Allocation Systems." Paper presented at annual meetings of the Western Agricultural Economics Association, Honolulu HI, June 1984.

Halverson, P. "Economic Impact of Interruptible Water Markets on Columbia Basin Project Irrigated Agriculture." Unpub. Ph.D. diss., Dept. of Agr. Econ., Washington State University, Pullman, 1990.

Hamilton, J., D. Reading, and N. K. Whittlesey. "Cost of Using Water from the Snake River Basin to Augment Flows for Endangered Salmon." Work. pap. developed for use by the National Marine Fisheries Service, Dept. of Agr. Econ. and Rural Soc., University of Idaho, Moscow, 1995.

Hamilton, J., and N. K. Whittlesey. "Contingent Water Markets for Salmon Recovery." Work. pap., Dept. of Agr. Econ., Washington State University, Pullman, 1992.

———. "Cost of Using Water from the Snake River Basin to Augment Flows for Endangered Salmon." Paper presented at winter meeting of the Western Regional Science Association, Napa CA, February 1996.

Hamilton, J., N. K. Whittlesey, and P. Halverson." Interruptible Water Markets in the Pacific Northwest." *Amer. J. Agr. Econ.* 71,1(1989):63–75.

Huffaker, R. G., and N. K. Whittlesey. "Agricultural Water Conservation Legislation: Will It Save Water?" *Choices* (4th Quarter 1995):24–28.

Huppert, D. D., D. L. Fluharty, and E. S. Kenney. "Economic Effects of Management Measures Within the Range of Potential Critical Habitat for Snake River Endangered and Threatened Salmon Species." Work. pap., Dept. of Econ., University of Washington, Seattle, 1992.

Middaugh, J. "Water Marketing: Promise or Peril?" *Northwest Energy News* 14,3(1995):4–5.

National Marine Fisheries Service (NMFS). "Consultation: Biological Opinion, Operation of the Federal Columbia River Power System and Juvenile Transportation Program in 1995 and Future Years." *Endangered Species Act*, Sec. 7. Seattle WA, 1995.

Northwest Power Planning Council (NPPC). *Columbia River Basin Fish and Wildlife Program.* Portland OR, 1994.

———. *Columbia River Basin Fish and Wildlife Program.* Resident Fish and Wildlife Amendments. Portland OR, 1995.

Peterson, S., J. R. Hamilton, and N. K. Whittlesey. "What Role Can Idaho Water Play in Salmon Recovery?" Pub. No. PNW 462, Pacific Northwest Ext., Moscow ID, 1995.

Sims, L., and V. Ossiender. "Snake River Water Routing Study, Vol. I: Technical Support." Report to the National Marine Fisheries Service, Hydrosphere, Denver, 1992a.

———. "Snake River Water Routing Study, Vol. II: Appendices." Report to the National Marine Fisheries Service, Hydrosphere, Denver, 1992b.

———. "Water Supplies to Promote Juvenile Anadromous Fish Migration in the Snake River Basin." Report to the National Marine Fisheries Service, Hydrosphere, Denver, 1991.

Sommers, C. Unpub. memo to Daniel D. Huppert, Chairman, Economics Technical Committee, National Marine Fisheries Service. ERO Resources Corp., Denver, April 1992.

U.S. Army Corps of Engineers. "Lower Snake River Juvenile Salmon Migration Feasibility Study: Draft Economic Analysis Project Study Plan." Walla Walla District Office, Walla Walla WA, April 1997.

Wernstedt, K., J. B. Hyman, and C. M. Paulsen. "Evaluating Alternatives for Increasing Fish Stocks in the Columbia River Basin." *Resources*, No. 109(Fall 1992):10–16.

Whittlesey, N. K., J. Hamilton, and P. Halverson. "An Economic Study of the Potential for Water Markets in Idaho." Report prepared by Idaho Water Resources Research Institute, University of Idaho, for the Snake River Studies Advisory Committee, Office of the Governor, State of Idaho, 1986.

Willis, D. "Modeling Economic Effects of Stochastic Water Supply and Demand on Minimum Stream Flow Requirements." Unpub. Ph.D. diss., Dept. of Agr. Econ., Washington State University, Pullman, 1993.

257–68
[99]

[18]

WATER RESOURCES RESEARCH, VOL. 35, NO. 4, PAGES 1257–1268, APRIL 1999

Limiting pumping from the Edwards Aquifer: An economic investigation of proposals, water markets, and spring flow guarantees

Bruce A. McCarl,[1] Carl R. Dillon,[2] Keith O. Keplinger,[3] and R. Lynn Williams[4]

Abstract. The Edwards Aquifer, near San Antonio, Texas, is an important water source for both pumping and spring flow, which in turn provides water for recreation and habitat for several endangered species. A management authority is charged with aquifer management and is mandated to reduce pumping, facilitate water markets, protect agricultural rights, and protect the species habitat. This paper examines the economic dimensions of authority duties. A combined hydrologic-economic model is used in the investigation. The results indicate that proposed pumping limits are shown to have large consequences for agricultural usage and to decrease the welfare of current aquifer pumping users. However, the spring flow habitat is found to be protected, and the gains from that protection would have to exceed pumping user losses in order for the protection measures to increase regional economic welfare. Agricultural guarantees are shown to cause use value differences, indicating the opportunity for emergence of an active water market. Fixed quantity pumping limits are found to be an expensive way of insuring adequate spring flow.

1. Introduction

The Edwards Aquifer (EA), near San Antonio, Texas, is an important water source for agricultural, industrial, municipal, ecological, and recreational uses. The San Antonio municipal area almost exclusively relies on EA water, with 1995–1997 annual municipal and industrial usage averaging close to 300,000 acre-feet (af) (370 million m^3). Agricultural use, west of San Antonio, averages about 180,000 af (222 million m^3) [*USGS*, 1997]. The EA also supports springs at San Marcos and New Braunfels (Comal Springs) which provide habitat for endangered species [*Longley*, 1992]. In turn, the springflow supports recreation and downstream water usage. Average annual recharge is 637,000 af (786 million m^3), and mid-1990s' average pumping is 480,000 af (592 million m^3). The level of total pumping leaves only 150,000 af (185 million m^3) of water to support spring flow that is much smaller than the historic 50-year average spring flow (350,000 af, or 432 million m^3). In recent years, pumping has frequently exceeded recharge [*USGS*, 1997].

The EA is a fractured limestone formation that recharges quickly. Pumping has grown at about 1.1% per year during the last 40 years [*Collinge et al.*, 1993]. With increased pumping, the spring flow share of recharge has fallen, and spring flow has twice been close to cessation during the last 5 years. Meanwhile, aquifer level fluctuations have increased [*Collinge et al.*, 1993].

There has been considerable concern regarding EA management (see the *Water Strategist* [1996] or the San Antonio Water System web page, available at http://www.saws.org/other/htm, for a more detailed history). In the late 1950s the Edwards Underground Water District was formed to manage the EA. In the late 1980s the western agricultural counties seceded from the district because of disagreements about drought management plans. In the early 1990s lawsuits were filed asserting that the EA should be declared an underground river and that endangered species in the springs should be protected by maintaining spring flow. The Texas Water Commission declared the EA an underground river subject to surface water law during mid-1992, but this declaration was overturned by the courts during the fall of 1992. In early 1993 the district federal court upheld the endangered species lawsuit and ordered that pumping limits be imposed to protect spring flow. Texas Senate Bill 1477 [*Texas Legislature*, 1993] (SB1477) (1) establishes the Edwards Aquifer Authority (EAA) to manage the aquifer; (2) requires the EAA to reduce pumping to 450,000 af (560 million m^3) in the near future and to 400,000 af (490 million m^3) by 2008; (3) mandates establishment of water rights; (4) provides for water sales and leases; (5) guarantees that the agricultural share will be a proportional share of historic use and a minimum of 2 af (2468 m^3) per acre (0.405 ha); (6) limits off-farm water leasing so that 1 af (1234 m^3) per acre must be retained for use in irrigation; and (7) charges the EAA to "protect terrestrial and aquatic life, domestic and municipal water supplies, the operation of existing industries and the economic development of the state."

The EAA formally began operation in fall 1996 and, as of this writing, is expending substantial efforts on water rights establishment. (EAA activities and charter are set out on the EAA home page, available at http://www.e-aquifer.com).

This paper provides results from an analysis of issues regarding EAA duties. We report an economic evaluation of the

[1]Department of Agricultural Economics, Texas A&M University, College Station.
[2]Department of Agricultural Economics, University of Kentucky, Lexington.
[3]Texas Institute for Applied Environmental Research, Tarleton State University, Stephenville, Texas.
[4]California State University, Fresno.

Paper number 1998WR900116.
0043-1397/99/1998WR900116$09.00

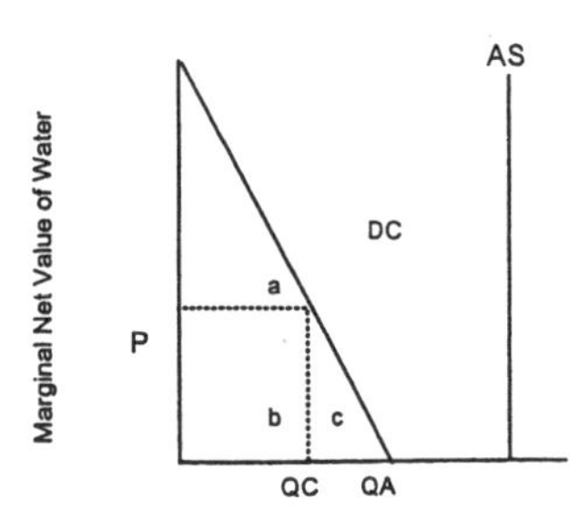

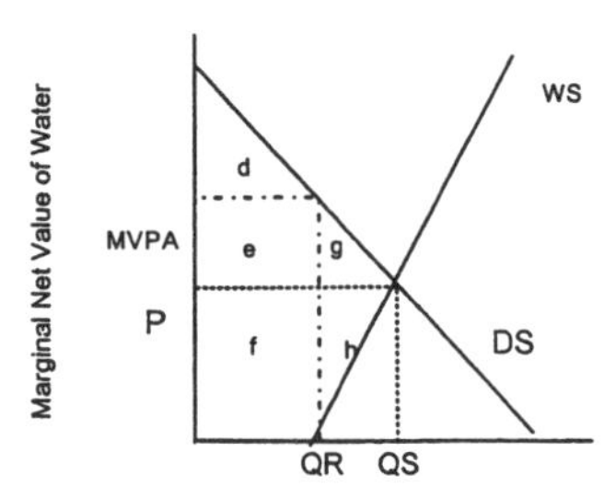

Figure 1. Residual versus cooperative use.

450,000 and 400,000 af (560 million and 490 million m^3) pumping limits, agricultural guarantees, and the associated imposition of water markets. Investigations are also performed on the costs of guaranteeing spring flows to protect aquatic life.

The contribution of this paper is best viewed in terms of the total literature. Fundamentally, the paper will build upon prior economic analyses of water allocation and water markets involving multiple users as previously studied in the surface water context by authors such as *Vaux and Howitt* [1982], *Howe et al.* [1986], *McCarl and Parandvash* [1988], *Michelsen and Young* [1993], and *Ward and Lynch* [1997] and extend the analysis to simultaneously treat multiple users, uncertainty, and a groundwater case. Second, the model used unifies results from a grid cell–based groundwater hydrology aquifer simulation (using the model described by *Thorkildson and McElhaney* [1992], which is similar in structure to that of *Gharbi and Peralta* [1994]) into an economic framework by using a regression summary of the groundwater model results. Third, the case study provides information on the trade-offs and considerations in the interesting EA case which involves, to mention a few salient characteristics, groundwater pumping by three economic sectors, endangered species, rapid recharge, spring flow, groundwater pumping rights, water markets, and agricultural use guarantees (see the web page by G. A. Eckhardt, available at http://www.txdirect.net/users/eckhardt, for a wealth of material on the aquifer). Fourth, this paper is an outgrowth of 10 years of work by the authors on EA issues. Compared to other journal articles involving the team, this is the first to treat nonagricultural and aquifer elevation–spring flow endogenously (*Keplinger et al.* [1998] and G. D. Schaible et al. (The Edwards Aquifer's water resource conflict: Do USDA farm programs increase irrigation water-use?, draft manuscript, 1998) limit treatment to the agricultural sector). This paper unifies work of the underlying dissertations [*Dillon*, 1991; *Williams*, 1996; *Keplinger*, 1996]. It extends an earlier bulletin on pumping limits [*McCarl et al.*, 1993], analyzing new issues raised by the most recent legislation regarding EA and using an improved analytical framework.

2. Water Use and Benefits With and Without Pumping Limits

Before reporting empirical findings, a graphical economic exploration of pumping limits is presented. In particular, we consider three questions: (1) Why have pumping limits?, (2) Should pumping limits be independent of water availability?, and (3) What are the welfare effects of pumping limits?

2.1. Why Have Pumping Limits?

A fundamental EA problem regarding efficient water use is the lack of transferrable water rights. (The current lack of property rights inhibits a market solution also rendering EA a common property pool. Provisions in SB1477 [*Texas Legislature*, 1993] provide a way for pumping interests to acquire and trade rights but do not provide a way for spring flow interests to acquire rights. Rather, a reduction in total use is envisioned leaving the residual for spring flow users.) Currently, individuals can use as much water as they can pump from beneath their land, although actions by EAA will soon limit usage. However, EA is an atypical aquifer. Water recharge is rapid, as is flow. Users jointly determine aquifer elevation, which determines pumping cost. Spring flow users are residual claimants obtaining water left over after pumping usage. The basic economics of such a case are depicted in Figure 1, where (1) the curve DC gives the demand for consumptive use by pumping, after water lifting cost has been paid, by a party that can pump as much as desired; (2) the supply curve AS is the aquifer supply of water; and (3) the demand curve DS is the demand for spring flow (arising through demands for both endangered species existence and instream flows, which in turn affect river ecology, recreation/tourism, and downstream water users).

This graphic format can be used to evaluate the effects of treating spring flow as a residual (as has been the case) versus actively considering spring flow–related demand. In the residual case, since available water (AS) is greater than the maximum pumping demand, pumping users withdraw water until the marginal value product (MVP) of water use is zero (or the MVP at the surface is just equal to water lifting cost). Thus pumping use is QA and the net water value is zero. In turn, unused water (AS − QA = QR) goes into spring flow, and that market yields a water value of MVPA. Such a situation yields a disparity between the value of water in spring flow and pumping usage arising because of a classic market failure [*Baumol and Oates*, 1975]. If somehow a market or other institutional mechanism confronted the value of displaced spring flow on pumping user decisions, with WS being the excess supply curve of water after pumping use, then the market would clear

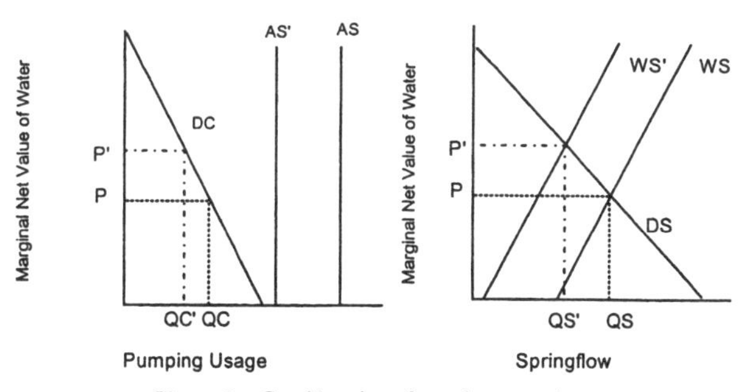

Figure 2. Consideration of supply uncertainty.

at price P with QC consumed by pumping and QS being the amount of spring flow.

This graphical analysis provides the basic rationale for SB1477 pumping limits. Namely, in the absence of a market that reflects spring flow value there is a less than economically efficient amount of water left after pumping use for spring flow. One way of making sure that enough water is left for spring flow is to impose a pumping limit at QC. The parties crafting SB1477 must have felt that an appropriate amount of pumping is somewhere around 400,000 af (490 million m^3). Such a level of pumping leaves about 100,000 af (120 million m^3) more for spring flow than is currently the case.

2.2. Should Pumping Limits Be Independent of Water Availability?

EA recharge in the last 10 years has varied from 240,000 to 2.4 million af (300 million m^3 to 3 billion m^3), while averaging about 630,000 af (780 million m^3) [*USGS*, 1998]. Figure 2 contains a second water supply/recharge level, AS′, which is considerably smaller than the original and leads to a different excess water supply curve (WS′). Under AS′ price P′ should be charged and consumptive use should be QC′, whereas under "normal" recharge (AS) quantity QC should be pumped. Thus a fixed pumping limit may not be appropriate as recharge varies.

In the empirical work we evaluate the effects of constant 400,000 or 450,000 af (490 million or 560 million m^3) pumping limits and of requiring spring flow to exceed various critical levels. No attempt is made herein (nor do we know of any attempt that has ever been made) to specify a spring flow demand equation and measure economic benefits as they relate to spring flow due to immense data gaps.

2.3. What Are the Welfare Effects of Pumping Limits?

To examine this we return to our certainty world of Figure 1. Pumping user welfare without pumping limits is the sum of the areas marked a, b, and c, while spring flow welfare is d + e + f. However, if the spring flow and pumping markets are in equilibrium, then pumping welfare equals a, while welfare area b arises because of the pumping limit and accrues to water rights and/or whatever agency is charging water rates while area c no longer accrues in the pumping sector. (A large portion of the water rights is likely to be assigned to agencies such as the San Antonio Water System, which in turn would need to develop a pricing scheme that would cause an appropriate level of water use. That might lead to such agencies accruing significant amounts of funds, which would need to be redistributed since they are public utilities. In the rest of this paper this component of welfare will be referred to as welfare to water rights and agencies.) Simultaneously, spring flow welfare increases by areas g + h. Thus the overall net welfare gain would be g + h − c. However, if the analysis is here, h = c limited to the pumping sector (as we do below), then there will be a loss of welfare equaling area c, which is offset by unmeasured gains in the spring flow sector. We develop estimates of area c that are a lower bound on what the water has to be worth when it is released as spring flow if potential pareto optimality is to be achieved. Thus welfare accruing to pumpers will fall under pumping limits. In Figure 1 pumpers lose areas b and, c while b accrues to the water rights holders and water agencies, which in turn, may redistribute that in various forms to those in the region.

3. The Modeling Framework

The empirical counterpart of the Figure 1 analysis was implemented using an equilibrium year economic and hydrologic aquifer simulation model, herein named EDSIM. EDSIM is the unification of cumulative developments by *Dillon* [1991], *Williams* [1996], *McCarl et al.* [1993], *Lacewell and McCarl* [1995]; *Keplinger* [1996], and *Keplinger et al.* [1995], with this being the first comprehensive publication. EDSIM depicts pumping use by the agricultural, industrial, and municipal sectors while simultaneously calculating pumping lift, ending elevation, and spring flow. EDSIM simulates choice of regional water use, irrigated versus dryland production, and irrigation delivery system (sprinkler or furrow) such that overall regional economic value is maximized. Regional value is derived from a combination of perfectly elastic demand for agricultural products, agricultural production costs, price elastic municipal demand, price elastic industrial demand, and lift sensitive pumping costs. The municipal demand elasticity is drawn from *Griffin and Chang* [1991], while the industrial elasticity is from *Renzetti* [1988]. The quantity demanded by municipal users depends upon rainfall and climatic conditions following *Griffin and Chang* [1991]. Agricultural water use dependency on climate is developed using EPIC [*Williams et al.*, 1989].

An algebraic representation of the fundamental relationships in EDSIM is presented below. A complete model specification is available in the source GAMS code. All variables are typed in upper case, while all the parameters are typed in lower case.

The unifying force in EDSIM is the objective function. EDSIM is a two-stage stochastic program with recourse model [*Dantzig*, 1955]. The model is solved as one simultaneous model but includes variables at two "stages" of uncertainty. The first (stage 1) set of variables depicts decisions on irrigation investment and crop mix and are constant across all states of nature chosen on the basis of average irrigation returns before anything is known about the weather event. The second (stage 2) set of variables are chosen with knowledge of state of nature (irrigation scheduling, crop sale, and nonagricultural water use). Thus, following Dantzig, the objective function involves two types of terms, one where certain costs are borne regardless of the uncertain outcome and the other where stochastically based decisions are weighted by their probability.

The first stage which is constant across all stochastic outcomes appears in the first three lines of the objective function, equation (1), and contains (1) the differential costs of irrigation development (costirr) by lift zone type (z) times acres developed (IRRLAND) in a place (p) for lift zone z and (2) the cost of establishing the crop mix times the acres in the mix (IRRMIX, DRYMIX) by place and irrigated or dryland crop (irracrecost, dryacrecost).

The second stage is defined for each state of nature (r) depicting alternative weather, recharge, and irrigation demand conditions and is weighted by probability (prob) as derived from historical observations. The second stage objective terms include the following state of nature dependent components: (1) irrigation and dryland net income (irrincome, dryincome) by place, lift zone, crop (c) and irrigation strategy (s) times acres produced (IRRPROD, DRYPROD); agricultural pumping cost involving per unit cost (AGPUMPCOST) times volume pumped (AGWATER) by month (m); integrals under the municipal and industrial demand curves (the terms with MUN, IND) by place; and municipal and industrial pumping costs defined in an analogous manner to the above agricultural pump cost term (MIPCOST).

3.1. Objective Function

Algebraically, the objective function is as follows:

$$
\begin{aligned}
\text{maximize} & - \sum_p \sum_z \text{costirr}_z \, \text{IRRLAND}_{pz} \\
& - \sum_p \sum_k \sum_z \text{irracrecost}_{pk} \, \text{IRRMIX}_{pzk} \\
& - \sum_p \sum_k \sum_z \text{dryacrecost}_{pk} \, \text{DRYMIX}_{pzk} \\
& + \sum_r \text{prob}_r \Bigg(\sum_p \sum_z \sum_c \sum_s \text{irrincome}_{rcs} \, \text{IRRPROD}_{pzrcs} \\
& + \sum_p \sum_c \text{dryincome}_{rc} \, \text{DRYPROD}_{prc} \\
& - \sum_p \sum_z \sum_m \text{AGPCOST}_{pzr} \, \text{AGWATER}_{pzrm} \\
& + \sum_p \sum_m \int \text{mprc}_{prm}(\text{MUN}_{prm}) \, d\text{MUN}_{prm} \\
& + \sum_p \sum_m \int \text{iprc}_{prm}(\text{IND}_{prm}) \, d\text{IND}_{prm} \\
& - \sum_p \sum_z \text{MIPCOST}_{pr}(\text{MUN}_{pr} + \text{IND}_{pr}) \Bigg) \qquad (1)
\end{aligned}
$$

where dMUN and dIND indicate the variables being integrated over.

The above objective function is maximized subject to the constraints in equations (2)–(17), which link the stage 1 and stage 2 variables and constrain the stage 2 variables to be conditioned by the stage 1 variables (e.g., allowing only as much irrigation as the amount of installed irrigation equipment and requiring the irrigation schedule to be consistent with the irrigated crop mix).

3.2. Irrigation Water Use Accounting

Irrigation water use is added up into the variable AGWATER by taking the water use per acre by each strategy times the acres produced (IRRPROD) by place (p), lift zone (z), state of nature (r), and irrigation strategy (s).

$$\sum_c \sum_s \text{wateruse}_{prcsm} \, \text{IRRPROD}_{pzrcs} - \text{AGWATER}_{pzrm} \le 0$$

$$\text{for all } p, z, r, m \qquad (2)$$

The AGWATER variable is used in the hydrologically based accounting of aquifer elevation, pump lift, and spring flow.

3.3. Total Farm Land Availability

Total acreage irrigated (IRRLAND) plus that converted to dryland use (MAKEDRY) cannot exceed the total land historically irrigated by place p and lift zone z.

$$+ \text{MAKEDRY}_{pz} + \text{IRRLAND}_{pz} \le \text{landavail}_{pz}$$

$$\text{for all } p, z \qquad (3)$$

3.4. Irrigated Land Availability

Total acreage irrigated (IRRLAND) in a place and lift zone is set equal to irrigated production by crop and irrigation strategy (IRRPROD) in that place and lift zone. The land to be irrigated is the same across all states of nature (note that IRRLAND does not have an r subscript), but the land can be assigned to different irrigation strategies depending on state of nature. Irrigated crop choice is constrained by (6) and (7).

$$\sum_c \sum_s \text{IRRPROD}_{pzrcs} - \text{IRRLAND}_{pz} \le 0 \qquad (4)$$

$$\text{for all } r, p, z$$

3.5. Dryland Availability

Acres in each place converted to dryland (MAKEDRY) summed across lift zones are set equal to dry production by crop (DRYPROD). The acreage converted is a stage 1 variable and is therefore equal across all states of nature. Initially, dryland production is zero since we are modeling only the irrigated portion of the region. This equation, however, allows conversion to dryland if the cost of water makes the conversion profitable.

$$\sum_c \text{DRYPROD}_{prc} - \sum_z \text{MAKEDRY}_{pz} \le 0 \qquad (5)$$

$$\text{for all } r, p$$

The acres converted to dryland are summed across the lift zones (since no pumping is done and thus lift zone is not an a cost factor for dryland production) and are set equal to total dryland cropping across all crops (c) for a place (p) and state of nature (r). The choice of dryland crop mix is (8) and (9).

3.6. Irrigated Crop Mix Restriction

These constraints require that the irrigated crop production for a place and lift zone be a convex combination of prespecified allowable crop mixes (where IRRMIX gives the weight in the combination and selects from k multicrop mix possibilities) following *McCarl* [1982]. The crop mix variables are stage 1 activities and do not differ by state of nature. The constraints require that the crops in each stage 2 state of nature summed over irrigation schedule(s) equal the stage 1 crop mix chosen. Thus the model can adjust the water use strategy to the climate, but the crop mix is chosen before exact weather conditions are known. The mixes chosen can differ by place and lift zone. Constraint (6) insures that the acres chosen by crop across all irrigation strategies s for a state of nature r are a weighted average of the relative frequency of that crop in that region and lift zone where IRRMIX is the weights in the average. Equation (7) forces the acres in the average to equal the acres farmed.

$$\sum_s \text{IRRPROD}_{pzrcs} - \sum_k \text{imix}_{pck}\,\text{IRRMIX}_{pzk} \leq 0 \quad (6)$$

for all p, z, r, c

$$\sum_c \sum_s \text{IRRPROD}_{pzrcs} - \sum_c \sum_k \text{imix}_{pck}\,\text{IRRMIX}_{pzk} = 0 \quad (7)$$

for all p, z, r

The mix data include historical and survey-based mixes. The survey-based mixes arose from a farm survey [*Schiable*, 1996] that asked irrigators what crop mix changes they would employ if the farm program were eliminated. These constraints cause the solution to cause realistic crop mixes without requiring modeling of detailed farm level resource allocation as argued by *McCarl* [1982].

3.7. Dryland Crop Mix Restriction

This is the dryland counterpart of (6) and (7) and requires that dryland production falls into a convex combination of previously observed dryland crop mixes by place. The DRYMIX variables are the weights in the convex combinations and multiply observed mixes. Constraint (9) requires that the total dryland acres in the mix equal the total dryland acres farmed. The dryland crop mixes are developed from historical dryland acreage statistics for the region.

$$\text{DRYPROD}_{pzrc} - \sum_k \text{dmix}_{pck}\,\text{DRYMIX}_{pzrk} \leq 0 \quad (8)$$

for all p, z, r, c

$$\sum_c \text{DRYPROD}_{prc} - \sum_c \sum_k \text{dmix}_{pck}\,\text{DRYMIX}_{pk} = 0 \quad (9)$$

for all p, r

3.8. Agricultural Pumping Cost Determination

Per-acre-foot agricultural pumping cost (AGPCOST) is set equal to a regression estimated linear function of lift (AGLIFT) by place, lift zone, and state of nature where agcosti is the intercept and agcostl is the slope.

$$-\ \text{AGPCOST}_{pzr} + \text{agcosti} + \text{agcostl AGLIFT}_{pzr} = 0 \quad (10)$$

for all p, z, r

3.9. Regional Base Pump Lift Equation

Base regional pumping lift is determined based on ending elevation. The base aquifer lift (LIFT) in a region is set equal to the difference between ending elevation (ENDWATER) and a zero lift level for each state of nature (r) and region (w).

$$-\ \text{LIFT}_{wr} + \text{ENDWATER}_{wr} = \text{zero lift}_w \quad \text{for all w, r} \quad (11)$$

3.10. Agricultural Lift Determination

The agricultural lift in a place and lift zone is set equal to the overall regional lift plus the agricultural lift difference (agdiff) for each state and lift zone.

$$-\ \text{AGLIFT}_{pzr} + \text{LIFT}_{wr} = -\text{agdiff}_{pz} \quad (12)$$

for all r, z, w, p ∈ reg(w)

The agdiff was calculated as a weighted average of the agricultural pumplift differential from the eastern area average lift in the GWSIMIV input data [*Thorkildsen and McElhaney*, 1992] for lands falling in the three lift zones.

3.11. Nonagricultural Pumping Cost Calculation

The per-acre-foot municipal and industrial pumping cost (MIPCOST) is set equal to a regression estimated linear function of the calculated M and I lift (MILIFT) by place and state of nature where micosti is the intercept and micostl is the slope.

$$-\ \text{MIPCOST}_{pr} + \text{micosti} + \text{micostl MILIFT}_{pr} = 0 \quad (13)$$

for all p, r

3.12. Nonagricultural Lift Determination

The municipal and industrial lift in place p falling in region w is set equal to the overall regional lift level for state r plus the place dependent municipal and industrial lift difference.

$$-\ \text{MILIFT}_{pr} + \text{LIFT}_{wr} = -\text{midiff}_p \quad (14)$$

for all r, w, p ∈ reg(w)

The midiff was calculated as a weighted average of the nonagricultural pumplift differential from the eastern area average lift in the GWSIMIV input data [*Thorkildsen and McElhaney*, 1992] for pumping falling in a county.

3.13. Regional Ending Elevation Determination

The ending aquifer elevation by region (ENDWAT) is computed through a linear equation that includes an intercept term (rendi), a recharge parameter (rendr) times the state dependent exogenous level of recharge(rech), an initial water level parameter (rende) times the endogenous initial water level (INITWAT) term, and a water-use-by-region parameter (rendu) times summed municipal, industrial, and agricultural use. Initial water level in both this and the adjacent eastern or western region affects this region's ending water level. Thus the subscript w2 is used to sum across both regions. The same is true for usage. The rend terms in the equation are regression response surface estimates over the entire set of results from a wide variety of aquifer hydrology model runs as described in section 5.

$$\begin{aligned} ENDWAT_{wr} = {} & rendi_w + \sum_m rendr_w \, rech_{rm} \\ & + \sum_{w_2} rende_{ww_2} \, INITWAT_{w_2} + \sum_{w_2} rendu_{ww2} \\ & \cdot \sum_{p \in reg(w_2)} \sum_m \Bigg(MUN_{prm} + IND_{prm} \\ & + \sum_z AGWATER_{przm} \Bigg) \qquad \text{for all w, r} \end{aligned} \tag{15}$$

3.14. Initial Elevation Balance

Initial elevation is set equal to the probability weighted average of ending elevation by region.

$$-INITWAT_w + \sum_r prob_r \, ENDWATER_{wr} = 0 \qquad \text{for all w} \tag{16}$$

3.15. Spring Flow Equation

Flow for the two springs is predicted from a regression-based forecast of similar structure to that used in the ending water level equation (15). The regression-based forecast considers only the cumulative use and recharge summed over months m* which proceed a particular month (m). Thus the regression equation for August will consider the initial water level and all pumping use in recharge from January through August. A linear equation is used that includes an intercept term (rsprni), a recharge parameter (rsprnr) times the state dependent exogenous level of recharge (rech), an initial water level parameter (rsprne) times the endogenous initial water level (INITWAT) term, and a water-use-by-region parameter (rsprnu) times summed municipal, industrial, and agricultural use. This equation is defined for each spring during each month for each state of nature. Equation estimation and resultant parameters is described in section 5.

$$\begin{aligned} SPRNFLO_{srm} = {} & rsprni_{sm} + \sum_{m^* \le m} rsprnr_{smm^*} \, rech_{rm^*} \\ & + \sum_w rsprne_{smw} \, INITWAT_w + \sum_w \sum_{p \in reg(w)} \sum_{m^* \le m} rsprnu_{smwm^*} \\ & \cdot \left(MUN_{prm^*} + IND_{prm^*} + \sum_z AGWATER_{przm^*} \right) \qquad \text{for all s, r, m} \end{aligned} \tag{17}$$

4. Model Component Elaboration

There are several key characteristics of the EDSIM framework which merit discussion. First, EDSIM is a price endogenous optimization model following the work of *McCarl and Spreen* [1980]. Water is allocated to the highest and best use in terms of generating greatest net economic value. Thus EDSIM is not constrained to simulate current use, but rather it simulates best use in an economic sense. However, when it ran without pumping limits under current water demand, the EDSIM water use solution corresponds closely to water use in the current unrestricted pumping environment where the marginal water value is basically driven to zero, as in Figure 1. When EDSIM is executed with the pumping or spring flow limits imposed, the results simulate the "best" total regional economic outcome under that limitation as well as permitting comparison with the existing situation.

Second, EDSIM incorporates uncertainty. The uncertain phenomena involves recharge and associated climate. The handling of uncertainty in EDSIM is based upon discrete stochastic programming or stochastic programming with recourse [*Dantzig*, 1955; *McCarl and Parandvash*, 1988; *Ziari et al.*, 1995]. Decision making is modeled as a two-stage process. In stage 1, decisions on irrigated acreage, furrow versus sprinkler irrigation, and crop mix are made which are state of nature independent. In stage 2, water use decisions are made which depend upon the state of nature. EDSIM maximizes average regional welfare over the recharge events and their probabilities. (The recharge distribution used herein is a nine-event representation of the recharge and climate distribution observed from 1934 to 1992. *Dillon* [1991] discusses its initial development and statistical characteristics.) This uncertainty model depicts an important fact coloring the agricultural production environment. Namely, the amount of agricultural irrigated acreage, the choice of furrow verus sprinkler, and the crop mix are generally chosen before the weather is revealed and persist as fixed decisions once the weather is known. However, the use of water is dependent on recharge and climate after their characteristics become known. Thus municipal demand, industrial demand, and choice of irrigation strategy depend upon water available.

Third, EDSIM incorporates hydrological processes based upon a regression summary of the Texas Water Development Board's EA simulation model [*Thorkildson and McElhaney*, 1992]. The estimated regression equations are directly incorporated as EDSIM equations in (15) and (17). Regression equations were estimated for monthly spring flow and ending elevation at two wells (one in Uvalde county and one in San Antonio) during the year at hand.

Fourth, EDSIM depicts economic competitive equilibrium water use by municipal, industrial, and agricultural interests. The agricultural submodel assumes farmers are profit maximizers choosing between dryland and irrigated cropping under 1996 commodity prices. The irrigation strategy depends on the recharge/weather situation, pumping lift, crop mix, and installed irrigation system. Three pumping lift zones and two irrigation delivery systems (furrow and sprinkler) are considered. The municipal and industrial submodels derive an economic equilibrium by intersecting explicit demand curves for water with their water supply prices. The supply price equals the pumping cost plus any water opportunity cost stimulated by pumping or spring flow restrictions. Thus EDSIM allocates water among sectors so that to the extent allowed by the scenario, marginal productivity is equalized and the overall level of economic activity is maximized.

Fifth, EDSIM is run under year-2000 demand projections. The year-2000 conditions assume that municipal and industrial demands expand according to growth-based, regional forecasts by the Texas Water Development Board.

Sixth, the EDSIM data evolved over time from efforts by *Dillon* [1991], *McCarl et al.* [1993], *Williams* [1996], *Lacewell and McCarl* [1995], and *Keplinger* [1996]. The agricultural part was largely specified using EPIC and extension service regional budgets [*Lacewell and McCarl*, 1995; *Keplinger*, 1996].

Table 1. Regression Coefficients for Annual Comal and San Marcos Spring Flow

	Dependent Variable			
	Spring Flow, Acre-Feet		Elevation, Feet Above Sea Level	
Parameter	Comal	San Marcos	J17 Ending	Sabinal Ending
J17 starting elevation, feet above sea level	2,651	412	0.542	0.348
Sabinal starting elevation, feet above sea level	551	0.0	0.155	0.583
Annual recharge, acre-feet	0.080	0.024	0.000019	0.000023
Western pumping, acre-feet	−0.04	−0.0005	−0.000028	−0.000091
Eastern pumping, acre-feet	−0.28	−0.025	−0.000136	−0.000059
Intercept	−1924677	−203976	225.41	102.22
R^2	0.93	0.77	0.95	0.96

One acre-foot equals 1234 m^3; 1 foot equals 0.3048 m.

Seventh, EDSIM is a single-equilibrium-year model. It starts from a single initial elevation across all recharge states. This initial elevation is set equal to the probabilistic weighted average of the ending elevations via (16). This means that the model always returns to an average initial elevation and does not account for decision making that would occur in a period of multiyear drought.

5. Hydrologic Regressions

The hydrologic regressions have large influences on the results and thus merit discussion. Equations were estimated that predict monthly spring flow at two springs and ending elevation at two wells. The monthly spring flow equations predict flow in month m during this year as a simple linear function of beginning year (January) well elevations, water pumping between January and month m, and recharge in months up to month m following a linear functional form. The same functional form was used for the ending elevations, but only a end of December equation was estimated. The response function equations were estimated from the results of 136,800 monthly observations from the single layer, porous medium, aquifer simulation model documented by *Thorkildson and McElhaney* [1992]. These observations arose from model runs under all combinations of 57 recharge states, 25 pumping alternatives, and 8 initial water levels observing 12 monthly results for each case. Annual versions of all equations are given in Table 1. Note that because the simulation model was the data source, we do not have a true underlying random distribution. Thus t statistic significance levels are not presented as they cannot be interpreted as valid statistical tests. We do however provide R^2 as a measure of fit.

The equations fit the simulated data well and contain expected results in terms of pumping, elevation, and recharge effects on the spring flow and ending elevation. Two features of the regression results merit discussion. First, western pumping and elevation have much smaller effects than their eastern counterparts. This is due to a granite intrusion (called the Knippa Gap) that separates the east and west EA regions, restricting flows and hydraulic pressure transmission. This finding manifests itself in later results. (These results have large implications for regional water use when spring flow is to be protected. Thus, we decided to try to verify the results using historical (rather that simulated) data. Annual regressions over historical data yielded essentially the same east/west results although multicollinearity did not permit estimation of the exact same equation. (See work by *Keplinger and McCarl* [1995] for details)). Second, the spring flow regression results do not fit the flows at San Marcos Springs as well as at Comal Springs but Comal is the critical spring. (Also as a reviewer pointed out, the GWBSIM model has more difficulty in predicting San Marcos spring flows.)

6. Base Model Results

The base model results appear in Table 2. Agricultural water use averages about 170,000 af (210 million m^3) with nonagricultural water use averaging around 330,000 af (410 million m^3). Maximum agricultural water use is about 190,000 af (230 million m^3), and nonagricultural water use is about 350,000 af (430 million m^3), for a total of 540,000 af (670 million m^3). This corresponds closely to historic maximum water use [*USGS*, 1997]. The smallest monthly spring flow at Comal, the most sensitive spring, is zero, an expected result given that a very dry year is in the data set. In terms of welfare, agricultural net income averages \$6.4 million with a municipal consumers' surplus of about \$471 million (being so large since it comes from a constant elasticity demand curve which goes asymptotic to the axis) and industrial surplus about \$2.8 million. Rents to water rights/agencies (area b from above) are around \$7 million. Total average welfare, the model maximand, is \$487 million across all sectors. All available acreage is irrigated (only acres irrigated in 1992 are included as land available in the model). Agricultural income has a 44.55% coefficient of variation. The optimal beginning (January) elevation for the J17 reference well in San Antonio is 642 feet (196 m). However, under dry conditions the ending elevation is as low as 626 feet (191 m).

7. Model-Based Policy Analysis

We are now in a position to investigate issues regarding EAA duties. We begin by examining the implications of the two SB1477 pumping limits, then follow that with an investigation of the pumping levels needed to maintain selected spring flow levels. Last, we turn attention to the implications of agricultural guarantees and associated water marketing.

7.1. SB1477 Pumping Limits

As stated above, SB1477 imposes pumping limits of 450,000 af (560 million m^3) in the near future and 400,000 af (490 million m^3) in the longer term. We simulate this by bounding the sum of the IND, MUN, and AGWATER variables to not exceed the pumping limit. The second and third columns of

Table 2. Comparison of Welfare Effects of Alternative Water Management Plans

		Change From Base Scenario					
		Pumping Limits, Acre-Feet		Spring Flow Limits, Cubic Feet per Second			
	Base	450,000	400,000	50	100	150	200
		Average Welfare Measures					
Agriculture income, 10^6 $	6.36	−0.40	−1.19	−0.08	−0.24	−0.27	−0.49
Percent change	...	−6.33	−18.69	−1.20	−3.82	−4.28	−7.73
Municipal surplus, 10^6 $	471.20	−1.90	−4.36	−0.26	−0.43	−1.90	−3.16
Percent change	...	−0.40	−0.93	−0.06	−0.09	−0.40	−0.67
Industry surplus, 10^6 $	2.77	−0.06	−0.13	0.00	0.00	−0.05	−0.08
Percent change	...	−1.99	−4.61	−0.22	−0.36	−1.67	−2.78
Authority surplus, 10^6 $	6.99	2.06	4.55	0.32	0.61	2.03	3.26
Percent change	...	29.48	65.10	4.63	8.74	29.08	46.70
Total surplus, 10^6 $	487.32	−0.30	−1.13	−0.02	−0.07	−0.19	−0.47
Percent change	...	−0.06	−0.23	−0.005	−0.01	−0.04	−0.10
		Agricultural Activity Measures					
Irrigated land, 10^3 acre-feet	79.89	−13.89	−26.98	−4.16	−11.19	−13.89	−18.64
Dryland usage, 10^3 acre-feet	0	13.89	26.98	4.16	11.19	13.89	18.64
Agricultural income coefficient of variation, %	44.56	−1.84	−0.93	−1.41	−3.74	−4.36	−4.57
		Water Use Measures					
East agricultural, 10^3 acre-feet	81.85	−32.85	−63.41	−9.35	−24.81	−32.55	−44.23
West agricultural, 10^3 acre-feet	87.41	−8.82	−14.08	0.03	0.79	0.37	−1.33
East nonagricultural, 10^3 acre-feet	322.00	−9.92	−22.49	−1.47	−2.42	−10.50	−16.89
West nonagricultural, 10^3 acre-feet	9.73	−0.48	−1.01	−0.01	−0.01	−0.12	−0.20
Total use, 10^3 acre-feet	500.99	−52.08	−100.99	−10.80	−26.44	−42.80	−62.65
		Hydrologic Result					
Comal spring flow, 10^3 acre-feet	97.56	93.12	183.91	21.83	54.49	86.17	123.55
San Marcos spring flow, 10^3 acre-feet	64.16	9.88	19.56	2.34	5.85	9.25	13.25
J17 Well End Elevation, feet	641.83	22.25	44.01	5.18	12.90	20.58	29.57
Minimum Comal Spring flow, cfs	0	139.72	250.56	50.00	100.00	150.00	200.00

One acre-foot equals 1234 m^3; 1 cubic foot equals 0.028 m^3; 1 foot equals 0.3048 m.

Table 2 give results from EDSIM under those pumping limits. The total loss in regional welfare (equivalent to area c in Figure 1) is $300,000 per year under the 450,000 af limit and $1.1 million (0.23%) under the 400,000 af limit. Under the 400,000 af limit agriculture loses $1.2 million, or 18.7% of base income level while municipal surplus is reduced by $4.3 million (about 1%). (Note that municipal surplus percent change is small because we are dealing with a constant elasticity demand curve.) Industrial surplus falls by $130,000, or 4.6%. Simultaneously, an additional $4.55 million a year accrues to water agencies or water rights holders. Thus the pumpers lose $5.68 million a year (areas b and c, above) whereas the agencies and rights holders gain $4.55 million, leaving a net regional loss of $1.13 million. As shown in the theoretical section above, for society to gain as a whole, this $1.13 million must be recouped through the value of the additional activities stimulated by the increased spring flow. Thus, if total net benefit is the standard for making decisions, then at least $1.13 million must be gained annually through the value of the continued existence of the endangered species, the increased aquifer elevation, and the benefits of increased spring flows, including the improved ecological characteristics in the rivers fed by spring flows, the increased recreational and other social values stimulated by expanded in-stream flows, and the value of additional downstream water consumption permitted by the increased spring flows.

The results show, under the 400,000 af (490 million m^3) limit, that average Comal spring flow almost triples, increasing by 184,000 af (230 million m^3) from the base level of 97,000 af (120 million m^3). Further, the smallest monthly Comal spring flow under the worst case recharge (approximately 50,000 af, or 68 million m^3) has risen from zero to 250 cubic feet per second (cfs; 7 m^3 s^{-1}), which is above the U.S. Fish and Wildlife Service (USFWS) take and jeopardy levels, which are 200 and 150 cfs (6 and 4 m^3 s^{-1}), respectively, as described in the next section [*USFWS*, 1995].

Irrigated acres are reduced under the 400,000 af (490 million m^3) pumping limit by 27,000 acres (11,000 ha; almost 33%). Eastern agricultural water use falls by over 75% whereas western use falls by about 17% due to higher pump lifts. Municipal use is reduced by 1000 af (1.2 million m^3) in the west and 22,000 af (27 million m^3) in the east. Total average water use falls from 501,000 af (62 million m^3) down to the 400,000 af limit. The elevation of the San Antonio reference well rises by 44 feet (14 m) in the typical year.

The pumping limit causes agriculture relative to municipal and industrial uses to experience larger water and percentage welfare adjustments, particularly in the east. This occurs because (1) the regression shows that eastern usage has more profound implications for eastern pumping lifts, (2) a less profitable overall crop mix is used in the east (which is congruent with historic observation), and (3) agricultural use values are smaller than municipal and industrial use values on the margin. Annual welfare reductions to the pumping users are almost a million dollars less under the 450,000 af (560 million m^3) limit than the 400,000 af (490 million m^3) limit. Considerable revenues could accrue to those allocating the rights

Table 3. Total Pumping Usage under Alternative Scenarios

Typical Weather Year*	Probability	Recharge, Thousands of Acre-Feet	Base Usage, Thousands of Acre-Feet	450,000 Acre-Foot Pumping Limit, Thousands of Acre-Feet	Usage When Spring Flow ≥200 cfs, Thousands of Acre-Feet
1956	0.018	43.7	537.8	450.0	401.8
1951	0.018	140.1	535.1	450.0	433.6
1963	0.089	170.8	526.9	450.0	442.4
1989	0.143	214.5	521.0	450.0	440.0
1980	0.214	406.3	519.6	450.0	461.5
1974	0.214	658.4	497.8	450.0	445.5
1976	0.214	894.1	463.9	445.8	412.2
1958	0.071	1701.2	487.7	450.0	434.9
1987	0.018	2003.6	454.3	439.8	409.8
Average	...	626.9	501.0	448.9	438.3

One acre-foot equals 1234 m^3; 1 cubic foot equals 0.028 m^3.
*These weather years provide the states of nature used in the model.

under the pumping limit, which might need to be dissipated if a public utility were involved.

7.2. Consideration of Spring Flow Limits

The EAA is charged with maintaining spring flow. The USFWS has estimated that Comal flows less than 200 cfs (6 m^3 s^{-1}), the "take" level, will result in unsuitable habitat for the endangered fountain darter species, while finding that 150 cfs (4 m^3 s^{-1}) is the Comal "jeopardy" level. In order to investigate the implications of maintaining spring flow we ran four minimum spring flow scenarios placing a lower bound on the monthly SPRNFLOW variable from (17). These scenarios require 50, 100, 150, and 200 cfs (1, 3, 4, 6 m^3 s^{-1}) of Comal flow during each and every month. Some of these levels are underneath the USFWS take and jeopardy levels. However, during August of 1996 spring flow fell to 79 cfs (2 m^3 s^{-1}) even though the USFWS levels had been announced.

The results show that relatively small adjustments can guarantee the lower spring flow minima. However, the 200 cfs (6 m^3 s^{-1}) minimum requires relatively larger adjustments. The welfare lost by pumping users to achieve this level of spring flow is only about half as much as that implied under the 400,000 af (490 million m^3) limit because the use of water generating spring flows above those needed is not allowed even if available. In particular, a 200 cfs limit reduces pumping user welfare by about a half million dollars a year. This rises through welfare losses of about \$0.5 million to agriculture, \$3 million to municipal interests, and \$80,000 to industrial interests, but a \$3.2 million increase in the welfare account accruing to water rights and agencies. In terms of irrigation, irrigated acreage falls by about 25%. Again the most substantial adjustment is in eastern water use by agriculture and municipal interests again because of the Knippa gap as discussed above.

These adjustments are stimulated by a pumping pattern which adjusts to climate, as implied in Figure 2. Table 3 details water use under the 200 cfs (6 m^3 s^{-1}) and 450,000 af (560 million m^3) scenarios. The 200 cfs column shows that when recharge is low, usage approaches 400,000 af (490 million m^3). However, as recharge grows, so initially does usage until at high levels demand is reduced by abundant rainfall. The reductions in usage relative to the base are greater the dryer the year. This result implies that pumping limit or water cost policies, which vary with aquifer water supply, would be desirable to take advantage of plentiful or scarce water.

7.3. Agricultural Water Use Guarantees

The above results arise from a scheme which allocates limited EA water so as to maximize total regional welfare. The EDSIM structure allocates water in a manner that implies that agriculture would forego pumping to allow higher-valued users access to water. In practice, that would be unlikely without compensation. Further, most of the agricultural use is west of the municipal use, and EA water flows from west to east. Thus agriculture generally has first access to the water.

Here we explore the implications of agriculture being guaranteed the amounts suggested in SB1477. Namely, agricultural water use will be no less than either (1) the base unrestricted level of usage adjusted down proportionally or (2) 2 af per acre (2500 m^3 per 0.4 ha, or 6800 m^3 per hectare) irrigated. These minima are imposed on all acres in all counties in all lift zones as a lower bound on the AGWATER variable. Also for now we do not allow agriculture to sell or lease water.

Table 4 presents the agricultural guarantee results under the 400,000 af (490 million m^3) limit. The optimal column gives EDSIM results under the pumping limit when agricultural water use is not guaranteed. The next two columns give the results under the proportional and 2 af (2500 m^3) guarantees. (The two market columns will be discussed in the next section.)

The results demonstrate water use and welfare trade-offs between sectors. Under guarantees, average agricultural welfare is 22–35% higher. These welfare gains occur due to 15–50% higher agricultural water use associated with the guarantee particularly in the east. However, total welfare is reduced by \$0.3 million to \$3 million with the agricultural gains achieved at the expense of nonagricultural users. Equivalently, without an agricultural guarantee, the gains by nonagricultural users are achieved at agriculture's expense (in the absence of compensation for reduced water use).

The results show the choice of guarantees involves more than a million dollars annually at a 400,000 af (490 million m^3) limit. There the 2 af (2500 m^3) guarantee has a million dollar greater annual welfare effect.

7.4. Water Markets

The results under the agricultural pumping guarantees show there is room for the establishment of water markets. EDSIM, in effect, simulates water use under an idealized water market with no transactions cost. We examine the effects of not having a water market by imposing water usage minimums by parties

Table 4. Welfare Effects of Agricultural Guarantees and Water Markets: 400,000 Acre-Foot Pumping Limit

		Guarantee, Change From Optimal			
				With Market	
	Optimal	Proportion	2 Acre-Feet	Proportion	Proportion/ 1 Acre-Foot
	Average Welfare Measures				
Agriculture income, 10^6 $	5.17	1.14	1.82	0.18	0.62
Percent change	...	22.04	35.13	3.46	12.05
Municipal surplus, 10^6 $	466.84	−9.89	−16.57	−1.18	−5.03
Percent change	...	−2.12	−3.55	−0.25	−1.08
Industry surplus, 10^6 $	2.64	−0.26	−0.43	−0.03	−0.13
Percent change	...	−9.83	−16.41	−1.07	−5.09
Authority surplus, 10^6 $	11.54	7.92	13.09	0.97	4.08
Percent change	...	68.66	113.46	8.38	35.34
Total surplus, 10^6 $	486.20	−1.09	−2.09	−0.06	−0.47
Percent change	...	−0.22	−0.43	−0.01	−0.10
	Agricultural Measures				
Irrigation development, 10^3 acres	52.91	17.39	26.31	2.76	14.65
Dryland usage, 10^3 acres	26.98	−17.39	−26.31	−2.76	−14.65
	Water Use				
East agricultural, 10^3 acre-feet	18.44	42.79	55.59	6.20	28.55
West agricultural, 10^3 acre-feet	73.33	−3.50	4.64	−0.92	−6.79
East nonagricultural, 10^3 acre-feet	299.51	−38.04	−58.31	−5.13	−21.08
West nonagricultural, 10^3 acre-feet	8.72	−1.25	−1.92	−0.15	−0.69
	Hydrological Measures				
Comal spring flow, 10^3 acre-feet	281.47	−9.16	−1.79	−1.82	−11.16
San Marcos spring flow, 10^3 acre-feet	83.72	−0.97	−0.06	−0.20	−1.26
J17 well ending elevation, feet	685.84	−1.38	0.79	−0.31	−2.17
Minimum Comal spring flow, cfs	250.56	−5.31	1.85	−1.06	−7.64

Note the scenario definitions are as follows: "Proportion" means agriculture gets its proportional share; "2 acre-feet" indicates no more than 2 acre-feet of water can be used per acre (as opposed to 2.12 in the base scenario); "with market" means that permanent sale of water is allowed; and "proportion/1 acre-foot" means that agricultural gets its proportional share but must retain use of at least 1 acre-foot per acre. One acre-foot equals 1234 m^3; 1 cubic foot equals 0.028 m^3; 1 foot equals 0.3048 m.

who do not freely participate in water sales by placing lower bounds on the AGWATER variable. Then through comparison with the unrestricted model, we evaluate the economic implications of market presence.

Two market forms are considered: temporary (lease) and permanent (sale) markets. We examined potential trades between agricultural and non agricultural interests under the 400,000 af (490 million m^3) limit. In particular, we develop our discussion from results of four water marketing related scenarios under the 400,000 af limit.

1. Agriculture does not trade any water and is guaranteed its proportional share (the second column in Table 4).
2. Agriculture is deeded its proportional share and can sell water to nonagricultural interests, but the same amount of water is required to be transferred across all states of nature simulating a permanent water sale of a given number of acre-feet (the next to last column in Table 4).
3. Agriculture realizes the guarantees of case 2 but must retain usage of 1 af per acre (2500 m^3 per 0.4 ha, or 6800 m^3 per hectare) following a restriction appearing in SB1477 (the last column in Table 4).
4. Agriculture sells whatever it wants with the volume sold varying by state of nature (the first column in Table 4).

Table 5 presents results on water marginal value product (MVP), derived as a weighted average of monthly and recharge-dependent shadow prices for water less the pumping cost under these scenarios. In the absence of a water transfer mechanism, there is about a $70 per acre-foot ($170 per hectare) difference in MVPs. Allowing permanent sales reduces the MVP disparity to about $4 ($41.64 to $37.55) per acre-foot ($10 per hectare) in the east or about $5 ($12 per hectare) in the west. When leasing is allowed, the MVPs are almost equal with slight differences due to seasonality in water use intensities. (Note that the MVPs are usage-weighted averages across

Table 5. Marginal Value of Water under Three Water Marketing Scenarios at the 400,000 Acre-Foot Limit

	Base 400–No Guarantee (Leasing)	No Market	Permanent Sale
	East		
Agriculture water value	38.23	26.15	37.55
Nonagriculture water value	37.66	99.93	41.64
	West		
Agriculture water value	30.21	21.23	28.02
Nonagriculture water value	30.28	90.93	33.73

Values given in dollars per acre-foot. One acre-foot equals 1234 m^3. Note that the base 400–no guarantee scenario allows agriculture to vary water sold across state of nature while the permanent sale scenario depicts the same amount of water sold by agriculture in each and every state of nature.

all states of nature and months. So in cases differences arise between agricultural and nonagricultural values due to the fact that these are a composite of 108 differentially weighted shadow prices across months and recharge years. The water value differs slightly owing to place and time of diversion.) The data in Table 4 for the water market scenarios show that most of the losses in the municipal and industrial areas can be mitigated by water markets. However, these results do not consider the magnitude of the transaction costs involved in the parties finding each other or factor in the agricultural and municipal welfare implications of water payments. Hence we conclude that the emergence of an active market is likely.

8. Concluding Comments

The economic impacts of Texas Senate Bill 1477 provisions which deal with management of the Edwards Aquifer were investigated using a multiuser, stochastic, linked economic-hydrologic simulation model. The results indicate that the near term consequences of a 450,000 af (560 million m^3) total pumping limit plan are not large (around $300,000 per year). The simulations show that the 400,000 af (490 million m^3) limit will cause the springs to flow above the endangered species critical limits. However, as the limit is reduced to 400,000 af some users show almost a 75% drop in water use, particularly eastern irrigators. This reduces total agricultural income by 18.7%. Simultaneously, the plan decreases municipal and industrial sector welfare by about $4.6 million (1%), but generates a $4.5 million return to water rights, agencies or permit holders. Overall, a 400,000 af pumping limit reduces pumping user welfare by $1.1 million per year. Such losses would be offset through gains from increased spring flows at Comal and San Marcos Springs and higher ending aquifer elevation. For the spring flow limitation to the yield and economic gain for society the gains from these other sources would have to exceed the $1.1 million annual loss borne by Edwards Aquifer pumping users.

The results also indicate that an active water market is likely to arise under the tighter pumping limits. For example, when historic levels of agricultural use are guaranteed, then the study results show water use value differences of more than $70 per acre-foot ($170 per hectare). An active water market can reduce this difference thereby benefitting both sectors.

Finally, the results indicate that minimum spring flow can be achieved at a higher level of pumping than the 400,000 af (490 million m^3) limit, but that total allowed water use needs to be sensitive to weather and recharge with less water consumed under the dryer events. This would require implementation of a more complex water use management regime such as the seniority system used for western water rights.

Acknowledgments. B. A. McCarl is Professor of Agricultural Economics, Texas A&M University; C. R. Dillon is Associate Professor of Agricultural Economics, University of Kentucky; K. O. Keplinger is Research Economist, Tarlton State University; and R. L. Williams is Assistant Professor, California State University, Fresno. Thanks to Ron Griffin, Ron Lacewell, Manzoor Chowdhury, Perry New, Wayne Jordan, and two anonymous reviewers for contributions and comments. This research project was sponsored by the Texas Water Resources Institute and the Texas Agricultural Experiment Station. This manuscript was done in association with the USDA NRCS cooperative agreement and with Texas A&M University entitled Natural Resource Modeling and Policy Analysis. B. A. McCarl is senior author. The other authors all contributed equally.

References

Baumol W. J., and W. E. Oates, *The Theory of Environmental Policy: Externalities, Public Outlays, and the Quality of Life*, Prentice-Hall, Englewood Cliffs, N. J., 1975.

Collinge, R., P. Emerson, R. C. Griffin, B. A. McCarl, and J. Merrifield, The Edwards Aquifer: An economic perspective, *TR-159*, Tex. Water Resour. Inst., Tex. A&M Univ., College Station, 1993.

Dantzig, G. B., Linear programming under uncertainty, *Manage. Sci.*, *1*, 197–206, 1955.

Dillon, C. R., An economic analysis of Edwards Aquifer water management, Ph.D. dissertation, Tex. A&M Univ., College Station, 1991.

Gharbi, A., and R. C. Peralta, Integrated embedding optimization applied to Salt Lake Valley aquifers, *Water Resour. Res.*, *30*, 817–832, 1994.

Griffin, R. C., and C. Chang, Seasonality in community water demand, *West. J. Agric. Econ.*, *16*, 207–217, 1991.

Howe, C. W., D. R. Schurmeier, and W. D. Shaw Jr., Innovative approaches to water allocation: The potential for water markets, *Water Resour. Res.*, *22*, 439–445, 1986.

Keplinger, K. O., An investigation of dry year options for the Edwards Aquifer, Ph.D. dissertation, Tex. A&M Univ., College Station, 1996.

Keplinger, K. O., and B. A. McCarl, Regression based investigation of pumping limits on springflow within the Edwards Aquifer, Dep. of Agric. Econ., Tex. A&M Univ., College Station, 1995.

Keplinger, K., B. McCarl, M. Chowdhury, and R. Lacewell, Economic and hydrologic implications of suspending irrigation in dry years, *J. Agric. Resour. Econ.*, *23*, 191–205, 1998.

Lacewell, R. D., and B. A. McCarl, Estimated effect of USDA commodity programs on annual pumpage from the Edwards Aquifer, Final report submitted to Nat. Resour. Conserv. Serv., U.S. Dep. of Agric., Temple, Tex., 1995.

Longley, G., The subterranean aquatic ecosystem of the Balcones fault zone Edwards Aquifer in Texas—Threats from overpumping, paper presented at First International Conference on Ground Water Ecology, U.S. Environ. Prot. Agency, Tampa, Florida, April 26–29, 1992.

McCarl, B. A., Cropping activities in agricultural sector models: A methodological proposal, *Am. J. Agric. Econ.*, *64*, 768–772, 1982.

McCarl, B. A., and G. H. Parandvash, Irrigation development versus hydroelectric generation: Can interruptible irrigation play a role?, *West. J. Agric. Econ.*, *13*, 267–276, 1988.

McCarl, B. A., and T. H. Spreen, Price endogenous mathematical programming as a tool for sector analysis, *Am. J. Agric. Econ.*, *62*, 87–102, 1980.

McCarl, B. A., W. R. Jordan, R. L. Williams, L. L. Jones, and C. R. Dillon, Economic and hydrologic implications of proposed Edwards Aquifer management plans, *TR-158*, Tex. Water Resour. Inst., Tex. A&M Univ., College Station, 1993.

Michelsen, A. R., and R. A. Young, Optioning agricultural water rights for urban water supplies during drought, *Am. J. Agric. Econ.*, *75*, 1010–1020, 1993.

Renzetti, S., An economic study of industrial water demands in British Columbia, Canada, *Water Resour. Res.*, *24*, 1569–1573, 1988.

Schaible, G. D., Summary/interpretation: Three county irrigation survey of irrigators in the EA area, in *Biological Evaluation of USDA Farmer Assistance and Rural Development Programs Implemented in Bexar, Medina, and Uvalde Counties, Texas on EA Threatened and Endangered Aquatic Species*, appendix II, Nat. Resour. Conserv. Serv., U.S. Dep. of Agric., Temple, Tex., 1996.

Texas Legislature, *Senate Bill 1477*, 73rd session, Austin, May 1993.

Thorkildsen, D., and P. D. McElhaney, Model refinement and applications for the Edwards (Balconies Fault Zone) Aquifer in the San Antonio region, Texas, *Rep. 340*, Tex. Water Dev. Board, Austin, 1992.

U.S. Fish and Wildlife Service, San Marcos/Comal recovery plan, Albequerque, N. M., 1995.

U.S. Geological Survey, Recharge to and discharge from the Edwards Aquifer in the San Antonio Area, Texas, Austin, Tex., 1997. (Available at http://tx.usgs.gov/reports/district/98/01/index.html.)

Vaux, H. J., and R. E. Howitt, Managing water scarcity: An evaluation of interregional transfers, *Water Resour. Res.*, *20*, 785–792, 1984.

Ward, F. A., and T. P. Lynch, Is dominant use management compatible with basin-wide economic efficiency?, *Water Resour. Res.*, *33*, 1165–1170, 1997.

Water Strategist, On groundwater control and markets: Managing the Edwards Aquifer, newslet., vol. 10, no. 3, 1996.

Williams, R. L., Drought management and the Edwards Aquifer: An economic inquiry, Ph.D. dissertation, Tex. A&M Univ., College Station, 1996.
Williams, J. R., C. A. Jones, J. R. Kiniry, and D. A. Spaniel, The EPIC crop growth model, *Trans. ASAE*, *32*, 497–511, 1989.
Ziari, H. A., B. A. McCarl, and C. A. Stockle, Nonlinear mixed integer program model for evaluating runoff impoundments for supplemental irrigation, *Water Resour. Res.*, *31*, 1585–1594, 1995.

C. R. Dillon, Department of Agricultural Economics, University of Kentucky, Ag Engineering Bldg., Lexington, KY 40506-0276.

K. O. Keplinger, TIAER, Tarleton State University, Box 10410, Tarelton Station, Stephenville, TX 76402.

B. A. McCarl, Department of Agricultural Economics, Texas A&M University, College Station, TX 77843-2124. (mccarl@ranger.tamu.edu

R. L. Williams, California State University, 5245 N. Backer Avenue, M/SPB 101, Fresno, CA 93740-8001.

(Received June 1, 1998; revised November 30, 1998; accepted December 11, 1998.)

Name Index